DATE			

D1402448

BAKER & TAYLOR

YOURDON™ SYSTEMS METHOD

YOURDON™ SYSTEMS METHOD

Model-Driven Systems Development

YOURDON Inc.

YOURDON PRESS
Prentice Hall Building
Englewood Cliffs, New Jersey 07632

Yourdon Systems Method: model-driven systems development
    by YOURDON Inc.
        p.   cm. --(Yourdon Press computing series)
    Includes bibliographical references and index.
    ISBN 0-13-045162-2
    1. System design.   2. System analysis.
I. YOURDON, Inc.   II. Series.
QA76.9.S88Y6 1993   92-44387
004.2'1--dc20   CIP

Editorial/production supervision: *Harriet Tellem*
Cover design: *Jerry Votta*
Buyer: *Mary Elizabeth McCartney*
Acquisitions editor: *Paul W. Becker*
Editorial assistant: *Noreen Regina*
YOURDON, Inc. coordinator: *Julian Morgan*

Published by PTR Prentice Hall
A Simon & Schuster Company
Englewood Cliffs, New Jersey 07632:

The publisher offers discounts on this book when ordered in bulk quantities.
For more information, contact:
    Corporate Sales Department
    PTR Prentice Hall
    113 Sylvan Avenue
    Englewood Cliffs, NJ 07632

    Phone: 201-592-2863
    FAX: 201-592-2249

ISBN 0-13-045162-2

Prentice-Hall International (UK) Limited, *London*
Prentice-Hall of Australia Pty. Limited, *Sydney*
Prentice-Hall Canada Inc., *Toronto*
Prentice-Hall Hispanoamericana, S.A., *Mexico*
Prentice-Hall of India Private Limited, *New Delhi*
Prentice-Hall of Japan, Inc., *Tokyo*
Simon & Schuster Asia Pte. Ltd., *Singapore*
Editora Prentice-Hall do Brasil, Ltd., *Rio de Janeiro*

Dear Yourdon Reference Manual User,

Effective software engineering methods such as YSM are an evolving technology. Critical to this process of evolution is feedback, especially when related to the practical application of YSM in the field as experienced by the system engineer using the method.

Please find attached to this letter a Yourdon Reference Manual Improvement Form. You may use it to report any features of the manual you think might benefit from change or enhancement. The same form may also be used to make suggestions to enhance the method.

We will review all suggestions and incorporate those changes that will benefit the majority of our users and give you appropriate credit.

Sincerely

Julian L. Morgan
Director

Contents

Preface

1 Purpose of this manual

This reference manual describes "The Yourdon System Method" (or YSM) and how it can be used for systems development and support. It provides a definitive statement of what constitutes YSM in terms of models, tools and methods.

1.1 What this manual is not

This manual is not intended to act as a textbook or training document, nor would it be suitable for either.

It is not intended to be read sequentially, but as a reference on individual topics such as a specific modelling tool, model or technique. It is organised for minimal redundancy (as far as we can determine).

1.2 This version

This is the first part of the YSM reference manual. It covers analysis of system requirements and real-world information. Additional parts will cover implementation models, strategic planning, testing, quality assurance.

At present, there is no intention to define the YSM position on all topics of relevance to system development and maintenance. This manual is intended to consolidate the ideas used in teaching and consultancy.

2 Coverage

YSM covers:

- Enterprise activities and resource management. An enterprise is any economic unit that is resourced and managed as a unit. An enterprise could constitute the whole company, a division of the company, or a specific managed account. YSM provides models to capture and understand enterprise requirements.

- System modelling. A system is part of the enterprise. Systems usually have shorter life cycles than the enterprise. They are built, modified and eventually replaced. YSM provides system models that can be used to produce quality systems on time and to budget.

A project is an activity that is mandated over a fixed time period to achieve a specific goal. It may be to provide an improvement in enterprise resources or to build a given system.

The current release of YSM is mainly concerned with models, modelling tools and how to use them. Topics such as cost-justification, sizing of requirements, testing and project management are not covered by this release of YSM.

3 Changes/additions to existing methods

3.1 Upwards compatibility

YSM as described in this manual, should be regarded as a consolidation, rather than replacement of previous versions of the method taught and used. Most of the techniques used in the past still have their place within YSM. They will be described at the appropriate place in this manual (or later versions).

3.2 Models

Yourdon continues to be a model-oriented company and indeed, is even more so than before. As already marketed in the Strategic Information Modelling course, there is a distinction between enterprise and system models. This is true of information, but also many other "shared resources". YSM allows cases where the system is the enterprise. (This is achieved by combining the enterprise and system model, or "folding over" the enterprise model into the system model.)

The Hardware Implementation Model is new to the method. This is built to model how analog devices are used (if they are used).

The Manual Implementation Model is also new. It is, in effect, the user manual. It is included in the method to allow traceability of system documentation back to the earlier models.

3.3 Additional features

The main change has been a more rigorous theory underlying the method. This will allow better quality assurance and support by CASE tools. As far as possible, any changes have been hidden from the practitioner of YSM. However, where there were ambiguities in what the method consisted of, additions were made. Generally speaking, information modelling is now firmly based on set theory, as used in other formal approaches to data modelling and rule-based systems. Text specifications now use well-defined grammars.

Two new features of the method are entity life-cycles and abstract data types. These are well-known in the industry and their addition should enhance the method.

Another important change is in the area of text specifications. Rather than trying to define a 'data dictionary' notation to cover all types of specification, each model component has it's own type of specification. These are given in a 'frame' type format, but there are guidelines for use of alternative notations.

In the important case of minispecs, there is a proposed grammar. This may be relaxed, as required. However, some formalisation in this area is required for automated tools.

In the design version of the reference, a very significant development will be the tying of services on technology-level diagrams to specific service types. This is termed 'strong typing' of the implementation models.

The general approach has been to try and 'tie down' the concepts, so that the 'relaxed' form of YSM does not look very different from what they have used in the past. However, by basing the method on more formal principles, at the lowest level, a more formal approach may be used when needed.

An additional consideration has been the requirement that any grammar must be sufficiently well-defined to be automatable.

3.4 Rationalisation of terms

Some simplifications in terms of names, etc. has been made to reflect current practice. The terminology used in the real-time and commercial versions of YSM has been brought into alignment.

One significant change of this type is the return of 'minispec'. The terms 'object type' (or object) has been replaced by 'entity' (this avoids confusion when using YSM with object-oriented additions).

4 Organisation of YSM reference manual

The analysis reference manual is divided into the following chapters:

1. **Introduction**: this gives a historical background and description of the main features of YSM.

2. **Modelling Tools**: this describes the graphic and textual modelling tools used to build models. These may be system or enterprise models. Notation and rules are discussed, with quality guidelines.

3. **The Enterprise Essential Model**: this describes how enterprise requirements are modelled.

4. **The Relationship Between Systems and the Enterprise**: this describes how system projects make use of enterprise resources. It also describes how enterprise resources are put in place to support systems. This version only deals with general principles and the relationship between the information aspects of the Enterprise Essential Model and System Essential Models.

5. **The System Essential Model**: this describes how an Essential Model is built to capture system requirements. It also takes account of the relationship between the system and the rest of the enterprise.

6. **Appendices:** this contains reference material to back up the main chapters. This material provides additional rigour to the main sections.

There is also a bibliography, which gives a consolidated list of all references.

5 Cross-referencing

The following conventions are used to help cross-referencing:

5.1 Page identification

1. For the analysis part of the reference, there are 6 chapters. The number and name of each chapter appears on the left page, below the page number.

2. Each chapter consists of a number of sections, numbered 1, 2, ... within that chapter. For example, within chapter 2, the sections are numbered 2.1, 2.2, 2.3 ... .

3. The section number and name appear on the right page below the page number. The section corresponds to the section starting on that page, or the continuing section if it began on a previous page and continues to subsequent pages.

An example is shown below.

52	**53**
2:Modelling Tools	2.3: The Entity Relationship Diagram

Pages 52 and 53 are shown. They are from the "Modelling Tools" chapter, section 3. The section "The Entity Relationship Diagram" commences on, or before the right hand page.

5.2 References to other parts of the manual

1. A reference to another part of the manual is always given by quoting the section number. For example:

   ```
   See §3.2
   ```

refers the reader to section 2 in chapter 3. This reference is given to the most specific point possible. In the above example, the indication is that all of section 2.3 is relevant. This might be quite long. More specific references, such as:

```
See §3.2.3.2.2
```

are also used to give more exact identification of smaller sections of the reference. Although this means that some of the section numbers are quite long, it is easier to use the manual as a reference.

5.3 References to other sources

These are given as, for example "[DeM78]". All such references are collected together in the bibliography. The index shows which pages the references occur on.

6 Typography etc.

The following conventions have been used in this release of the manual:

- Model components are in double quotes — for example, "Scheduled course"

- Technical terms, slang, etc. are in single quotes — for example, 'outside–in'

- Examples are shown in shaded boxes for text fragments of models — for example:

> **Pre-condition**
> Matching <Drug> exists

- Larger examples of specific modelling tools are given in doubly-boxed frames:

- Algorithms, heuristics and techniques are shown in (roughly) algorithmic form within double frames:

6.1 Spelling

This release is in UK spelling — hence modelling, behavioural, etc, are correct. The punctuation is also UK version and therefore full stops, commas, etc. appear *after* quotes, not inside them.

6.2 Capitalisation

This manual has moved away from the idea of capitalising nearly every noun in sight. The only capitalisation that is used is for the names of specific YSM models. Generic model descriptions are given lower case names.

Chapter and highest-level section headings within the chapter are fully capitalised. Other section headings are not capitalised.

Names of model components are capitalised as they would be in the real-world. Thus entities, terminators and data processes are usually given a capital letter for the first word only. Most other model components, (e.g. names of data flows, states, etc.) are lower-case throughout.

7 History and acknowledgments

This Yourdon Reference Manual (YRM) is the result of many months of work and extensive contributions from a team of professionals at both Yourdon, Inc. in the USA and Yourdon International, Ltd. in London, United Kingdom. Following is a partial list of those individuals who have provided inputs to the YRM in recent years:

> Dr. Sami AlBanna, Paul Allen, John Baker, Steve Bodenheimer, Roger Boudreau, John Bowen, Dr. Adrian Bowles, Corrine Brandi, Mike Brough, Dave Bulman, Peter Davis, Keith Edwards, Roger Holmes, Viv Lawrence, Dr. John Leung, Nick Mandato, Rick Marden, Nan Matzke, Scott McBride, Jim McCallum, Julian Morgan, Edward O'Connor, Lew Sherry, Chris Spruyt, Dennis Stipe, Phil Sully, and Edward Yourdon.

The YSM method has been in continuous evolution since its initiation in 1974. It evolved in response to the contributions of numerous organizations, projects, professionals, and reviewers. Many authors and researchers chronicled the evolution of the method after participating in developing its concepts, tools, and techniques. These works remain part of the heritage of the method. Listed here are but a sample of these contributors:

> Paul Allen, Larry Costantine, Tom DeMarco, Matt Flavin, Chris Gane, Timothy Lister, Steven McMenamin, Stephen Mellor, Meilir Page-Jones, John Palmer, Sally Shlaer, Trish Sarson, Dr. Steve Weiss, Dr. Paul Ward, and Edward Yourdon.

1: Introduction

Table of Contents

1.1 Historical Background

1.1.1 Stages in system development methods

The evolution of system development methods has been gradual, with many people contributing to their improvement. The three stages in the evolution of system development methods may be identified:

1. **First generation methods.** Development of system modelling techniques for a varied set of problems and concerns. However, most people still relied heavily on one technique and modelling tool.

2. **Second generation methods.** A wider view of system behaviour was seen to be of importance and system developers started to use a range of modelling tools. However, the relationship between these tools and their use at different stages of the system life cycle was rather ill-defined. Some automated tools were developed, but they were still based on the approaches developed for manual application.

3. **Third generation methods.** During the third generation, a more integrated approach to system modelling is used. A range of modelling tools are still used to deal with the multi-dimensional nature of systems, but they are supported by more rigorous modelling principles. Models can be checked for quality, simulations and calculations carried out. These models use an underlying proven theory.

At the present time, most people are using techniques from the first or second generations. YSM is a third generation method.

1.1.2 First generation methods

First generation methods can be equated with the various 'structured techniques' developed during the late 1960s and 1970s. Structured techniques break down a complex problem into smaller components, with well defined inter-relationships between the components. The recognition that this is fundamental to the construction of complex software may be traced back to a paper by Dijkstra [Dij65]. In it he said:

> *The technique of mastering complexity has been known since ancient times: Divide et impera (Divide and rule).*

He justified this by stating:

> *I have only a very small head and must learn to live with it.*

G. Miller, an experimental psychologist, had earlier recognised just how few factors can be considered at once when making decisions [Mil56]. He suggested that the limit for understanding is about seven items at once.

1.1.2.1 Structured programming

1.1.2.1.1 Sequence, selection, iteration and avoiding 'GOTOs'
The first structured techniques helped organise source code. These techniques use standard constructs that can be used to implement any algorithm [Böh66]. These program 'building blocks' are 'sequence', 'selection' and 'iteration'.

At the time these standard constructs were recognised, not all the programming languages in use provided support for them. However, it was understood that reliable, maintainable code could only be produced by using this disciplined approach — ad-hoc methods of organising code were no longer acceptable [Dij68a], [Dij69]. Modern programming languages provide direct support for these constructs and existing languages have been modified to provide them. It is now taken for granted that code should only use these constructs.

1.1.2.1.2 Modular design and structure charts

In addition to these constructs, many programming languages allowed groups of statements to be reused. Such a group of statements is called a 'program module' and their use is referred to as 'modular programming' or 'modular design'.

Structure charts were used to show the way the modules connected to form the program. Structure charts were developed out of organisation hierarchy charts. Their adoption was largely due to the work of Constantine [Ste74]. YSM still uses structure charts to model source code organisation.

1.1.2.1.3 Programming style

Even using the standard constructs, code could be complex, 'clever' and difficult to understand, or simple, and easy to understand. This is a style issue and in the early 1970s, it was realised that maintainable code also required style guidelines. Kernighan and Plauger had a major influence in helping to define what is good programming style [Ker74].

1.1.2.2 Data structures

The data design techniques in use in the 1970s were largely pragmatic ways of building data structures and files to support programs. The CODASYL organisation was an important influence in formulating ideas on database design during the period 1967 to 1972. In particular, it defined the abstract model for a network database [COD71]. (For a discussion of the work of the CODASYL group, see [Oll78].)

The data structure diagram is a graphic tool used to model data structures in the same way that structure charts are used to model program organisation. Data structure diagrams show how links provide navigation around a set of data structures. Their use can be traced to the pioneering work of Bachman [Bac69].

1.1.2.3 Structured design

1.1.2.3.1 Limited scope of structured programming

Structured programming techniques only addressed 'design in the small' — how individual statements should be organised within a module and what was an acceptable organisation of modules. Structured programming gave no guidelines on how to carry out 'design in the large' — what would be *good* modules to choose and how they should be organised.

'Structured design' techniques were developed to solve this problem. These techniques were used for single programs and also for very large software projects containing many components.

1.1.2.3.2 Successive refinement

The strategy of successive refinement (or top–down design) was a major design technique [Wir71]. A required system function was broken down into a small number of tasks that could be combined to accomplish the function. Each of these was then considered in turn, breaking it down into smaller subtasks. The process continued until each subtask constituted a unit that could be coded with little chance of error.

1.1.2.3.3 Abstraction
In 1967, Dijkstra presented a paper which had a profound impact on development methods [Dij68b]. In it, he described how an operating system was built using successive layers of architecture. At the lowest level, the hardware was controlled, with higher layers providing successively more application-oriented services.

When the lower-level hides detail about data from higher-level layers, this technique is called 'information hiding', as first described by Parnas [Par72].

Hiding of implementation details from higher levels is an important design principle. It led directly, or indirectly, to many important abstraction techniques. It is a special case of the lump law — see §1.2.1.4.3.

1.1.2.3.4 Techniques based on the semantics of the structure chart
As described in §1.1.2.1.2, the structure chart was used to model the way program modules inter-connected. This led to the development of techniques that looked at the structure chart and tried to improve the design of the program. The main criteria for improving program design were based upon:

- Coupling: this is a measure of the complexity of interfaces. If the interfaces between modules are simple, the design is described as having low coupling; if the interfaces are complex, the design is described as having high coupling.

- Cohesion: this is a measure of how 'single-minded' a module is. If a module carries out exactly one, well-defined function, it is described as having functional cohesion. A functional module is highly cohesive. Other modules might be less cohesive, carrying out several functions (or only part of a single function).

Although it was difficult to quantify the coupling and cohesion parameters, they formed a useful basis for program design techniques [Mye75], [You75]. A good design had low coupling and high cohesion.

These criteria are still useful guidelines in design, but their application is not the only design technique now used. Other techniques will be described in future releases of this manual.

Coupling and cohesion are also useful criteria when organising system requirements. They may be applied (for example) to 'levelling' of data flow diagrams and grouping of system events. (See §5.7.7.2 and §2.22.5.2.1.)

1.1.2.3.5 Data refinement techniques
Most 1970s data design techniques were based on the idea of a relational data model, which treats all data as 'tables' [Cod70]. Refinement criteria were used to check the quality of the design. Application of these criteria is called 'normalisation' — it is still a useful technique (see §3.5.8).

1.1.2.4 Structured analysis

1.1.2.4.1 Failure to identify user requirements
Even with appropriate system design principles, there was still a way in which a project could fail to meet the user requirement. This was that the requirement might be misunderstood by the system developer. Often, the user requirement was ambiguous, confused or inconsistent.

For small systems, the requirement *could* be understood by one person, who took the responsibility for ensuring that the requirement was met. Even then, there was no guarantee that the requirements were correctly understood. For larger systems, a team of developers was required. This gave more problems, because each developer had their own understanding of what was required.

With such problems, the failure or success of the project became a matter of chance.

The set of tools and techniques referred to as 'structured analysis' were developed to address the problem of identifying and stating the user requirement in an unambiguous, understandable form [You76]. These techniques were mainly 'process-oriented' (see §1.1.2.6.1), concentrating on required system functions.

1.1.2.4.2 Data flow diagrams
The main need was for modelling tools that would help to partition complex requirements into smaller units with a well-defined inter-relationship. As the subject matter knowledge generally resided with the users, and not computer professionals, these modelling tools had to be 'user-friendly', so that they could be used jointly by system developers and users.

The modelling tools used were:

- data flow diagrams to show the partitioning of the proposed system into smaller functional areas, or 'mini-systems'.

- minispecs to specify each 'mini-system'.

As both system developers and system users understood these tools, they could be used to capture the users' requirements. The use of review sessions, where part of the requirement was presented, was a key technique in capturing user requirements. These review sessions are termed 'structured walkthroughs' [You77].

In addition to the tools, techniques to use them correctly were required. These techniques were based on the use of data flow diagrams to partition the system requirements into different functional areas [Gan77], [Ros77], [DeM78].

1.1.2.4.3 Top-down functional decomposition
The partitioning of the requirement proceeded in a 'top–down' manner, applying the principles of successive refinement — see §1.1.2.3.2. Top–down functional decomposition is still a useful analysis technique in certain circumstances (see §5.7.4), but has been supplemented by other techniques (for example, event-partitioning — see §1.1.3.3).

1.1.2.4.4 Avoiding technological bias
Structured analysis techniques established the user requirements without any discussion about the technology to be used. Capture of user requirements involved discussion of system policy only. Discussion of how the technology would impact on the system was deferred to the design

stage. Previously, requirements analysis had been poorly distinguished from fact-gathering on one hand and design on the other. The development of structured analysis gave a clear separation of these two stages of system development.

The term 'logical model' was used to describe the statement of the user requirements in a form that was independent of the technology [DeM78]. (Because of ambiguity about the use of the term 'logical', this term was later replaced by 'essential' [McM84].) This concept of an implementation-free statement of requirements is an integral part of YSM — see §1.2.1.4.1.

1.1.2.4.5 Information modelling
At the same time as structured analysis techniques were being developed, a semantic approach to information modelling was formulated [Che76]. This was based on real-world entities and relationships. However, the approach of most data modellers in the 1970s was still technology-oriented and based on relational or network data structures. The semantic approach to information modelling was incorporated into second and third generation methods.

1.1.2.5 Problems with top–down functional decomposition
Top–down functional decomposition (TDFD) was carried out by trying to 'guess' what the main functional areas were. Each functional area was then examined in turn and successively broken down until the functions were simple enough to verify with the user. This led to several problems that lead to TDFD being supplemented or replaced in second generation methods:

1.1.2.5.1 Analysis-paralysis
Correct partitioning of the system could not be carried out until the system was understood by the analyst and the system could not be understood until it had been partitioned! This was termed 'analysis-paralysis'.

Because of this problem, an optimal statement and organisation of the user requirement was very time-consuming, requiring considerable reworking of models. Due to time pressure, analysts often had to compromise on the quality of the work they performed.

1.1.2.5.2 Analyst bias
Top–down functional decomposition is a creative technique and reflects the way the analyst perceives the system. Even though the intended technology is suppressed, the logical model tends to reflect the unconscious bias of the analyst towards a specific solution. There was a tendency for the analyst to 'design a system' using perfect technology. This was presented to the user, who was asked to verify that this was what they required. The inherent bias in the model was not easily removed.

1.1.2.5.3 Fragmentation of policy
Although there were guidelines that helped to identify the *best* organisation of the user requirements, it was sometimes found that functions which were widely separated in the model needed to be cross-checked. A specific problem related to identifying what the system did when an event occurred. There might be many parts of the system that needed to carry out some function at this time. To verify that together they gave the correct system policy, the analyst had to 'flip' from one part of the specification to another.

A good analyst had sufficient experience and common-sense to avoid the problem by a suitable organisation of the model. Unfortunately, this sense was not always very common! Techniques were needed to help organise the model in a way that avoided this problem.

1.1.2.6 Differing system development philosophies

1.1.2.6.1 Process and data-driven methods
During the first generation of structured methods, analysts and designers tended to be either:

- Process-oriented: concentrating on system functions and regarding the data as only being present to support system functions.

- Data-oriented: concentrating on information requirements, particularly in terms of identifying the data to be stored in databases. Systems functions were considered to be less important.

System development methods based on these two points of view are sometimes referred to as 'process-driven' and 'data-driven' methods respectively.

1.1.2.6.2 Phase-oriented and model-oriented development
Another categorisation of system development methods was:

- Phase-oriented: here, the system development life cycle was seen as a sequence of phases, each consisting of well-defined activities that had to be carried out. This was sometimes used as a 'cookbook' approach, in which a sequence of steps was slavishly followed.

- Model-oriented: here, the system development life cycle was seen as a sequence of models of the system requirements. Each model had a defined structure that could be checked. Heuristics and guidelines were available to help build the models, but they did not have to be followed in a mechanical way — the system developer could choose a sequence of building the model that suited the system characteristics and which parts of the model were easiest to build.

1.1.2.7 Acceptance of structured techniques
In spite of the evidence that control of large projects was very difficult [Bro75] and costly [Boe76], [Boe81] industry was slow to take up structured techniques and they did not come into general use until the 1980s. (There are still some companies who are not entirely converted to the use of structured techniques.)

1.1.3 Second generation methods
Second generation methods can be characterised by a more mature approach to system modelling. In first generation methods, developers were process-oriented, or data-oriented, and used modelling in a fairly informal way. In all second generation methods, the main stress is on the construction and checking of models.

1.1.3.1 Evolutionary model sequence
The first generation of structured techniques provided largely independent approaches for dealing with each of the stages of the system development life cycle. Second generation

methods provide a smoother development path from requirements analysis to the later design and implementation stages. This is achieved by use of a sequence of models, each addressing different stages in the system life cycle [War85].

The first model is used to capture system requirements in policy terms. Later models elaborate on how this model is mapped onto the available technology. The models are not independent, but each model 'evolves' into the next, taking into account another layer of technology. For this reason, these methods are described as being based on 'an evolutionary sequence of models'.

1.1.3.2 Viewpoint integration
In first generation methods, system developers tended to model the system from one viewpoint, with relatively poor modelling from other viewpoints (see §1.1.2.6.1). Second generation methods regard system functions and data as two equally important aspects of the same system.

This is now recognised by most authorities — Yourdon Inc. uses three viewpoints — functions, information and time [YOU84].

1.1.3.3 Externally oriented analysis
To avoid analyst bias (see §1.1.2.5.2), analysis techniques are based on real-world requirements of two types — to respond to real-world events and to store information about real-world entities.

These techniques provide ways of collecting system requirements that are independent of pre-conceived ideas of the analyst. They provide objective, rather than subjective techniques for identifying requirements. Because they help avoid analyst bias, these techniques are referred to as neutral techniques.

1.1.3.3.1 Real-world events
Time is more explicitly modelled in terms of events that occur in the environment of the system [McM84], [War85]. Systems behave as they do *only* because of these events that occur and required policy for responding to these events.

Breaking down the system requirements according to the parts needed to respond to a specific event is termed event-partitioning. These techniques are described in more detail in §5.7.3.

1.1.3.3.2 Real-world entities
By collecting information about real-world entities and the relationships between them, the analyst avoids bias towards any possible data storage organisation. This is an integration of the theoretical work of Chen into practical analysis methods [Che76]. These techniques are described in more detail in §5.

1.1.3.4 Computer Aided Software Engineering
The use of computer software to help build systems was not common until the 1980s, although such tools became available from about 1975 [Rei75]. During the latter half of the 1980s, the market availability and demand for these tools greatly increased. Such products are now generally called Computer Aided Software Engineering (CASE) tools.

These tools reflect the currently used graphic modelling tools and the philosophy behind them — for some CASE products this is a very simplistic 'draw what you want' approach; others have a more well-defined grammar that the diagrams can be checked against.

1.1.3.4.1 Second generation CASE

Second generation techniques are mainly 'diagram-oriented'. In other words, the unit of modelling is a diagram. Diagrams are drawn and checked. Components of the diagrams are specified. Diagrams are checked against each other.

This way of thinking of models carries over into CASE tools designed to support these techniques. Typically, these tools allow the diagrams to be drawn and then cross-checked. This is a 'hang over' from when the methods were only supported with pencil and paper. The CASE tool allows diagrams and specifications to be stored and checked against each other. In effect, they provide an 'electronic pencil and paper' approach. An example is shown below, with four model components:

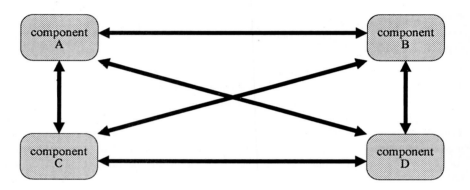

Such tools may offer an improvement over manual support, but they do not fully utilise the modelling capabilities of the computer. Because of the 'point to point' nature of checks, maintenance and cross-checking of large numbers of different specifications may become very complex. Model components appear in many specifications and this redundancy is difficult to control. The cross-checking rules also become very complex and unwieldy as the number of diagrams and specifications increase.

1.1.4 Reasons for third generation methods

The second generation view of models is rather low-level. It deals with individual diagrams, rather than the larger issues of how individual analysis and design units fit together and interact. For CASE to achieve its full potential, a change in the way models are thought of is required. Third generation methods will be distinguished by a philosophy that is more concerned with the whole, rather than the parts. They will continue to use graphic techniques, but these diagrams will be more a way of capturing and verifying the model, rather than the model itself. This is the fundamental shift that constitutes third generation methods. It allows third generation methods to break away from the 'pencil and paper' paradigm.

Cost-effective use of third generation methods will require software tools that can edit and store the models, carry out simulations, strategic studies and produce delivered systems with data and code generators. Manual involvement will be in terms of such activities as 'model input', 'phase management', 'check initiation', and integrity control.

The user and analyst will not 'see' all the technical details of the underlying models, but will interact with the system in terms of defining what they need.

1.1.4.1 YSM as a third-generation method

YSM is a third generation method. For the remainder of this reference manual, any reference to YSM will usually be assumed, rather than explicitly stated. This manual will describe how to use YSM, and not how to use any general or alternative approach.

1.2 Overview of YSM

YSM should be regarded as a consolidation, rather than a replacement of earlier methods used and taught by Yourdon consultants. Many of the techniques described in the preceding section still have their place within YSM. They will be described at the appropriate place in this manual (or in later versions).

The main change that this version of YSM introduces is a more rigorous theory underlying the definition of the models. This will allow better quality assurance and support by CASE tools. YSM continues to be a flexible approach to system modelling (see §1.4.6).

1.2.1 Model-oriented nature of YSM

During the development of a system, many decisions have to be made about what the user needs and how these requirements would be best met. If these decisions are not modelled carefully, requirements are missed or misunderstood. Sometimes, systems are built with functions that are not actually required. Any of these failures are potentially costly and damaging.

Models allow the system builder to understand these decisions and the effect of choosing particular options. These effects may be in terms of cost, quality, or performance.

1.2.1.1 Use of the term 'Model' in YSM

The models used by YSM are the foundation of the method. In a general sense, a model is 'a simplified representation that helps simplify calculations or predictions'. The model abstracts the main features of what is under examination and presents them in a form that is more usable than the real thing. On the basis of results from the model, decisions can be taken. This is shown below:

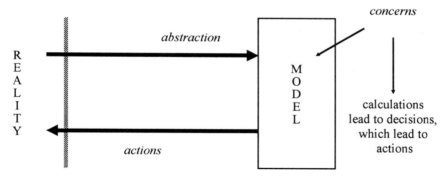

Models have been used to support many human activities — architecture and engineering are two obvious examples. YSM reserves the term 'model' for quite specific structures that are complete and checkable (see § 1.3.1).

1.2.1.2 The enterprise, systems and projects

An enterprise is any economic unit that is resourced and managed as a unit. An enterprise could be the whole company, a division of the company, or a specific managed account.

A system is a collection of information and operations, organised to meet a specific purpose. It may include computers, programs, files, manual operations, and associated support facilities. Although the enterprise has inputs and outputs that it processes according to prescribed policy, it differs from a system in the following ways:

1. The enterprise has a longer life than any one system. In a typical enterprise, a system may be developed, operated, maintained, enhanced and eventually retired (and possibly replaced by a new system).

2. The enterprise is usually made up of many systems grouped together to achieve a strategic goal of the business. The enterprise also contains an infrastructure that allows these systems to work together. This infrastructure contains information, functions and resources that are used in many systems. Examples of resources include provision of office space, personnel to carry out development work, cabling. More specific computer examples include program libraries, and software packages.

A project is an activity that is mandated to achieve a specific goal. It has a limited duration with a start time to initiate activities and a target time to conclude all activities and produce its deliverables. Projects may provide an improvement in enterprise resources or build a given system.

No project should be viewed in isolation, but, rather, as part of the larger picture of how it impacts on the operation of a company or enterprise. Strategic planning is required to ensure that each project contributes to the long term goal of the company.

1.2.1.3 Charter of YSM models
The distinction between the enterprise and individual systems is recognised within YSM by having system *and* enterprise models.

1.2.1.3.1 Enterprise models
YSM provides an enterprise model for each shared resources such as information. This allows sharing of resources by several systems to be checked. These models also support strategic planning. Enterprise models are discussed in more detail in §1.3.2.

1.2.1.3.2 System models
Each system has a set of models. Because there are different issues that need to be addressed at different times in the life cycle of a system, YSM uses several system models, each chosen to highlight one set of issues. The effect of decisions about those issues is modelled explicitly and verified before consideration of more detailed technical issues. System models are discussed in more detail in §1.3.3.

1.2.1.3.3 When the system is the enterprise
When the system scope is the whole enterprise, the system is 'stand–alone'. In this case, YSM allows enterprise models to be combined with the corresponding system models. The same model acts as both a system model *and* an enterprise model.

1.2.1.4 Models to show policy and implementation

Models either show policy or technology. Those that show policy are described as being 'essential' models; those that show technology are described as being 'implementation' models.

There is a clear separation of concerns between the two types of model — essential models show policy and no technology; implementation models show the effect of technology on the policy.

1.2.1.4.1 Models to show policy

A model of the essential type does not show or assume that any specific technology will be used. It should not be possible to infer that a particular technology is to be used from the way the model is organised.

Both the Enterprise Essential Model and the System Essential Model are essential. The System Essential Model shows system policy; the Enterprise Essential Model shows enterprise information, operations and events.

1.2.1.4.2 Implementation models

Implementation models are used to show how the policy is allocated to technology. System implementation models are used to show allocation of system requirements to processor, run-time and source language technologies. Enterprise implementation models show how the technology used by several systems 'fits' together.

1.2.1.4.3 Law of deferment

One of the main reasons for building essential models before implementation models was well stated by Tom DeMarco [DeM78]. He summed up this principle in the law of deferment:

Defer anything that you can get away with.

This is a specific case of Weinberg's lump law [Wei75]:

If we want to learn anything, we musn't try to learn everything.

The law of deferment indicates that the most important decisions should be made first, then less important decisions, and so on. Finally, the most detailed parts of the problem are addressed. Models are used to capture policy requirements before using models to visualise how this policy is mapped onto the technology. The mapping onto the technology proceeds in layers, with the most important design decisions being taken first.

1.2.2 Model structure

Each model has a well-defined structure to allow checking for completeness and consistency. Systems are complex and even a model of a system may be difficult to understand. Correctness is established by providing 'user-friendly' views of the model, so that subject matter experts are able to verify the applicability to the system's environment.

1.2.2.1 The view concept

A view is a diagram, table or other specification used to highlight an area of concern. Each view shows part of the model and is used to discuss one set of issues. By choosing the correct set of overlapping views, it is possible to completely cover all of the system requirements, without any one single view being beyond the comprehension of a user.

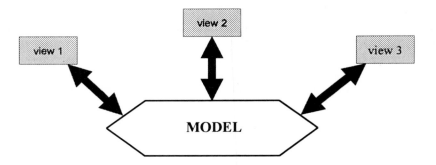

Examples of views include: a context diagram (to highlight information flows into and out of the system), an event list (to highlight real-world events that the system must respond to), and a minispec (to highlight a single function performed by the system).

This description seems very much like that given for a second generation model, where a model was defined as 'a collection of views'.

The difference is that:

- for second generation methods, the model *is equal to* the collection of views.

- for third generation methods, the views are *derived from* the model. The model is now 'internal'.

To the user of the method, this distinction may seem a fine point, but it has important consequences in terms of the power of the method.

There can be many views into the internal model. Each shows part of the total model in a user-friendly format. This is shown (for four views) below:

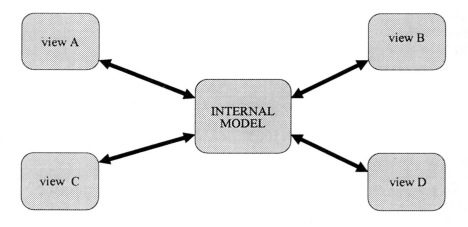

1.2.2.2 Internal model

Each standard model (for example, the System Essential Model) has an internal structure which is a representation of all the information in that model (for example, a specific system's essential requirements). It is 'minimal', in the sense that there is no repeated information in this model.

The internal model is defined using abstract mathematical techniques. This internal model is formal (and not very 'user-friendly'), the formulation being chosen to provide rigour, completeness and precision, rather than any ease of understanding. It provides support for cross-checking, simulation, system generation, traceability etc.

The internal model is not 'visible' to system modellers, who only interact with it by means of views, as defined below.

1.2.2.3 Views into the model

The modeller interacts by means of views into the model. Each view shows part of the internal model, selected as being visible in that view. To change any part of the model, the system modeller modifies the view. Any changes made are recorded in the internal model.[1] This is shown below:

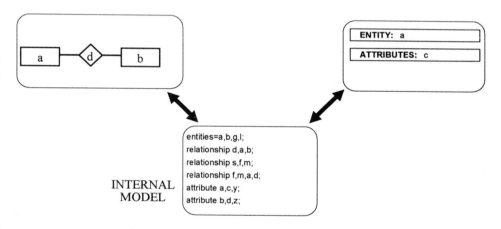

The relationship between the internal model and the components shown in the views is defined by YSM. These rules are transparent to the user of the method, but would be of concern to a CASE developer. In the above diagram, the internal model contains entities such as a, b; relationships such as d; attributes such as c. The two views shown are generated by using two standard modelling tools – the entity relationship diagram and entity specification.

---

1 There is also information about each view, such as the position of a model component on a graphic view. This is stored with the view, rather than with the internal model.

The entity relationship diagram shows an entity as a named rectangle and a relationship as a named diamond linking the entities to which it refers. With a given internal model, this allows an entity relationship diagram of selected components to be generated automatically.

Note that if a component appears in several views, then modifying it in one view may change it in other views. The views are, in a sense, 'active' windows into the model.

1.2.2.3.1 *Standard views for each model*

For each of the models, YSM provides a set of standard views which will assure the quality of the internal model when used correctly. Each such view provides a presentation of some area of concern to the user or analyst. For example, the required views that must be captured and verified for a System Essential Model consist of a statement of purpose, context diagram, event list, ….

1.2.2.4 Changes from second generation methods

Using the method in terms of views does not seem very different from before. The system modeller still builds diagrams and other specifications. However, the separation of the specification components of the model from the views that are used to capture and evaluate it has the following advantages:

- The internal model can be organised in a rigorous, mathematical way. This facilitates checking, maintenance, traceability and such features as 'running the model'.

- The external views can be chosen for their user-friendliness. Any ambiguity should be resolved by the CASE tool as it imports the information in the view into the internal model.

By carefully controlling which of these external views are used (and how they are used), the method guarantees to provide full rigour, without the internal model being visible directly. The benefits of using formal techniques (including checking, sizing, and control) are therefore achieved without losing 'user-friendliness'.

This is a more precise, yet flexible way of thinking about modelling. The remainder of this reference will assume this approach. The formal mathematical model will be referred to as 'the internal model' and the views as 'views into the model' or just 'model views.'

1.2.2.5 Consistency of views

Because many of the views can contain elements in common, it is important to ensure that there is no contradiction between them. The way this is ensured depends on the technical support environment — see §1.4.4.

1.2.2.6 YSM without the internal model

If the internal model is not supported, then YSM reduces to a second generation method. The effect of the technical support environment is discussed in §1.4.4.

It is important to realise that YSM is a third generation method and can be used with sophisticated CASE support. It may also be used without CASE support. A second generation method could only be used manually or with 'electronic pencil and paper' CASE tools.

1.2.3 Modelling tools and presentations

1.2.3.1 Types of modelling tools

There are several types of modelling tool used in YSM.

1.2.3.1.1 Graphic modelling tools

Because humans understand pictures best, the views are mainly in terms of different types of diagrams. The data flow diagram is an example of a graphic tool — used to show a group of functions and how they fit together and use stored information.

Because different issues need to be dealt with at different times, and discussed with different audiences, a range of different graphic modelling tools are used. Each graphic tool has its own area of application. When used together in the correct way, it is possible to capture different aspects (see §1.2.5.2) of the requirements and how they can be achieved.

Although graphic techniques are easier to understand than text, it is important to have guidelines to help organise them to aid communication. These guidelines will be discussed for each tool in §2.

Section 2.1.4 includes a more detailed discussion of graphic modelling principles.

1.2.3.1.2 Text support

Not all relevant facts about a system can (or should be) modelled graphically. For example, names of system functions, cost of processors and many other aspects of concern are best modelled by text or numbers. These specifications are required parts of the model. They are needed to allow checking and traceability.

These text specifications are organised by using the graphic views to provide a 'map' of model components. At one level of presentation, a high-level diagram shows the system components and their inter-connections; at a lower level of detail, the detailed characteristics of any specific components are described in terms of such text. Textual modelling tools are described in §2.1.7.

1.2.3.1.3 Use between different types of modelling tool

It is important to realise that a complete model can only be built by using both graphical and textual specification techniques. Graphic modelling tools are used to aid understanding of the system. They suppress the detail, which is then specified by the supporting text.

1.2.3.1.4 Other types of modelling tool

As an alternative to graphic modelling tools, tables may be used. These are discussed in §2.1.5.

'Frame' specifications are used when standard items need to be given values. These are discussed in §2.1.6.

1.2.3.2 Presentations

Even for one specific view, there several modelling tools might be used to present the information. For example, in a view showing system inputs and outputs, the information could be presented as a context diagram, or in a table. In either case, the logical content of the view is the same, but the modelling tool is different. These are referred to as different presentations of the same view.

Within YSM, each view has a recommended modelling tool. For example, to show system inputs and outputs, the recommended modelling tool is the context diagram. Some views have alternative modelling tools supported by YSM. For example, an entity life-cycle can be modelled graphically by means of an entity state transition diagram, or using a pair of tables, as described in §2.1.5. These are different (but equivalent) presentations of the same view.

There are not many examples where YSM offers this choice and, generally speaking, each view has only one recommended modelling tool.

1.2.3.2.1 Different dialects

For textual tools, an equivalent concept is the ability to define different syntaxes for the same underlying grammar. These are referred to as different dialects of the grammar. The appropriate dialect to use will depend on the audience and technical support environment. A one appropriate to user review is not necessarily one for which 'code generation' tools are available. YSM permits different dialects of the same operation. With suitable automated support, one dialect may even be automatically translated into another.

1.2.3.3 Choice of modelling tools

In general, it is advisable to restrict the number of modelling tools and dialects used within an organisation. However, because of factors such as different audiences, project types, and technical environments, there may be a need to use more than one type of presentation of a particular view. YSM has tried to anticipate this, wherever possible, by describing a suitable range of presentations. However, if no presentation tool is suitable for a specific view of a model, the creation of a different presentation is allowed. The method is mainly concerned with defining the required views of a particular model and what should be shown in those views. The modelling tool chosen should clearly represent the components modelled in that view.

1.2.4 Three main viewpoints

As mentioned, different points of view are addressed by the use of different graphic modelling tools, together with supporting textual specifications. It is helpful to think of there being three main 'viewpoints' from which the system can be viewed:

1. Function. Here we are concerned with "what does the system *do*?".

2. Time. Here we are concerned with "what happens *when*?".

3. Information. Here the main concern is "what *information* is used by the system?".

In a sense, each of these questions can be regarded as being independent of any question in one of the other two groups. Systems may therefore be thought of as 'three-dimensional':

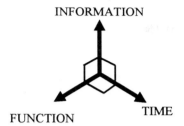

INFORMATION

FUNCTION   TIME

1.2.4.1 Relating modelling tools to the three viewpoints
The above leads to a convenient way of thinking about the modelling tools. There are modelling tools whose main accent is on each of the primary dimensions. For the three principle dimensions, the primary modelling tool is:

1. Time dimension. The event list is used to show anything that happens that the system needs to respond to.

2. Information dimension. An entity relationship diagram is used to declare entities and show the relationships between them.

3. Function dimension. A data flow diagram is used to declare system functions and their interfaces.

1.2.4.1.1 Modelling tools that link dimensions
There are also modelling tools for the 'planes' between the primary dimensions:

1. Time–information plane. An entity state transition diagram (or an equivalent tabular presentation) is used to show the effect of time on one entity. An entity–event table is used to examine the relationship between a group of entities and relationships and a set of events.

2. Time–function plane. A behavioural state transition diagram (or tabular equivalent) is used to show the effect of time on a group of system functions.

3. Function–information plane. The major modelling tool for the use of information by functions is the data flow diagram. For a higher-level view of the use of information by a group of functions, a function–entity table may be used.

Each of these are used to provide views that show model components. Each component has a more detailed specification view, which uses a different modelling tool. These supporting views often use textual modelling tools, which are sometimes referred to 'textual support tools'.

This is in accordance with the lump law (see §1.2.1.4.3) — the major tool declares the components and the interfaces between them; the support tool fills in the detail. For example, each entity on an entity relationship diagram has a corresponding entity specification; each data flow seen on a data flow diagram has a corresponding data flow specification; each event shown in an event list has a corresponding event specification.

1.2.4.2 Characteristic systems

Each system has a certain complexity in terms of its 'functionality', 'behaviour over time' and 'information usage'. This gives a 'system profile' that could be obtained by 'plotting' the complexity for each dimension and then defining a tetrahedron using these three points and the origin. Two such profiles are shown below:

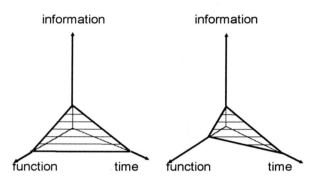

Note: It is difficult to find a metric that assigns these measures to a system with correct weights for each dimension. Although theoretically possible to do so, a reliable profile would only be available when a system model was complete. Nevertheless, this concept is useful in an informal way, to determine the type of system being built and the best strategy for building it. The adjectives 'event-driven', 'information-driven' and 'function-driven' are sometimes applied to systems whose greatest complexity is in the corresponding dimension.

1.2.4.2.1 Changes of system behaviour

Some systems have a constant policy to which they operate at all points in time. Other systems (for example, a lift control system) change behaviour over time. In the specific example chosen, 'requests' are dealt with in different ways, depending on whether the lift is moving or not.

For systems that change behaviour over time, a complete model requires one or more views showing their changes of behaviour. Systems that do not change behaviour, do not require this type of view. This will be discussed in §5.7.1.1.3.

1.2.5 Standard models

To allow all issues of importance to be checked, each YSM model has a standard structure. These structures are defined by Yourdon Inc. Different models built for different requirements by different people will all have this generic structure, although the subject matter specifics will be different. For example, each Essential Model contains a statement of purpose, context diagram, event list, ….

The way this statement is interpreted depends on whether the model is maintained:

- manually (or equivalently, using a CASE tool that keep a repository of diagrams and text specifications and then 'cross-check' these against each other)

- using a CASE tool that fully supports third generation methods

For most system developers, the distinction between these two cases will be of no concern. For method specialists and anyone developing CASE tools, the existence of the internal model will be significant — see §1.2.2.2.

1.2.5.1 Benefits of standard model structure
By using the standard YSM models, with the standard views, the system developer can succeed in building quality models that are supported by all the rigour of the underlying method.

This relies on the framework of model structure and methods of building models that compose the method. YSM guarantees that these methods work — the practitioner does not have to prove the underlying theory of these methods over and over again from first principles.

1.2.5.2 Aspects
There are many concerns that need to be addressed by the system developer. Each of these is dealt with by providing a set of views into the system. For example, when aiming to verify that all expected system outputs are captured, a set of views to highlight system inputs and outputs is chosen. On the other hand, if the current concern is identifying the information used by the system, a different collection of views would be chosen.

YSM uses the concepts of aspects to formalise this concept of concerns. An aspect is a collection of views, organised together to highlight a specific concern. Two named aspects of a model are shown in the diagram below:

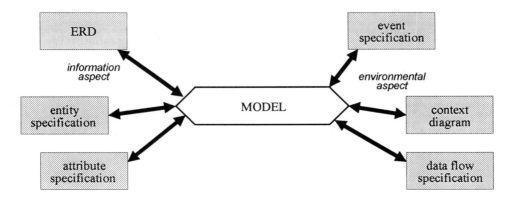

These are, in fact, specific aspects of the System Essential Model. The environmental aspect shows the interface of the system to its environment; the information aspect shows the information used by the system.

A model may have several aspects, depending on how it is developed, verified and used. There are four standard aspects to the System Essential Model.

1.2.5.3 Choice of views

Even for the same model, there might be several different ways of grouping and organising views into it. This is not a problem; in fact, different views might well be chosen for different types of system (or different technical audiences). It is the responsibility of the YSM practitioner to choose the correct views (although guidelines are contained in this reference manual); for example, if the system does not have any changes of behaviour, it is not sensible to draw behavioural state transition diagrams.

1.3 YSM Models

1.3.1 Completeness of models

As a guideline, any model should be complete enough to be realised in some conceivable architecture. If the model is not sufficiently complete to contemplate this, then further work is required. For each of the models, YSM provides rules and guidelines to ensure their completeness, consistency and correctness.

1.3.2 Enterprise models

Enterprise models are concerned with global or company issues and are used in strategic planning and co-ordination. Enterprise activities are generally 'open ended' — they do not have a specific point in time at which they are 'finished'. Enterprise activities provide an infra-structure within which more specific, goal-oriented activities may be supported. The only enterprise model covered in this release of YSM is the Enterprise Essential Model (EEM), but later releases of YSM will define enterprise implementation models.

In addition, YSM supports the concept of a 'resource library'. This is a collection of items that may be used to build systems. They may be re-used in several systems, even if each system is stand-alone. This collection of re-usable items is held in the Enterprise Resource Library.

1.3.2.1 Enterprise Essential Model

This model provides a long-term stable reference model of the way requirements for different systems fit together. This release of YSM mainly deals with the enterprise information aspect (EIA). This models the information requirements of the enterprise.

The EIA is free of any potential implementation details and is organised around real-world facts. It uses both graphic and textual specification tools to capture the total information requirements.

1.3.2.2 Enterprise Resource Library

The Enterprise Resource Library (ERL) is a repository of all re-usable items. It includes:

- conceptual or essential items such as operations, abstract data types and objects (not covered in this release of YSM).

- implementation items, such as subroutines, processor types, data base management system types, objects (not covered in this release of YSM).

The ERL is not a model — it does not satisfy the criterion of running in a specific architecture. It is more a collection of things that can be used to construct models.

1.3.3 System models

System models directly address the issues relevant to a specific project. The project is usually not 'open ended' — it has a specific initiation point and a target date for delivery.

1.3.3.1 Evolutionary sequence of models

The system life cycle involves a sequence of decisions. Each set of decisions is captured in the form of a model. The first model is the Essential Model, which captures the user policy requirements without any technical implementation details. The later models each elaborate on how the organisation is affected by successive layers of implementation technology. These technology layers include processors and software and language architecture for automated processors.

The models may be regarded as an 'evolutionary' sequence of models — one is converted into the next by a process of elaboration. In practice, there is some iteration back to preceding steps and 'look-ahead' to future models, particularly in critical system aspects. As a first approximation, however, the diagram below gives a good overview:

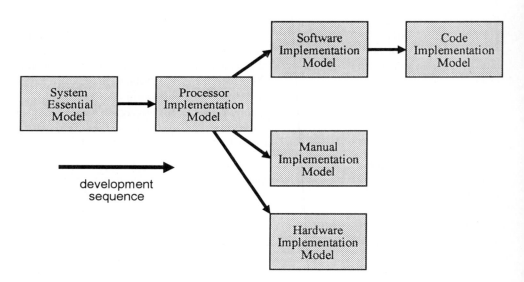

1.3.3.2 System Essential Model

The System Essential Model captures the user requirements in a form that is not dependent on any specific technology that may be provided to meet those requirements. The structure of the model is well-defined and correct use of the suggested rules and guidelines should ensure that this model is a faithful representation of what the users actually require. To deal with different types of systems and the existence (or not) of existing systems, YSM provides techniques for identifying those requirements in the most cost-effective way.

In third generation methods, the main accent is on the development of models of systems requirements. These models can be checked for completeness and internal consistency but, more importantly, they can be verified against user requirements.

This model deliberately suppresses the effect of technology on the requirements. It may be considered to run in a 'perfect conceptual processor', with zero cost and instruction time, and infinite memory and mean time between failures.

1.3.3.3 System implementation models

There are several system implementation models, each dealing with a different layer of technology:

1.3.3.3.1 System Processor Implementation Model

This model specifically addresses the problem of how system requirements are mapped onto processors that will support that system. Processors may be automated (both digital and analog devices), or they may be human. The main concern addressed by this model is the ability of available processors to meet the system requirements taking into account the capabilities of processors and the essential requirements that have been allocated to them.

Building the System Processor Implementation Model may be described as mapping the System Essential Model onto the hardware platforms to be used. When the System Processor Implementation Model is built, YSM requires that processor capabilities and capacities are carefully examined to ensure that the model meets performance requirements. Communications between processors must be provided by available inter-processor communications protocols.

This model may be imagined to be running in one or more processors, each with finite performance and an associated cost.

1.3.3.3.2 System Software Implementation Model

This model shows how the System Processor Implementation Model is mapped onto the available run-time software architectures. For each digital processor, the effect of choices of operating system, DBMS, off-the-shelf packages, etc., are explicitly modelled. The conceptual processors in this model correspond to the run-time software architectures of each of the digital processors.

Specific detail in the model shows the use of the enterprise software resources..

Note that this model shows processors that are not digital computers. However, the 'internals' of such processors is not explicitly shown in this model (see §1.3.3.3.4 and §1.3.3.3.5).

1.3.3.3.3 System Code Implementation Model

For each software unit identified in the System Software Implementation Model, the System Code Implementation Model shows how the source code is organised to build that unit. The detailed design of each delivered software unit will therefore take account of the mechanisms available in the source programming language.

1.3.3.3.4 Hardware Implementation Model

This model elaborates and provides more detailed design for any analog processors that were identified in the Processor Implementation Model. Not all systems will include such analog processors.

There is no prescribed format for the Hardware Implementation Model, although it is likely that the standard YSM modelling tools will be useful for the higher levels of hardware design. At the lower levels, circuit diagrams and standard electronic design techniques will be used. In effect, these would replace the minispecs that would be used in a digital processor.

The components of this model must be traceable back to the Processor Implementation Model. The Hardware Implementation Model should be in a form that allows it to be checked against the inputs and outputs of the corresponding analog processors in the Software Implementation Model.

1.3.3.3.5 *Manual Implementation Model*

This model elaborates and documents the essential functions that are people. Each person or organisation acts as a processor. There is no prescribed format for the Manual Implementation Model, but its components must be traceable back to the Processor Implementation Model. Note that the inputs and outputs to this model act as an interface to the Software and Code Implementation Models.

Although there is no prescribed format for this model, it may correspond to the delivered user manual for the system.

1.3.3.4 Strong typing of implementation models

Each component of an implementation model must correspond to an identifiable unit or feature in the technology architecture. For example, on a processor diagram, each processor is associated to a specific physical processor that is owned by the enterprise. Each enterprise processor is of a given processor type. This processor type has given characteristics, which may be checked as being suitable for the functions that are allocated to the specific processor. This association of the implementation model component with available resources is termed 'strong typing' of implementation models and will be discussed in later versions of this manual.

1.3.3.5 Traceability

There are two types of traceability supported by YSM:

It is important that the requirements, once captured, are not lost in the process of building the system. Traceability of a delivered system back to the user requirements, as originally captured, is built into YSM by:

1. Within essential specifications, there are ways of tracing back the specification to user documents, interviews, etc. This allows the capturing of user requirements to be controlled. Quality assurance procedures are used to ensure that all relevant sources have been examined for the requirements.

2. In implementation models, it is important that the policy requirements captured in essential models are correctly mapped onto the technology. This is achieved by the use of 'traceability tables', showing how components of one model is mapped onto the chosen technology of the next.

Traceability is also an important aspect of version control and will be discussed in more detail in later versions of this manual.

1.4 Using YSM

1.4.1 Strategic planning

One of the ongoing enterprise activities is strategic planning. This examines current and future practices and tries to evaluate cost-effective strategies for achieving the long-term goal or mission of the enterprise. In doing so, many different types of project may be initiated. Such projects may include:

- procurement of new equipment, sites, or even other companies;
- improvement of company infrastructure, information accessibility, etc.;
- provision of new (or replacement of existing) systems that are provided for a specific reason.

This version of YSM mainly deals with the third type of project, which uses system models to represent the system and enterprise models to show the way the project relates to other enterprise resources.

Strategic improvements to infrastructure will be dealt with in later versions of this reference.

1.4.2 Enterprise support for system projects

A project to build a system may be said to be supported by ongoing enterprise activities. It receives support from the information, databases and other resources in different ways throughout its life cycle, as shown below:

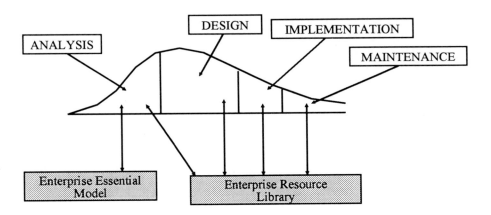

This enterprise support is provided for multiple system projects, each of which is likely to be in a different stage of its life cycle:

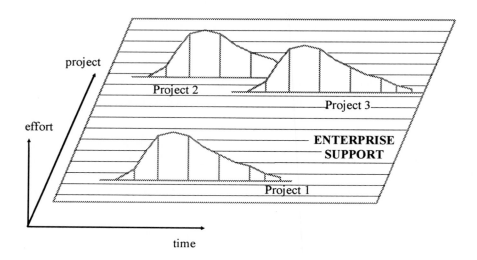

1.4.3 Identifying enterprise support for systems

Enterprise activities support systems, as described above. When projects to build systems are initiated or more detailed system modelling is carried out, requirements for a further new resource (such as hardware, or information) is often identified. These project requirements are 'fed back into' the enterprise support activity. The enterprise resource management activity then plans and provides extensions to ensure these are available to meet project time scales. This inter-relationship between systems and the enterprise is shown below:

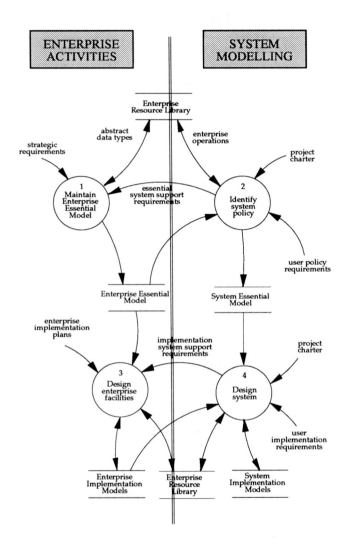

1.4.4 Technical support environment

As far as possible, the tools and techniques described in this reference are independent of the technical support environment. Where there are likely to be differences for different technical environments, this is specifically mentioned at the appropriate point.

For example, if a model is maintained manually, it is important to reduce the redundancy between different views, because of the possible inconsistencies that may arise. For automated tools there is, in principle, much less of a problem about information appearing on several views.

1.4.4.1 'Pencil and paper' environments

The term 'pencil and paper' will be used where diagrams and supporting text are maintained manually. Software tools that allow diagrams to be drawn and text specifications to be entered, edited, and cross-checked should also be regarded as 'pencil and paper' tools in this respect.

1.4.4.2 'Automated' environments

The term 'automated' will be restricted to an environment in which each of the diagrams, tables and text specifications are regarded as views into an underlying model. Changing the information in a view automatically changes the corresponding component in the internal model. This change would also be automatically reflected in other views. In some tools, multiple active views into the internal model would be actively displayed on a screen.

See §1.2.2.2 for a discussion of this internal model.

1.4.4.3 View consistency

Because the information in views may overlap, it is important to ensure that overlapping views are consistent. For automated support environments, this consistency depends on the link between the view and the internal reference model. For pencil and paper environments, consistency requires specific cross-checking rules between different views.

In either case, these rules should be enforced, either by software or manual cross-checking. It is important that models must be built to a certain quality, not just built.

1.4.4.4 Model change control

Control of the way models are stored and changed over time is outside the scope of this manual. However, it is important that suitable version and change control is in place. This could be manual change control and review procedures, or built into the technical support environment for automated support tools.Concerns such as who last modified a component of a model, who reviewed it, what the changes were, when they were carried out etc., are all important aspects of successful resource and system management.

YSM is independent of this layer of system development and control, although it cannot be used successfully without such protocols, procedures and/or support tools being properly used.

1.4.4.5 Using YSM without software support

The most cost-effective system development strategy incorporates the use of software tools to carry out model maintenance, checking, simulation, production, etc. However, YSM has been developed so that it can be used in environments that do not offer this level of automated support. Indeed, YSM can be used in a 'pencil and paper' environment. The method is intuitive so that it is possible to capture the systems requirements — these always come from human sources. It does not rely on a specific technical tool for support. In fact, the 'pencil and paper' method, although workable, involves a great deal of manual effort. Cross-checking models by hand is difficult, time-consuming and costly, except for small systems. Use of automated support tools greatly increases ease of application of YSM and improves productivity.

1.4.5 Appropriate rigour

While YSM is capable of being applied in a very rigorous manner, the appropriate level of rigour depends on the system being built.

For some systems, such as aircraft control systems, the costs associated with proving that the models are correct and complete are easily justified, even if this increases the cost of the models.

For other types of system, the same level of rigour might not be appropriate. It would be too costly to justify and there might be problems in communicating with end-users. In these cases, a more relaxed, informal approach may be used. Formal methods do not have to be used to benefit from using YSM.

YSM can be used in a variety of situations. It has sufficient formal mathematical foundations to be used in a very formal rigorous way. It also has user-friendly views to allow it to be used in a more digestible form. With an automatic support environment, this would not be an exclusive choice — there can be rigorous modelling *and* user-friendly views.

1.4.6 Flexibility of method

One of the attractions of using YSM is its flexibility. Yourdon modelling techniques can be applied to many types of problem that cannot be handled within the strict formats and procedures of other methods.

YSM provides this flexibility in several ways:

- Names of models could be changed without compromising the method. It is unlikely that many companies would want to do this, but it is allowed.

- Names of views and components of views can be changed. Again, this is not likely to be a common requirement.

- New views may be defined into models. This is an important feature of the method, although the reference manual describes sufficient views to build and verify a model. With automated support, creating additional views could be very useful; without automated support, it is less likely to be cost-effective.

- Notation of diagrams and syntax of text specifications may be changed, demonstrating the flexibility that comes with the presentation concept.

- The sequence of building model views may be varied. Although some guidelines and heuristics for choosing a sequence are provided in the reference manual, this order may be varied, depending on circumstances. There is no suggestion that there 'is only one true way'.

1.4.6.1 Extensibility of method

The above gives an idea of the type of flexibility that should be allowed. It should also be noted that the current version of the reference manual deals with certain types of system only. Other types of system (for example, those including material flows) will be covered by later versions of the manual. Modelling techniques may still be applied in those areas; they are still 'Yourdon techniques'; they are just not covered in this version of the reference manual. In these cases, however, there is no guarantee that there is a 'standard' Yourdon way of modelling that situation. Providing that is accepted, pragmatic extension of the techniques is allowed.

The above may be summed up by the motto:

> *Adapt, don't adopt.*

1.4.6.2 Standardisation and local style guidelines

Although the method is flexible, this does not mean that each individual within an organisation should change the method at whim. Changes to model, view or component names should only be carried out at an organisation (enterprise) level. The same applies to modelling tool notation — although it is possible to choose a different set of graphic icons for the data flow diagram (for example), correct communication requires some standardisation.

Choices of style for minispecs might depend on the target audience and other factors, so guidelines for minispec style probably vary from project to project.

It may also be helpful to develop a 'house style', both for diagrams and textual specifications. For example, Yourdon has adopted certain house style guidelines for diagrams used in courses. These relate to such things as layout of diagrams, capitalisation of component names.

This manual was also produced with style guidelines of this type, so that diagrams have a standard 'look and feel'. However, it should not be taken to mean that this is part of the method — it is much more a presentation issue. In particular, the manual has probably erred on a more pedantic, formal textual style than would be appropriate in most applications. This is deliberate. The intention is that the more dry, rigorous style can always be relaxed in the field. The alternative, with a set of informal textual specifications, would probably lead to problems where they needed to be 'tightened up'.

2: Modelling Tools

Table of Contents

2.1 Introduction

2.1.1 Intended audience

2.1.2 Organisation

There is no obvious way in which the tools should be organised — probably the most logical organisation would be in alphabetical order, but that would still mean the table of contents would be required to find which tools were covered (and under what name).

The rough progression is:

- tools for modelling information — §2.2 onwards;
- tools for modelling functions — §2.11 onwards;
- tools for modelling time and dynamics — §2.22 onwards.

This organisation is neither strict, nor optimal. For example, eSTDs and entity–event tables are grouped under information modelling tools.

2.1.3 Types of modelling tools

Four main types of modelling tool are used in YSM — graphic, tabular, frame and textual. These are discussed below, with some general comments on each type.

2.1.4 Graphic tools

YSM uses graphic modelling tools to show high-level components of a particular aspect of a model. Graphic tools are the preferred type of modelling tool when the connection between model components is important.

Each graphic modelling tool has a set of icons that may be used, each icon being used to represent a specific model component. The icons may be connected in ways which have a well-defined meaning for that specific diagram type. For example, an entity relationship diagram is used to declare entities and relationships between them. (See §2.2.2.)

2.1.4.1 Semantic nature of graphic tools used in essential modelling

Graphic tools used in essential modelling are mainly of a semantic nature, highlighting the meaning of the requirements. The effect of the notation chosen on the way of thinking about the system requirements is subtle, yet important. The communication grammar, in other words, the graphic conventions, predetermines the way of thinking about the problem. This was summed up in a saying of Marshall McLuhan [McL67]:

> *The medium is the message*

In the specific example of the entity relationship diagram, the choice of notation presupposes a way of thinking that is in terms of entities, relationships, subtypes, etc.

Because of the semantic nature of these modelling tools, it is very important to carry out extensive peer reviews using walkthroughs etc. For essential models, this will require the involvement of subject-matter experts, as well as modelling specialists.

2.1.4.2 Structural nature of graphic tools for implementation modelling

For implementation models, there are graphic modelling tools to show the actual implementation units (this will be covered by later releases of YSM). These tools are more concerned with the *actual* units chosen and how they are connected.

2.1.4.3 Distinction between essential and implementation modelling tools

The essential modelling tools are mainly concerned with meaning, or semantics; the implementation modelling tools are more concerned with structure, or syntax.

2.1.4.4 Universal guidelines for graphic tools

For graphic views, there are certain guidelines that are always relevant, irrespective of the tool used. Mostly, these are self-evident, for example:

> *All text used on a hard copy diagram must be legible to a person with normal eyesight.*

> *Diagrams should not be over commented. Too many comments are a sign of a poor model. It should be revised, to avoid the need for comments. Sometimes these comments can be moved to the supporting textual specifications.*

2.1.5 Tabular tools

Some information is usefully laid out in a tabular form. For example, the relationship between entities and events may be visualised in an entity–event table (see §2.4.2).

Some graphic modelling tools have an alternative tabular presentation. This may be useful in some circumstances, particularly if the subject-matter expert finds the graphic modelling tool difficult to relate to. For example, a state machine is usually modelled using a bSTD, but may also be modelled using state transition and action tables (see §2.24.4.

Generally speaking, all YSM tables are all in 'spread sheet' format, with a single entry, or list of entries in each cell.

2.1.6 Frame specifications

The term 'frame' is used informally for a certain type of specification tool. Frame specifications are used to specify all relevant information about a model component that has been declared on a diagram or another frame. Each such frame has a standard layout, with 'cells' into which values can be entered. For example, see §2.5.2.2.

Sometimes each cell has a text grammar entry. This is discussed in §2.1.7.

Some frame specifications have variants. For example, an attribute has slightly different entries depending on whether its values are listed or defined in terms of an abstract data type §2.8.2.

2.1.6.1 Move away from complex data dictionary notation

YSM has avoided extending the DeMarco Data Dictionary notation [DeM78]. This was for several reasons:

1. The data dictionary notation was already becoming rather complex. It would have become even more complex if it had been extended to cover new areas of concern.

2. Data dictionary notation is textual and therefore not so easy to understand as other modelling tools. With the correct support, graphic and frame modelling tools are easier to understand.

3. In some cases, the data dictionary notation was ambiguous and open to personal interpretation. In these areas, the notation was replaced. Examples include a more clearly defined replacement to 'refs' (see §2.2.3.4) and explicit listing of multiple identifiers (see §2.5.2.3.4).

There is also a long term need to get away from the idea that the modeller 'writes the specification and then gets the software tool to check the syntax'. The reason that models are built is to capture the requirements and identify how these requirements will be met. Eventually, this will be much more interactive. The method therefore provides a more flexible way of providing specifications to model components declared in the graphical views. There will inevitably be a period of transition before such software tools are available, but the concept of frame specifications should provide a 'gentle' upwards progression.

Data dictionary notation has been retained for the composition of data flows (see §2.18). Even in this case, the method would allow an alternative presentation of the structure (see §2.18.3.4.6). This might be graphical, or some other technique, given the appropriate technical support environment.

2.1.6.2 Technical support for frame specifications

The use of frame specifications in different technical environments will vary. In automated environments that fully support the YSM view concept, the tool is likely to look rather like the examples given.

In 'pencil and paper' environments, or software environments that cross-check diagrams, pre-prepared, standard forms can be used, but this is not very popular with practitioners because of the large amounts of forms required. Alternatively, the use of comments in a more general specification format may be used.[1]

2.1.6.2.1 Stacking frame specifications

In either automated or pencil and paper environments, frames may be 'stacked' to reduce the amount of cross-referencing. The specifications given here are those derived from a minimal meta-model that was built on the principles described in this manual. Much of the required description that is entered as specifications becomes attributes of associative entities in this meta-model — these correspond to the 'frame' type of format. Although this is conceptually simple, particularly to automate, it may not be the most user-friendly way of organising the specifications. For example, the specification of each attribute (see §2.8.2) could be included in the specification of the entity (see §2.5.2.2).

For example, part of the specifications for the entity "Course" and one of its attributes are shown below in the form shown in this manual:

---

1 This is the approach to be adopted with the current version of the ADT™.

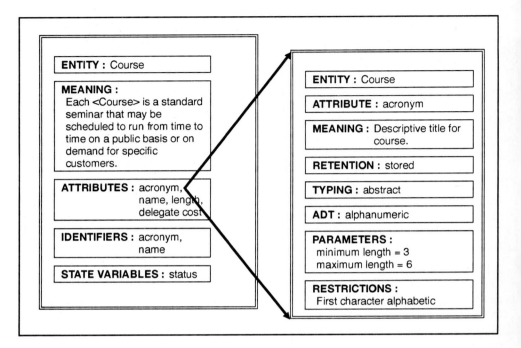

In the most obvious automated support, these might both be visible on the screen at the same time. In a more pencil and paper way of working, these could be combined into the single specification:

ENTITY : Course					
MEANING : Each <Course> is a standard seminar that may be scheduled to run from time to time on a public basis or on demand for specific customers.					
ATTRIBUTES					
name	**retention**	**meaning**	**typing**	**parameters**	**restrictions**
acronym	stored	a short description used within the organisation for identification purposes	adt: alphanumeric	minimum length = 3 maximum length = 6	first character alphabetic
name	stored	Descriptive title for course	adt: alphabetic	minimum length = maximum length =	
length	stored	number of days duration	adt: cardinal		
delegate cost	stored	charge for one delegate	adt: currency	unit = 0.01	

This latter approach be more convenient in true pencil and paper environments, as it reduces the total number of forms required.

2.1.6.3 Lists in frame specifications

Within this manual, the following convention is used:

1. When a list of items is entered in a single 'cell' of a frame specification, individual items are separated by commas. For example:

   > **ATTRIBUTES** : acronym, name, length, delegate cost

2. If a value includes a comma, the whole value is included between single quote characters. (This is a conventional way of showing 'literals'.) For example:

   > **VALUES** : 'Sorry, no such course', 'Sorry, course full'

   Note: these are two discrete values, not two strings.

3. If a value includes a quote character, it is shown with this quote repeated (this will be very rare). For example, in defining legal values for a punctuation mark ADT, one might have (enlarged for clarity):

   > **VALUES** : ',', ., :, ?, !, " , " , " , " , —

   In fact, this is a contrived example and this set of values would be better defined by:

   > **VALUES** : comma, full stop, colon, question mark, exclamation mark, open double quote, close double quote, open quote, close quote, dash

   This is equivalent and much more natural.

2.1.6.4 Null entries in cells

Within this manual, the convention of showing a '—' (dash) within a cell that has nor (or a null) value is used. This is to distinguish it from a cell that has not been 'filled in yet' (this would be blank). The '—' acts as a positive assertion that 'there is no value', rather than no value is known yet.

2.1.7 Textual tools

In order to completely specify enterprise and system behaviour, YSM does have to resort to textual grammars in some situations. Where this is required, the following principles have been followed:

1. The grammar was originally defined formally, using meta-language, or set theory. This grammar will usually only be released to CASE developers. See §6.1.6 for an example where set theory is used to define an abstract data type restriction.

2. It was then rewritten in ordinary language. This might be termed the relaxed form of the grammar. These versions of the grammar are given as appendices in this manual. See §6.5 for a slightly more user-friendly grammar.

3. Simple examples were given in the main body of the reference. Hopefully, YSM practitioners will be able to infer correct syntax from this and not need to use an appendix. See §2.8.3.10 for an example.

2.1.8 General view principles

2.1.8.1 View identification in this manual
This reference manual uses a simple 'low-tech' approach to the identification of views.

For example, each entity relationship diagram (ERD) has a name that identifies it. This name may either be shown on the diagram or stored as a 'key' to access the diagram. The manual does not assume that the name is 'hard-coded' on the diagram. The way the name is associated with the diagram is very dependent on the technical support environments.

Views also require identification of whether they relate to the enterprise as a whole or to one specific system. This is not dealt with in the manual. However, where a view relates to a system, the manual defines a 'meta-attribute, which is the name of the system that the view is for.

2.1.8.2 Version control
Most models will require version control. Often, there may be several versions of the same view. This can be dealt with in many ways, but as mentioned in §§1.4.4.4, YSM is independent of versioning. The manual does not address version-numbering of views.

All this is ignored here, but it is assumed that an automated tool would provide a standard way of achieving. For automated use, appropriate filing and diagram identification protocols must be in place.

2.1.8.3 Updating models in automated support environments
Dealing with information that appears in several views is very dependent on the technical support environment. For example, an entity appears in:

- Enterprise ERD: this shows the entity and any relationships or subtypings that it appears in;

- Enterprise entity specification: this defines such properties as the attributes, identifiers and state variables. If an entity is a subtype 'inherits' attributes from a supertype. (This is discussed in more detail in §2.5.3.6.)

These cannot be regarded as independent views — they are alternative views of the same entity. Changing the name of the entity (for example), is something that should be done as a single action. If all of the above views needed to be changed, then a 'pencil and paper' approach is being adopted.

In an environment that supports this multiple visibility in different views, there must be some control over which views 'own' the information and which merely display it. If the view owns the model component being shown, then that item may be changed by a person examining the view. Certain other criteria also have to be met, so the criteria for updating an item visible in a view are:

1. The model item must be owned by the view,

2. The user must have the correct access privileges,

3. The user must have opened a view showing the component for update (rather than just 'display'),

4. The attempted update must not compromise the integrity of the model of which the item is a component.

If all of the above are satisfied, then the model is updated. The change is propogated to all views in which that component is visible.

2.1.8.3.1 Pencil and paper methods
It is not feasible to support the above concepts using pencil and paper methods, except in a very informal way.

2.1.8.3.2 Interdependence of models
There are analogous ways in which one model 'imports' definitions and specifications from another. For example, the System Essential Model includes:

- System ERD: this defines which of the enterprise entities and relationships are used by the system;

- System entity specifications: these 'inherits' many properties from the corresponding enterprise entity specification. In fact, the specification merely defines which enterprise properties of the entity are visible;

- Minispecs: these define system functions in terms of connecting together pre-defined functional primitives, called operations. These operations may be re-used across different systems and are held as part of the Enterprise Resource Library.

Conceptually, each model may be regarded as complete and self-sufficient. However, where a model references a component owned by another model, an update can only be to the model owning the component. For example, it is not allowed for a system modeller to rename an enterprise entity.

2.1.8.4 Annotation of diagrams
The description of diagrams given in this volume is in terms of the minimum required for each view. It has been the practice to 'annotate' diagrams in certain ways, to highlight particular concerns. The method allows such annotation to be added as comments. However, because this information is generally recorded elsewhere, the use of such annotation should be carefully controlled.

In automated support environments, 'hard-coding' comments on the diagram should be avoided — this would still be a 'pencil and paper' approach. A better approach is to allow additional required information about a diagram to be presented in some other way. For example, additional information about a diagram could be 'popped-up', when required. This approach depends on suitable software support that implements the ideas of multiple views into the model. §2.6.2.3.6 discusses annotating entity relationship diagrams to show cardinality.

2.1.8.5 Cross-references

The modelling tools described in §2 are used to capture requirements and are part of well-defined models — see §3 and §5. It is important to provide references back to the original sources of information, including user interviews, text documents, government standards, and hardware specifications.

As these sources of information are not part of the model, the references back to them are 'outside the model'. However, it is sensible to have a standard way of dealing with these references, so that proposed changes to any model may be checked back against the original requirements in order to avoid compromising already agreed features.

In pencil and paper environments, references may be added as extra, unstructured text. In automated environments, each such reference would have a record referring to the model component, the external requirement, when it was identified, and who agreed to it.

2.2 Entity Relationship Diagram

2.2.1 Purpose

The entity relationship diagram (ERD) is a modelling tool is used to model particular roles of importance to the enterprise and the relationships between them. It is a semantic modelling tool and helps to clarify concepts.

The ERD is used to identify and organise information. It is used to organise all information used by the enterprise, not just stored information. ERDs may also be used as a tool for discovery of rules and events.

2.2.1.1 Real-world orientation of the entity relationship diagram

Each view shows a collection of real-world facts that are significant to the enterprise. These are aspects of reality that are not negotiable — they are not contingent on the modeller's point of view or interpretation. There is no assumed implementation bias in possible ways that this information might be stored or represented by a possible system.

2.2.1.2 Enterprise and system use of entity relationship diagrams

The Enterprise Essential Model uses ERDs to define the entities it uses and the relationships between them (see "The Enterprise Essential Model"). An entity relationship diagram used in this way is referred to as an enterprise entity relationship diagram, or enterprise ERD.

The System Essential Model uses ERDs to show the entities and relationships that the system has responsibility for collecting or using information about (see "The System Essential Model"). An ERD used in this way is referred to as a system entity relationship diagram or system ERD (see §2.2.4).

The prefix enterprise or system may be dropped if no ambiguity results.

In cases where no distinction between the enterprise and system is made, there is only one set of information views given by ERDs.

2.2.1.3 Supporting text specifications

For both the Enterprise Essential Model and the System Essential Model, the components shown on an ERD have corresponding specifications. These are covered by other sections in this chapter.

2.2.2 Example

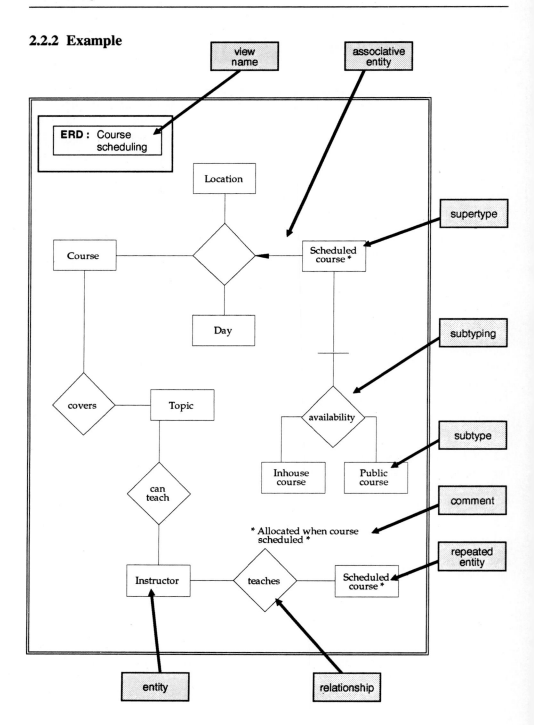

2.2.3 Components

2.2.3.1 Associative entity

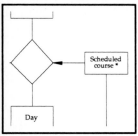

An associative entity acts as both a relationship and an entity. As a relationship it indicates that there is a group of real-world associations between entities (see §2.2.3.4, which discusses relationships). As with all relationships, an occurrence of the associative entity cannot exist without prior existence (or simultaneous creation) of occurrences of other entities.

A relationship should be replaced by an associative entity if the relationship has attributes or the relationship acts as an entity in other relationships. It is then both an relationship *and* an entity.

2.2.3.1.1 The associative entity as a relationship with attributes
Attributes of an associative entity do not describe the entities that participate in the relationship, but the occurrence of the association between them. For example, given the relationship:

the date on which any specific couple were married might be of importance. This is neither an attribute of "Man", nor of "Woman", but describes when the occurrence of the relationship was set up. It is modelled by replacing the relationship "is married to" by the associative entity "Marriage", with "date married" as an attribute.

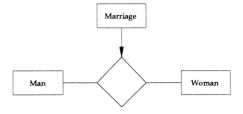

The associative entity retains its property of being a relationship — each occurrence of "Marriage" records the fact that a specific "Man" was married to a specific "Woman" on a specific date.

2.2.3.1.2 The associative entity as an entity in other relationships

an associative entity may participate in relationships with other entities. The associative entity acts as a relationship in the way it 'remembers' the original association and also an entity involved as a participant in the 'later' relationships. For example, the associative entity "Scheduled course" acts as a relationship between the "Course", "Location" and "Day" entities and also as an entity in the relationship "reserves place on":

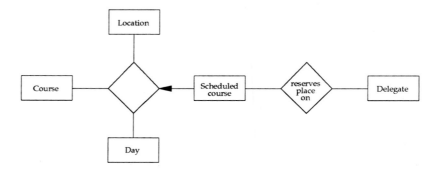

An occurrence of the associative entity must be created before it can participate in any other relationship itself. It does not make sense to think of having a "Delegate" reserve a place on a "Scheduled course" before that specific "Scheduled course" has been created by defining an association between "Course", "Location" and "Day".

The fact that an associative entity was created as a relationship between other entities may be of little significance to some later activities that relate to it. These may set up relationships in which the associative entity acts like an entity. For example, when discussing the reservation of places on a course, the entity that is of importance is "Scheduled course". Whether or not it is an associative entity is irrelevant in that context and it is shown as an entity.

2.2.3.1.3 Showing the dual aspect of associative entity on enterprise ERDs

Associative entities are modelled as an entity and a relationship with an arrow between them, as shown in the diagram. This is a single, indivisible icon. The relationship part is shown with lines to the entities to which it refers and the entity part is labelled with the name of the associative entity.

An associative entity appears in one view as a specific association between the entities to which it refers in its role as a relationship. It may appear on many other views as an entity. However, it still remains an associative entity and has an associative entity specification. For pencil and paper techniques it is only appropriate to show the associative entity as a relationship on one view because of the redundancy involved in repeating this on several views. If automated support is used, the associative entity may be shown on more than one view as a relationship.

2.2.3.1.4 Converting an entity into an associative entity

Sometimes an associative entity is initially identified and modelled as an entity. Only later may its behaviour as a relationship become evident later. The entity is then 'converted' into an associative entity in the model and the entity specification amended to an associative entity specification. This conversion, however, does not compromise the original insight, as captured and agreed in the first information view. Indeed, the associative entity is still modelled as an entity in that view. However, in the Enterprise Essential Model taken as a whole, it is as an associative entity.

2.2.3.1.5 Dependency of associative entities on other entities

Associative entities cannot exist independently of the other entities that they link as a relationship. For example, an occurrence of "Scheduled course" requires the prior existence of the occurrences of "Course", "Day" and "Location" that it links.

The associative object is sometimes described as being supported by the other entities. This dependency on the other entities is a good clue to an information model component being an associative entity.

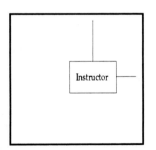

2.2.3.2 Comment

As with all diagrams, free-format comments may be added to the ERD to elucidate or stress any important points. See §2.1.8.4 for general discussion relating to comments.

Comments should be used sparingly — a diagram that is cluttered with too many comments is usually an indication that the view has been poorly chosen or that the model components are ill-defined (see the guidelines given in §2.2.6).

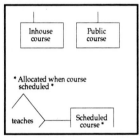

2.2.3.3 Entity

An entity is a class of real-world things whose role of interaction with the enterprise is well-defined. These things may be tangible, physical objects, or abstract concepts (see §2.2.3.3.4). The enterprise will use information about and interact with occurrences of the entity. Each entity has a unique name which should reflect the role that is played by that type of object.

Individual occurrences of the role will be identified by the enterprise and information stored about them. This information may be that the entity participates in relationships or that it has attributes (see §2.5 for a discussion of attributes of entities). Each occurrence of the entity must be distinct from, but fulfil the same role as other occurrences of the entity.

If there is an entity named "Instructor" in an information model, it should be read as:

> *There is a category of real-world people, with certain properties in common. The generic name used for any one of these is an Instructor. Several occurrences of Instructor exist, each uniquely identifiable. Certain facts about each Instructor (these correspond to*

values of attributes and the participation in relationships) are of importance to the enterprise.

It is an occurrence of the role, not the real-world thing that is referred to here — that same person could also correspond to an occurrence of another entity (for example, "Delegate" or "Salesperson").

2.2.3.3.1 Distinction between entity and occurrence of entity

The entity represents the whole set of such possible role occurrences.[1] For example, the entity "Course" might have many occurrences. Three such occurrences might be "Introduction to Russian", Navigation for master's certificate" and "Spherical astronomy".

2.2.3.3.2 References to entities in specifications

Because entities are an integral part of YSM, specifications such as minispecs, relationship specifications and rules of association refer to them in a standard way that can be checked for consistency.

Within model specifications, a specific occurrence of the entity is usually of concern. YSM uses a standard notation to represent this — "<Instructor>" is used to represent "an occurrence of Instructor" (for example). At any point where this is used, there must be an unambiguous way in which this specific occurrence can be identified. Taken out of context, this notation may seem rather strange, but in context, it is quite natural (see §2.2.3.4.2 for example). This notation allows the semantics of how the entity is used to be checked more formally by prescribing a specific syntax for a specific use of the entity.

2.2.3.3.3 References to entities in this manual

Where the entity as a whole needs to be referred to, the name of the entity is used. This is unusual in information models, but this reference manual often refers to them. To make it clear that a model component is being referred to, the manual includes the entity name in quotes, for example "Instructor".

2.2.3.3.4 Types of entities

Some entities will represent collections of tangible physical things that play particular roles in their interaction with the enterprise. These are generally fairly easy to identify and present little difficulty. For example, in an enterprise running courses for outside customers, the entities "Instructor" and "Delegate" are important.

Other entities have a more abstract nature. Examples include time period entities such as "Day" (see §3.3.4.1) and abstract concepts such as "hobby". Individual occurrences of "hobby" might be: "stamp collecting", "swimming", "knitting" ....

YSM does not have a categorisation into different types of entities such as 'characteristic entities' (see, for example, [Flav81]). Entities are allowed to represent both tangible and more abstract things.

---

1 Some authorities use the term entity for the single occurrence and entity-set for the whole collection of occurrences. Although this might be more theoretically defensible, YSM chose to go with the majority of information modellers. See [Kor86] for a treatment using the alternative terminology.

2.2.3.3.5 *An entity as a fact pattern*

If an entity "Instructor" is built into a model, it should be regarded as making the statement that there is a class of real-world things each of which is an occurrence of "Instructor". Each such occurrence could be referenced (for example, by pointing to it) and the statement: "This is an Instructor" made. This statement must be verifiable — there must be a way of determining whether it is true or false.

Although it is not possible to 'point to' abstract entities such as "hobby", it must still be possible to reference individual occurrences of them and declare them to be occurrences of that entity.

If a potential entity does not satisfy this criterion, it is not an entity.

2.2.3.4 Relationship

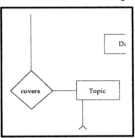

A relationship represents a possible association that may occur between occurrences of entities. Each occurrence of the relationship corresponds to specific occurrences of those entities being linked by that association. On the diagram, each relationship is shown linked by lines to the entities to which it refers.

The relationship may be regarded as a fact-pattern to which references to specific entities can be added to obtain specific facts about the real world. For example:

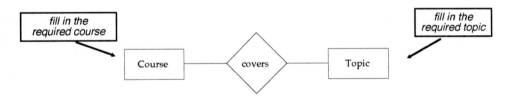

acts as a template into which occurrences of "Course" and "Topic" may entered to obtain the specific facts:

Introduction to Russian	covers	the Cyrillic alphabet
Navigation for master's certificate	covers	angular measurement
Spherical Astronomy	covers	angular measurement

Each occurrence of the relationship corresponds to an association of exactly one occurrence of each of the entities that participate in the relationship. This may be shown in an 'instance diagram':

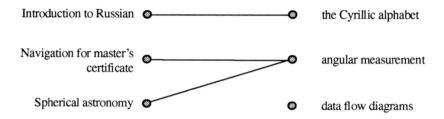

Introduction to Russian — the Cyrillic alphabet

Navigation for master's certificate — angular measurement

Spherical astronomy — angular measurement

data flow diagrams

2.2.3.4.1 Distinction between relationship and occurrence of relationship

YSM uses the term "occurrence of the relationship" to represent a single instance of the association between entities. The abstraction, or the whole collection of such instances is referred to as the "relationship". Care should be taken when reading other publications, which sometimes use the terms "relationship" for the single instance and "relationship-set" for the collection of instances. See §2.2.3.3.1 for references.

2.2.3.4.2 Relationship frame

The concept of a relationship frame formalises the above informal treatment. Each relationship has a relationship frame that includes the entities involved in the relationship with other text forming a complete sentence. For example, the above relationship has the relationship frame:

```
<Course> covers <Topic>
```

Each occurrence of the relationship corresponds to a specific occurrence of a "Course" and a specific occurrence of "Topic". <Course> should be read as "occurrence of Course" (see §2.2.3.3.2). To be well-defined, a relationship must have a frame that provides a verifiable statement for each possible substitution of entities in it. Each occurrence of the statement is either true or false. In an instance diagram, these two cases correspond to the existence or non-existence of a line between occurrences of the entities.

The relationship frame also removes ambiguity about which way the relationship "name" is read on the diagram. Any relationship on the ERD must have a corresponding frame (see §2.6.2.3.2 for how this relationship frame is specified). Note: there can be several relationships between the same entities. For example, "<Instructor> wrote <Course>" and "<Instructor> can teach <Course>".

2.2.3.4.3 Relationships as a 'trace' of an event or ongoing association

Each occurrence of a relationship shows that an event occurred and that event involved the specific occurrences of the entities linked by that occurrence of the relationship. Some relationships only remember that an event occurred (e.g. "<Person> married <Person>") — these are just a record of the event, or a 'trace' relationship. Other relationships reflect more of an ongoing association between the entities (e.g. "<Man> *is* married to <Woman>") — these are more of a continuing association.

YSM does not make any specific distinction between these two types of relationships, but there are certain heuristics that relate to them. For example, trace relationships are much more likely to have attributes and therefore be associative entities. Ongoing relationships sometimes do not have attributes.

2.2.3.4.4 Binary relationships
If a relationship has occurrences that each refer to two entity occurrences, the relationship is called a binary relationship. Binary relationships are shown on an ERD with two lines linked to the entities they refer to. Relationship frames for binary relationships always have two "<...>" entries.

2.2.3.4.5 Higher-order relationships
A relationship that involves more than two entity occurrences is referred to as a higher-order relationship. A third-order relationship appears on an ERD with three 'lines' attached to the 'diamond'. In the relationship frame for such a relationship, there must be three "<...>" entries (entity 'slots'). For example:

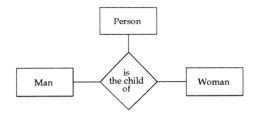

has a relationship frame of:

> <Person> is the child of <Woman> and <Man>

Note: YSM does not impose a limit on the order of relationships, although relationships of order greater than four are unusual.

2.2.3.4.6 Criterion for relationship to be well-defined
A relationship is defined by its relationship frame. This defines the entities that the relationship refers to. The number of entity slots is fixed — this is the order of the relationship. The entries in the entity slots are from given entities — the relationship cannot sometimes refer to one entity and sometimes to another. (Except that the relationship may refer to a supertype, which may be regarded as any of its subtypes, depending on the occurrence of the relationship.)

In addition, it must be clear what one occurrence of the relationship means. This corresponds to one occurrence of the relationship frame.

2.2.3.4.7 Recursive relationships
Sometimes the entities are not distinct, for example:

When the entity is repeated in the relationship, the relationship is referred to as a being 'recursive'.

In order to indicate which is the manager and who is reporting to whom, the relationship frame must distinguish between the two "Employee" occurrences. This could be done in several ways, but a suitable notation is:

    <Employee> reports to <Employee>(manager)

This allows each occurrence of the relationship to be for a pair of employees, one in the "manager" role and one in the "reporting" role. The same occurrence of the entity is allowed to play both roles. In this case, this is unlikely and would be specifically specified as a rule of association (see §2.6.2.3.7). The entity roles may be referred to as being 'qualified'.

2.2.3.4.8 Symmetric relationships

For most recursive relationships, the distinction between the multiple roles that the entity may play in the relationship must be clearly distinguished for the relationship to be well-defined.

For some recursive relationships, the position in the relationship frame is of no significance. These relationships are referred to as symmetric relationships. The relationship "is friend of" is an example of a symmetric relationship

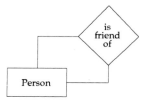

For symmetric relationships, the role for the entities does not have to be distinguished. For this example, the relationship frame for "is friend of" would be:

    <Person> is friend of <Person>

References to two different "Person"s can be filled in — it is not implied that "each person is friends only with themselves"![1]

Higher-order recursive relationships
As with binary relationships, entities are not required to be distinct. For example, another, equally valid way of modelling this third-order relationship is:

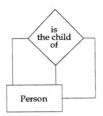

In this case, the relationship frame would be:

<Person> is the child of <Person>(mother) and <Person>(father)

The <...> entries are qualified to remove ambiguity about which occurrence of the "Person" is playing which role in the relationship — see §2.2.3.4.7.

2.2.3.5 Repeated Entities
An entity may be repeated on the same diagram. This may help to reduce complexity, particularly by avoiding crossing relationship references and allowing relationships to be placed 'near to' the entities that they refer to.

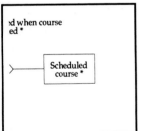

When an entity is repeated on a diagram, then it is helpful to indicate this in some way. In pencil and paper environments, it is conventional to 'tag' each repeat of the entity with a 'tagged' with a "*" symbol. In an automated support environments, any other suitable visual clue (e.g. colour, highlighting) may be used.

Note: if an entity appears on several ERDs, there is no special notation to indicate this. Automated tools may provide a view that highlights all relationships that refer to that entity, but this is not a necessary part of the method.

2.2.3.5.1 *Repeating associative entities*
If an associative entity is repeated on a diagram, only one instance of it shows the links to the entities to which it refers. The other instances are shown using the entity icon (possibly 'tagged' with a "*" symbol). This is dependent, to some extent, on the technical support environment. The main reason for this convention is a wish to avoid specifying the associative entity twice.

---

2 Where a recursive relationship *always* links an occurrence of the entity to itself, the relationship would be described as a reflexive relationship. Reflexive relationships are unusual in information systems, so the term is not generally used in information modelling.

2.2.3.6 Subtype

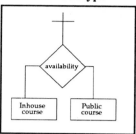

A subtype of an entity is a well-defined group of occurrences of the entity that may be regarded as an entity in its own right. As an example, consider the entity "Mammal". Individual occurrences of this entity correspond to Archimedes, John McEnroe, Tom (a well-known cat), Jerry (a mouse), etc.

One well-defined group is the entity "Human" containing many individual occurrences (two of which were explicitly mentioned above). The entity "Human" is said to be a subtype of the entity "Mammal". Other subtypes of "Mammal" are "Cat", "Mouse". These subtypes are shown below, as they would appear on an ERD:

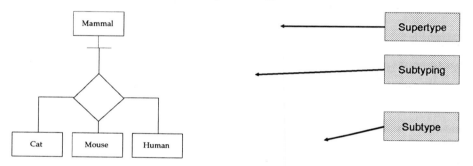

This subtyping may also be shown as a Venn diagram[1], which shows each entity as a planar area with occurrences of the entity 'inside' this area. (As is conventional in set theory, the sets are given plural names.):

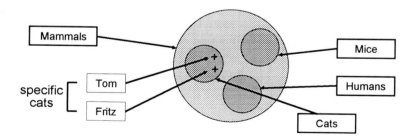

An occurrence of the subtype is always an occurrence of the supertype (for example, an occurrence of "Cat" is automatically an occurrence of a "Mammal"). As a consequence, any property that relates to the supertype is automatically a property of the subtype. The subtype may be said to "inherit" the properties of the supertype (see §2.2.3.8 for a further discussion of

---

1 For further discussion of Venn diagrams, see, for example [Sto61]. YSM does not use the Venn notation in modelling — in this section it is merely used to demonstrate a concept.

these common properties). In addition to properties "inherited" from the supertype, the subtype may have either or both (but never neither):

1. **Specific attributes:** some of the subtypes may have attributes that are not relevant to other subtypes. These attributes are declared in the specification for that subtype. This is contrasted with common attributes, which are declared for the supertype.

2. **Specific relationships:** some relationships may only refer to certain of the subtypes of an entity. For example, a delegate can only "reserve a place" on a "Public course" — "Inhouse course"s are not open for bookings of this type. Where the relationship refers to the subtype, it is shown linked to that subtype on the diagram, as below.

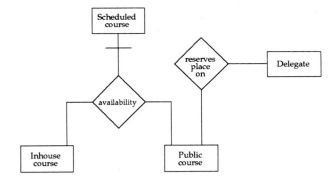

In the specification of the relationship "reserves place on", the reference is to the entity "Public course", rather than the supertype entity.

2.2.3.6.1 Multiple Supertypes
An entity may be a subtype of subtypings of several different entities. In this case it inherits any relationships and common attributes of **each** of those supertypes. For the example:

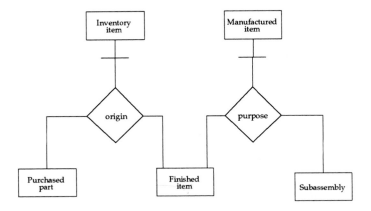

a "Finished item" is a type of "Manufactured item" and also a type of "Inventory item". This may be shown as a Venn diagram:

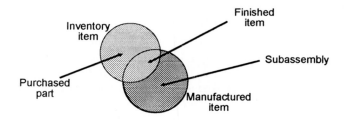

As a consequence, any occurrence of a "Finished item" is both a "Manufactured item" and also an "Inventory item" and thus has attributes inherited from both. This is discussed in more detail in §2.5.3.2.

2.2.3.7 Subtyping

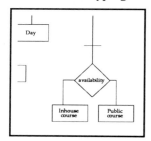

A subtyping indicates that the enterprise regards the entity as being made up of a number of distinct identifiable groups, each of which is referred to as a subtype. It is important to stress the semantic, rather than structural purpose of subtyping. It allows the modeller to think of general, high-level concepts (relationships, attributes) as being relevant to the supertype ; consideration of one subtype may involve more detail without being overwhelmed with the detail that is relevant to other subtypes.

As a simple example, consider the subtyping of the entity "Mammal" into "Cat", "Mouse", ... entities as discussed in §2.2.3.6. This subtyping makes a distinction between different "species" that each individual occurrence of a "Mammal" belongs to, so the subtyping could be given the name "species".[1] This example has three subtypes. In general, there may be any number of subtypes in a subtyping (see §2.9.3.5 for an example with a single subtype).

2.2.3.7.1 Subtyping is into mutually exclusive subtypes
Within a subtyping, there is no occurrence of the supertype that is simultaneously an occurrence of more than one subtype of that subtyping.

There are occasions when it may be useful to think of a specific occurrence of an entity as falling into more than one subtype. This can be dealt with by declaring more than one subtyping. This multiple subtyping is described in §2.2.3.7.4.

---

1 Note: as should be clear from this example, a subtyping is a very special kind of relationship. It is a "is same real-world thing" relationship defined on the supertype and subtype entities. This is the historical reason why the subtyping icon was chosen to be similar to the relationship icon. It should regarded as no more than of historical interest — regarding the subtyping as a relationship is not helpful and would only serve to confuse.

2.2.3.7.2 *Coverage of a subtyping*

A subtyping is said to be complete if any occurrence of the supertype is an occurrence of exactly one of its subtypes. As mentioned above, an occurrence of the entity can never be an occurrence of *more* than one of the subtypes.

Some subtypings have occurrences of the supertype that do not fall into any of the occurrences of the subtypes. These are described as partial subtypings. The subtyping of "mammal" into "cat", "mouse" and "human" is a partial subtyping. There are clearly occurrences of "mammal" that are not occurrences of any one of these subtypes.

2.2.3.7.3 *Repeated subtyping*

Where the number of subtypes in a particular subtyping is very large, it is permissible to repeat the subtyping on several views, with some of the subtypes shown on each views. A complete list of subtypes is given in the corresponding subtyping specification (subtyping specifications are covered in §2.9).

2.2.3.7.4 *Multiple subtypings*

There may be several ways of subtyping an particular entity, so each subtyping is given a name that is unique to the supertype entity and that subtyping.

For example, the entity "Person" might be subtyped by the subtyping "gender" into "Male" and "Female" and by the subtyping "seniority" into "Child" and "Adult":

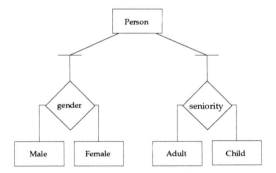

The above diagram should be read:

> Each "Person" may be either a "Male" or a "Female".
> The same person may be either an "Adult" or a "Child".

Both of these subtypings are complete — see §2.2.3.7.2 for a definition of the term "complete". As a consequence, the statement:

> Each "Person" *is* either a "Male" or a "Female".
> The same person *is* either an "Adult" or a "Child".

is also true.

2.2.3.7.5 Unnamed subtypings

If there is more than one subtyping of an entity, the subtyping name serves to distinguish them. A subtyping of an entity may be unnamed, particularly if the name does not aid the understanding of why the different subtypes are distinguished. For example, the following would be allowed:

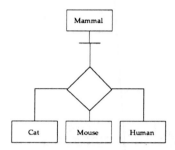

Any other subtyping of "Mammal" must be named to distinguish it from this subtyping. For any entity, only one subtyping of it is allowed to be un-named.

A subtyping is identified by the supertype name, together with the name of the subtyping. Many unnamed subtypings are allowed, provided they are of different supertypes.

2.2.3.7.6 Subtyping associative entities

If an associative entity is defined to have a subtyping, then each occurrence of one of the subtypes must be an occurrence of the supertype. Each occurrence of the subtype is thus also an associative entity, with references to the entities that the supertype references. On the ERD, these subtypes are shown as entity icons and the supertype as an associative entity. For example, the associative entity "Scheduled course" may be subtyped into "Inhouse course" and "Public course" on the basis of availability:

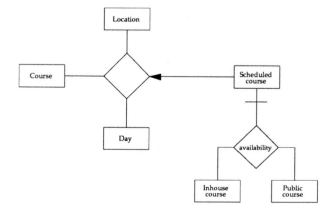

Supporting specifications are:

1. An associative entity specification for "Scheduled course". This would include references to the entities referred to in its role as a relationship.

2. Subtyping specification for "availability" is specified as a subtyping.

3. An associative entity specification for "Public course".

4. An associative entity specification for "Inhouse course".

2.2.3.8 Supertype

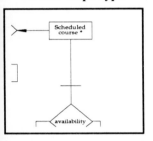

The supertype should be regarded as a general grouping of several entities. This general grouping may regarded as an entity in its own right. The entities that are grouped together into this more general entity are referred to as its subtypes.

When an entity is shown to contain one or more subtypes by means of a subtyping, there will be some general properties that are common to all the subtypes. In addition, there may be properties that are specific to each subtype (see §2.2.3.6).

The common properties are of two types — the supertype may:

1. **Participate in relationships**. If a relationship may involve *any* of the subtypes of an entity, it is described as "involving the supertype" and the relationship icon will be linked to the supertype entity. The relationship specification is in terms of the supertype entity. Each occurrence of such a relationship will involve an occurrence of one of the subtypes.[1]

2. **Have common attributes**. Common attributes are those attributes that are always found for all subtypes of an entity. In a sense, these attributes 'live' in the supertype icon on the diagram — they are defined in the specification of the supertype. Additionally there may be attributes that are specific to each subtype — these 'live' in the subtype icons — and are defined in the specification of the subtype.

2.2.3.8.1 Supertypes of associative entities

In some circumstances several associative entities may be quite similar and then it is useful to create a new entity of which these are subtypes. For example, both "Booking" and "Contract" are associative entities in their own right. On the other hand, there are many ways in which they are similar (bills need to be raised for them, etc.). A supertype entity "Service" may be created to deal with common relationships and attributes.

1. "Service" has an entity specification, which includes common attributes (such as "date provided").

2. "service type" has a subtyping specification.

3. "Booking" has an associative entity specification that refers to "Delegate", "Public course" and "Customer".

---

1 For a partial subtyping, one of the subtypes would have to be regarded as the 'anonymous' complement of the named subtypes.

4. "Contract" has an associative entity specification that describes the references to the entities "Inhouse course" and "Customer".:

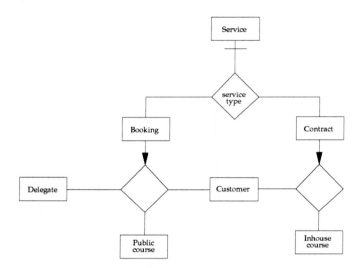

A supertype may be created for associative entities of different orders, as here.

2.2.3.8.2 *Identifiers of supertypes*
When a supertype is created for two (or more) existing entities, a new identifier usually needs to be created for the supertype. This is always true if the subtypes are entities or associative entities involving different entities. If the subtypes are associative entities that refer to the same entities, then a supertype entity could always be defined as the references to the entities ('reverse-inherited' from the subtypes), together with a 'tag' indicating which subtype the occurrence is of. The tag is necessary, because an occurrence of one subtype is not the same as an occurrence of another subtype, even if they refer to the same entity occurrences.

2.2.3.9 View name

Each view has a unique name which identifies it in an information model. The name should reflect the functional area shown by the view.

For example, "Course scheduling" shows all entities and relationships that are concerned with planning and scheduling new seminars including allocating staff to run them.

2.2.4 System entity relationship diagram

System ERDs show the part of the enterprise ERD (or ERDs) that are 'visible' to a system. They may be regarded as a partial copy of the enterprise ERD. However, information cannot be modified on the system ERD without changing the corresponding parts of the enterprise ERD. The system ERD is a read-only copy.

The relationship between system and enterprise ERDs defined by YSM is upwards-compatible with carrying out information modelling only on a system basis. In other words, the system models would not be affected if the Enterprise Essential Model were removed — the information aspect of the System Essential Model would become the sum total of the information used. The way this is achieved is given below — only the way in which enterprise subtypings are dealt with in terms of their visibility by the system is at all subtle.

2.2.4.1 System associative entities

If the system uses an enterprise associative entity as both a relationship and an entity, then it should be shown in a system ERD as an associative entity.

The associative entity is an entity and a relationship. However, if the system does not use it as a relationship, then, as far as the system is concerned, it is an 'ordinary' entity. The original references to the other entities are not visible to this system. It is therefore shown as an entity icon on all system ERDs for this system.

If the system uses an enterprise associative entity only in its role as a relationship, it should be shown on a system ERD as a relationship.

2.2.4.2 System entities

Any entity appearing on a system ERD is an entity that is accessed by the system. The access will be one or more of:

- creating, deleting, or determining whether specific occurrences of the entity exist;
- checking or changing the state of a state variable;
- reading or updating the values of its attributes;
- creating or deleting occurrences of a relationship that involves the entity;
- checking whether specific occurrences of a relationship that involves the entity exist.

2.2.4.3 System relationships

A system might be responsible for creating occurrences of a relationship; it might need to check whether occurrences of the relationship have been created elsewhere. In either situation, it should be shown on one system ERD as a relationship.

If the system uses an enterprise associative entity only in its role as a relationship, it should be shown on a system ERD as a relationship.

2.2.4.4 System use of subtypes

The system ERD only shows that part of the enterprise subtyping that is significant to the system.

2.2.4.4.1 *Subtype visibility*

If the system ever treats subtype entities in different ways, then the subtypes used by the entities, and *only* these subtypes, should appear on a system ERD. If the system always treats the subtypes of an entity in the same way, *none* of the subtypes appear.

2.2.4.4.2 *Supertype visibility*

If the system ever treats occurrences of the supertype in the same way, irrespective of which subtype they are occurrences of, then the supertype should be shown. If the system always treats occurrences of the supertype in a way that is specific to one of the subtypes, then the supertype does not appear.

2.2.4.4.3 *System view of subtyping*

If the supertype entity and one or more subtype entities appear on any system ERD, then at least one system ERD must show the subtyping of the supertype into the subtypes as a graphic icon.

2.2.4.5 **System view of supertypes**

If a system treats all occurrences of the supertype as equivalent in some respects, then the supertype should be shown on a system ERD. There may be relationships that the supertype participates in that are significant to the system, or attributes used by the system that are common to all of its subtypes.

If the system has no functions that it performs on all subtypes in the same way, but always regards them as separate entities, then the supertype should not appear on any system ERD.

2.2.4.6 **System views**

This gives the name of the view that is shown in the diagram. The system may use more than one such view. Each view has an identifying name that is unique within that system.

2.2.5 **Rules**

2.2.5.1 **View identification and completeness:**

1. Each view must be named.

2. Each view must contain at least one entity. This is a rather trivial rule. Unfortunately, it cannot be strengthened to "each view must contain at least one relationship", because there are (rare) examples of systems that do not deal with any relationships.

3. If a view contains a relationship then it must contain all entities that the relationship refers to. This is because:

   • the relationship cannot be understood, except as an association between entities;

   • the relationship cannot be created (or deleted or checked to exist) without checking (and sometimes deleting) for occurrences of the entities it refers to.

   The relationship cannot vary in its characteristics, as seen in different views. If it refers to specific entities, then it always refers to those entities. This is one of the consequences of demanding that the relationship be well-defined.

4. If a view shows a subtyping, then the supertype and at least one subtype must be shown.

5. A relationship must have at least two references to entities (not necessarily distinct).

6. An associative entity must have at least two references to other entities (not necessarily distinct).

2.2.5.2 Component identification:

1. Each entity and associative entity must be named.

2. Each relationship that is not an associative entity must be named. Associative entities are only named as entities. Note: associative entities also have a relationship frame (see §2.7.2.3.2).

2.2.5.3 View composition

1. If an associative entity is shown on an ERD, then all entities that it refers to must also be shown on that diagram with corresponding lines showing the references.

2. In pencil and paper support environments, if an associative entity is repeated on a diagram, then only one of the copies shows the references to the entities it associates.

3. An entity is not allowed to be a subtype of itself. In the case of repeated subtypings, this check must be applied 'all the way down'. An entity "X" is a subtype is an entity "A" if either of the following hold:

   - there is a subtyping of "A" showing "X" as a subtype of "A";

   - "X" is a subtype of "B" and "B" is a subtype of "A".

2.2.6 Guidelines

2.2.6.1 Information modelling by consensus

1. There are often several different ways in which a specific aspect of real-world information could be built into the model. Although there are some general criteria and guidelines about whether specific facts should be modelled as entities, relationships, or attributes, there is no unique 'correct' answer in many cases. It is important to obtain a consensus on the most appropriate model component to use. This is best achieved in walkthroughs and presentation sessions involving users and information modellers.

2.2.6.2 View definition

1. Each view should show a set of related facts — this is difficult to quantify, but the diagram should be capable of supporting a discussion of one functional area of the enterprises activities. The name chosen for the view should be an indication of this function. A subject-matter expert should be able to identify which entities and relationships appear on the view from that name.

2. Each view should be a well-defined set of entities and relationships supporting all or part of the enterprise functions. If an honest, precise name can be found for the diagram, then it is likely to contain related information. If this name is given to a subject-matter expert, they should be able to deduce which entities and relationships are on the diagram. If there is any difference between the list they produce and the components shown on the diagram, then the name or the grouping is not a good choice. Note: in carrying out this test, the diagram should not be described as concerning several unrelated areas, with "and"s (or similar) words. Such a group would not be cohesive.

3. For system ERDs, one way of choosing information views is to draw a system ERD to support each high-level function (as seen on the top-level data flow diagram of the System Essential Model).

2.2.6.3 Complexity issues

It is the nature of the human mind to only take in a certain, limited amount of information at once [Mil56]. In constructing an ERD this must be taken into account. ERDs are built to visualise the meaning of entities and the relationship between them. This is done within a certain context — there is a reason for wanting to review part of the enterprise information semantics at that point in time. Ideally, the selection of components in that view will meet that need, but not include any irrelevant 'noise', in terms of information that is not relevant.

Some possible reasons for constructing an ERD are:

1. **To highlight the meaning of a relationship**. All entities that participate in the relationship must be shown (this is an absolute rule). Any other relationships and entities that are needed to verify the relationship (when it is created, used, or destroyed) should also be shown. Sometimes this may involve understanding rules of association (see example in §6.3.4.2.1). All of the entities and relationships shown in the rule of association should be shown to understand the relationship in its context.

2. **To highlight an entity**. All relationships that the entity are involved could be shown. Sometimes there is a requirement that possible relationships be discussed and there are sometimes constraint rules of association that state "either one relationship, or the other, but not both" (in respect of a specific entity occurrence). These rules are best visualised by showing the entity and the relationships involved in this rule of association. If relationships that an entity participates are totally independent, though, little is to be gained from showing them in the same view.

3. **To highlight subtyping**. An entity relationship diagram also shows subtyping (although, in truth, it only works well for relatively simple cases). An entity relationship diagram constructed to show subtyping could be intended to highlight inheritance of attributes. In this case, it would be convenient to show the supertype and subtypes only, with no relationships. If the diagram were constructed to show relationships that are specific to subtypes, it is always important to show all the relationships that the supertype participates in.

4. **Review a data process**. If an entity relationship diagram is constructed to review a data process, all relationships and entities used by that data process (see §2.16.3.1) should be shown. Showing other information is irrelevant for the purposes of reviewing the data process.

The above list is not exhaustive, but it gives an idea of why an ERD might be constructed.

2.2.6.3.1 Selecting views in an automated support environment

In an ideal automated support environment, there could be many ERDs. Some would be constructed, but not saved — they would be created for a review session only. Others might be retained views. However, it is not required that any of these ERDs *must* be retained views — they can always be recovered unambiguously from relationship and associative entity specifications (from the frame formats). There is certainly no requirement that a single large ERD be verified (although it could be created automatically, if needed). The information model does not consist of a 'single large ERD, with supporting specifications', although it is a convenient conceptualisation to think it is.

2.2.6.3.2 Selecting views in a pencil and paper environment

In a pencil and paper environment, there are important trade-off decisions to be made. A small diagram is easy to maintain (it can be edited, or redrawn). However, there is an overhead in cross-checking diagrams (if an entity name is changed, for example). Many small views are also difficult to keep track of. Furthermore, it is sometimes the case that the exact collection of components (entities and relationships) required to verify part of the model are not available on any of the partial views. This would require a temporary ERD view to be constructed to show only those components.

It is tempting therefore to try to draw a single, large 'master' ERD, from which sections can be selected for review. Problems with large ERDs include:

1. The diagram cannot be read. Sometimes an Enterprise Essential Model might involve hundreds of entities.

2. There is a large inertia against redrawing the diagram. It is so complex, that changing even a small part of it involves considerable overhead (even for 'electronic pencil and paper' support, it takes a long time to lay it out properly). This is a very negative influence on the fundamental principle that modelling is an 'ongoing' activity, with little penalty in rebuilding a view when it is found to be in error. An important modelling motto is:

   Build one to throw away (describing a model).

   as attributed to F. Brooks [Bro75].

3. Relationships cannot be understood in context. Either there are many long, crossing relationship lines, or entities are repeated, making it difficult to understand the context of one entity — it is distributed 'all over the diagram'.

4. Parallel working becomes impossible. Each information modelling team could take their own copy of the diagram. However, when they modify it, there is a considerable integration problem involved in consolidating the changes and updating the master. In practice, when these techniques were used with no electronic support, this just was not dome at all. The master became history and there was no one, single information model.

As a compromise, therefore, the following are offered as general 'rules of thumb' for pencil and paper environments:

- Each diagram should be one that can be understood by a single user or subject-matter expert. If this is not the case, the view should be split into several smaller views that satisfy this criterion.

- A diagram should not contain too many components. Generally speaking, about ten components (each relationship, associative entity or entity counting as one) can easily be arranged into an understandable view, but more than thirty components is likely to lead to difficulty in relating different parts of the diagram to each other.

- Each diagram should not contain more than two levels of subtyping (very exceptionally three). Contravening this guideline leads to views that show high-level details in one area and very low-level details in another. If this problem is found in a view, then it should be split into two or more views.

- Where a subtyping has less than about seven relationships involving the subtypes, then all the subtypes and the relationships involving them should appear on one view. For more complex situations, the subtyping is repeated in several views.

2.2.6.4 Model interpretation

1. Both relationships and associative entities should be 'dependent' on the entities that they refer to. In other words, an occurrence of a relationship or associative entity should not be capable of existing without the prior existence of occurrences of each of the entities to which they refer.

2. Relationships and associative entities should reference entities that 'interact' together at a specific point in time. If they do not all interact at the same point in time, then new entities (probably associative entities) should be introduced so that the different interactions may be related.

3. If an entity has some attributes which are optional, or is involved in relationships with constraints that restrict the participation, then it is better to categorise the entity into subtypes.

4. If several entities have similar relationships or attributes of the same abstract data type, then it may be helpful to create a supertype entity of which they are subtypes. If there is no distinction made between the subtypes, the subtypes should be replaced by the supertype.

5. If an entity is used in different ways by different people and each uses different criteria for using that entity, then it may be helpful to subtype the entity. If there are no operations, relationships or attributes that are common to all the subtypes, replace the supertype by its subtypes.

2.2.6.5 Model component names

1. Names of pure entities should correspond to a role that real-world objects can play in interactions with the enterprise. In general, this will correspond to a noun phrase (often a common noun).

2. The name of the associative entity should be a noun phrase that reflects the name that is commonly used when referring to occurrences of this association. For example, the term "booking" would be used in everyday language by staff in an organisation that runs public courses. If the name is suggestive of real-world tangible objects, then it is generally better to model this as an entity, rather than an associative entity.

3. All relationships should be named in the 'active voice', rather than the 'passive voice'. Thus, for the relationship that links one employee to his superior, the name "reports to" is preferable to "is reported to".

4. Names of all relationships should allow a subject-matter expert to interpret what the relationship means without extensive explanations or reference to the relationship specification.

5. The relationship name should be chosen so that a subject-matter expert can immediately understand the significance of each occurrence of the relationship by looking at the ERD. Names of higher-order relationships should allow a simplified version of the relationship to be constructed. For example, "is married to" is still an acceptable name for the relationship that links the Man, Woman and Location entities (rather than the full form of the fact pattern, which would be " ... was married to ... at ... ").

6. Relationship names should not be longer than four or five words. If the need is felt to use a very long relationship name — resist it; it is a symptom of an ill-defined relationship, or possibly more than one relationship mis-recognised as one. Remember that the relationship name does not have to be unique to the enterprise, but only to the combination of entities that it associates.

2.2.6.6 Diagram aesthetics

The following could be interpreted as guidelines to follow for a modeller, or as for an automatic layout facility in an automated support environment.

1. Where two (or more) entities participate in a relationship, then they should be placed as close together as possible in such a way that the lines to the relationship icon are vertical or horizontal. If this is not done, then it will be difficult to understand the relationship in terms of the entities to which it refers.

2. Relationship lines should not cross as this also makes it difficult to understand the relationship in terms of just the entities to which it refers.

3. Because of the need to take in all of the entities and the relationship as a visual pattern, the lengths of the reference lines should be roughly equal for a given relationship. If this is not the case, the eye will need to read the entities in sequence, which obscures the fact that the relationship associates all of them equally and at the same time. For the same reason, the lines should not be too long, or the eye again has to search for the entities. The lines are better kept horizontal or vertical, with a minimum number of corners in them.

4. On the ERD, the 'link' lines may be shown connected to any part of the relationship icon, although they should go to distinct 'corners' wherever possible as this makes the identification of how many lines there are to be made more easily.

5. If any of the above are difficult to achieve, then one or more entities may be repeated, (possibly using a "＊" tag). However, repeating entities has a 'cost' in terms of complexity of the diagram and difficulty of understanding the entity in its context. Entities should not be repeated if there is a way of redrawing the diagram without repeats.

6. All entity and associative entity box icons should be the same size. Contravening this rule will makes it difficult to establish the entities on the diagram and some are given undue emphasis.

7. A large number of comments is a symptom that the view is a poor one to choose. Choosing another view, with more obviously related components, will remove the need for the comment.

2.3 Entity State Transition Diagram

2.3.1 Purpose

Entity state transition diagrams (eSTDs) are used to model significant changes in the properties of entities (including associative entities). These changes are significant in the sense that they affect the way the enterprise deals with them.

An entity may have one or more 'change patterns' called entity life-cycles. Each life-cycle is for a specific named characteristic, described as a state variable. The eSTD is used to model the different states that this state variable may take over time for one occurrence of the entity. (§2.5.2.3.5 shows how state variables are defined.)

Not all entities have state variables. Only those that have state variables will have eSTDs. There will be an eSTD for each state variable of that entity.

The eSTD shows how an occurrence of an entity changes its state when specific events occur. The access to the entity (create, read, update, delete, or match operations) may also be shown.

2.3.1.1 Enterprise entity state transition diagrams

Each entity that has state variables will have an entity life-cycle for that state variable. This eSTD is part of the Enterprise Essential Model. An eSTD used in this way is referred to as an enterprise entity state transition diagram, enterprise entity STD, or enterprise eSTD.

2.3.1.2 System entity state transition diagram

A system may deal with only some of the events that affect an entity. The system may not always be responsible for changing the state of the entity. An eSTD that only shows the events that are within the scope of a system is referred to as a system entity state transition diagram, system entity STD or system eSTD.

System eSTDs are discussed in §2.3.5.

2.3.1.3 Stand-alone systems

If a system is truly stand-alone and does not interact with other systems in the enterprise, then there is a single eSTD for each state variable. It acts as enterprise *and* system eSTD.

2.3.1.4 Alternative presentation tools

Because of the non-deterministic nature of eSTDs (see §2.3.3), an alternative tabular presentation is not recommended. (As bSTDs are deterministic, they do have an alternative tabular presentation — see §2.24.4)

2.3.2 Example

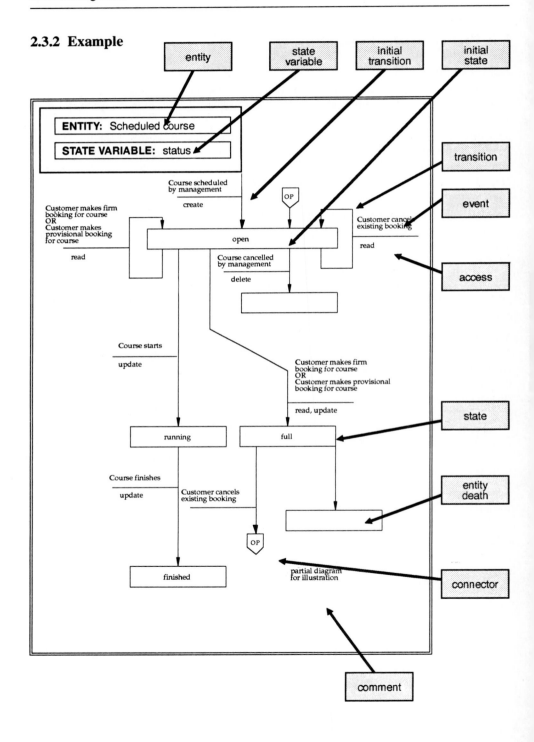

2.3.3 Non-deterministic nature of eSTD

The eSTD is non-deterministic. In other words, it describes *possible* changes of state when an event occurs. The change of state does not *always* occur, but it may occur. The responsibility for deciding whether there is a change of state lies in a minispec (or rarely, several minispecs).

Non-deterministic finite state modelling has been used in many other areas of system modelling for many years (for example, in lexical scanning parsers).

The non-determinism of eSTDs shows in two main ways:

- a transition is not always made when an event occurs,

- there may be two (or more) transitions from the same state for the same event.

In either case, the decision about change of state is made in a data process.

2.3.4 Components

2.3.4.1 Access

An access on an eSTD shows the possible access to the occurrence of the entity when the change of state occurs. Possible accesses are:

- **Create access**: a create access causes a new occurrence of the entity to exist. This always occurs for an initial transition. It cannot be used for any other transition. (See §2.3.4.7 for a description of the initial state and §2.3.4.8 for a description of initial transitions.)

- **Read access**: if an event occurs which requires some of the attributes to be used (but not necessarily changed), then this is referred to as a read access.

- **Update access**: if an event causes some of the attributes of an entity to be changed, then this is shown as a update access.

- **Delete access**: after a deletion the occurrence of the entity does not exist.

- **Match access**: a match access is a check on whether a specific occurrence of the entity exists. Showing "match" accesses is optional if there is a "read", "delete" or "update" on the same transition.

Note: the check and change access to state variables is not shown as an access in this diagram — it is implicit in the state and transition icons.

2.3.4.1.1 Showing accesses in different technical environments

For pencil and paper methods, the accesses may be shown as: "create", "update", "read", "delete" or "match", as in the example. For read and update, the attribute names could be shown after the keyword. Abbreviations for read and update are then conventional. For example: "R: start date" would indicate that the access reads one attribute "start date". Update can also be abbreviated, for example: "U: date closed" would indicate that the access changes (or allocate an initial value to) the attribute "date closed".The keywords create, delete and match are not usually abbreviated, but could be to "C", "D" and "M".

In an automated support environment, the states and transitions may be shown, with the accesses 'hidden', for example:

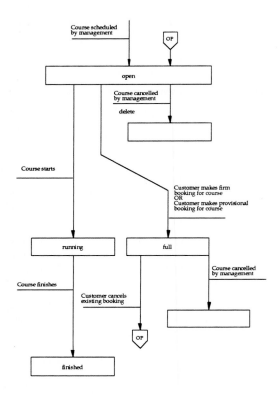

For any transition, the event causing the transition could be shown as required. Accesses could also be shown on demand, with lists of attributes for reads and updates.

2.3.4.2 Comment

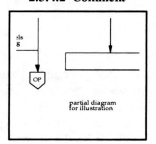

As with all diagrams, free-format comments may be added at any point of a diagram to elucidate or stress any important points.

Such comments should be used sparingly — a diagram that is cluttered with too many comments is usually an indication that the choice of names for the states and events has been poor (see the guidelines in §2.3.7).

2.3.4.3 Connector

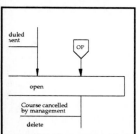

Where the diagram would be difficult to draw without crossing transition lines, a pair of connector symbols may be used. This is used to indicate a break in the transition line — the two ends of the line being identified by the 'name' given to the connector.

The technique is useful when there are several states with transitions to a given new state. This new state is given a single connector symbol, with a transition from the connector to the new state. No events or accesses are shown associated with this transition.

Each of the possible states that may make a transition to this new state are connected to the top of a connector symbol, with the event and access 'attached' to this transition. (Where there is more than one state that makes a transition to the same, new state, the events and accesses are not necessarily the same.)

On the diagram, the connector "OP" is used in this way.

2.3.4.4 Entity

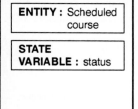

Each diagram is for one state variable of a specific entity. The example shown is for the associative entity "Scheduled course".

An entity may have several eSTDs, one for each of its state variables. It may have no eSTDs, if it has no state variables.

2.3.4.5 Entity death

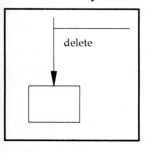

This is used to indicate that the occurrence of the entity no longer exists. Any transition with an associated "delete" must be to an entity death. After an occurrence of the entity has been deleted in this way, it no longer exists; it cannot be accessed.

Transitions to an entity death must include a delete access. They may also include read accesses (to produce a report of details about the occurrence deleted, for example).

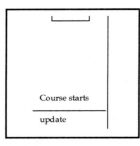

2.3.4.6 Event

The event is what causes the transition to take place. It is a real-world event, for example:

Course starts

When this event occurs, a change of state *may* occur for one occurrence of the entity. It does not *always* occur. The responsibility for determining whether or not it *does* occur resides in a data process. The minispec for this process defines the logic for determining whether there is a change of state or not. That data process will take account of the current state that the occurrence of the entity is in and other stored information (usually relating to other entities).

For example, when a "provisional booking" relating to a specific occurrence of "Scheduled course" occurs, the data process responsible for dealing with this event would check that there was a (matching) "Scheduled course" and that its "status" is "open". If this is not true, the booking will be rejected. If there is a corresponding "Scheduled course" that is open, the booking is recorded. The data process must also determine whether the course is still "open". This is dependent on attributes of other entities. For example: "<Course>.maximum number of delegates" is the maximum number of delegates that are ever accepted on that type of course. If required, the data process changes the "status" of the "Scheduled course" by:

<Scheduled course>.status := closed

This corresponds to the transition on the eSTD. This transition does not always take place. Only when the above checks are satisfied is there a change in state of the occurrence of "Scheduled course".

2.3.4.6.1 Multiple events for one transition

Where more than one event causes the same change of state, the events may be linked by "OR" operators. For example:

firm booking OR provisional booking

General expressions involving other stored information are not permitted. The only allowed format is a list of events, each separated by "OR".

2.3.4.6.2 Suppressing full event descriptions

Showing the full event names may lead to technical difficulties for some diagrams. There may be so many events that the diagram cannot be read. This is potentially more of a problem with pencil and paper environments. In these situations, an event identifier (e.g. the number) may be shown on the diagram. This may be linked to an enterprise or system event list, or the events can be added as an attachment to the diagram. An example is shown below:

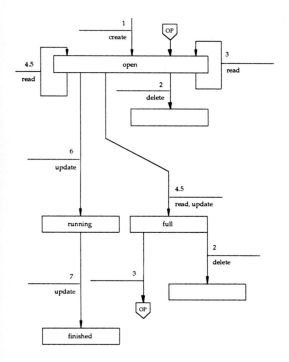

1:	Course scheduled by management
2:	Course cancelled by management
3:	Customer cancels existing booking
4:	Customer makes firm booking for course
5:	Customer makes provisional booking for course
6:	Course starts
7:	Course finishes

2.3.4.7 Initial State

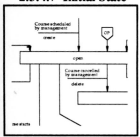

The initial state is the state of the occurrence of the entity when it is created. There may be more than one initial state, but there must be at least one. A transition to this initial state is referred to as an initial transition and must always include a "create".

2.3.4.8 Initial transition

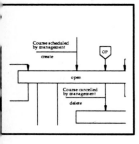

An initial transition is one that creates an occurrence of the entity. The transition is shown as a transition from 'nowhere' to the initial state. Before the event occurs, this occurrence of the entity does not exist; it is meaningless to refer to it. After the transition, the occurrence of the entity that was just created is in the initial state.

There may be several initial transitions to an initial state. Each corresponds to an event that could cause an occurrence of the entity to be created in that state.

Create accesses are only allowed on initial transitions. In addition, update accesses (setting values of some of the attributes) are allowed.

2.3.4.9 State

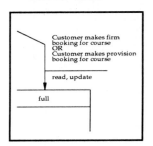

At a particular instant in time, each occurrence of the entity must have a well-defined state for a given state variable. For example, each "Scheduled course" may have status of "open", "running", "full" or "finished".

Different occurrences of the "Scheduled course" will be in different states at a specific point in time — the diagram does not make any statement about relative states of different occurrences of the entity.

2.3.4.9.1 Final State

A state from which there is no possible transition is termed a final state.

Retiring entities

It is quite unusual for entities to be deleted — more often they are 'retired' to a final state where archive information is still available, but that occurrence cannot be modified, nor take part in any new relationships. In the example "finished" is a state of this type. See also §2.3.4.5.

2.3.4.10 State variable

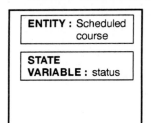

Each eSTD is for a specific state variable of the entity. Entities may have one or more state variables, each with a descriptive name (see §2.5.2.3.5). This state variable may be accessed in minispecs. The data access grammar for access to state variables is described in §6.10.16.

2.3.4.11 Transition

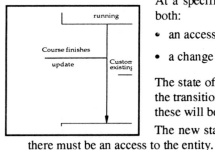

At a specific point in time, an event occurs which causes either or both:

- an access to the entity,

- a change of state of an occurrence of the entity.

The state of the entity before and after the event occurs is defined by the transition. (Where the role of these two states needs to be stressed, these will be referred to as 'current state' and 'new state').

The new state may be the same as the current state, but in this case there must be an access to the entity.

For example, a "Scheduled course" may change its state to "full" when "firm booking" occurs.

See §2.3.4.6 for a discussion of determining whether or not the transition takes place when the event occurs.

2.3.4.11.1 Choosing which transitions to show
Depending on the technical support environment, any of the following are allowed:

- showing all transitions;

- showing only those transitions that cause a change of state;

- showing all transitions that cause a change of state, with optional display of other transitions (this would only be possible in an automated support environment).

2.3.4.12 Description
The system eSTD is a partial copy of the enterprise eSTD for that entity. It is part of the information aspect of a System Essential Model.

Only the events and accesses that the system is responsible for are shown. This diagram may be considered to be a 'read-only' partial copy of the corresponding enterprise diagram.

Heuristics used in building the System Essential Model use these events to determine the systems response (see "The System Essential Model").

2.3.5 System entity state transition diagram

2.3.5.1 Example

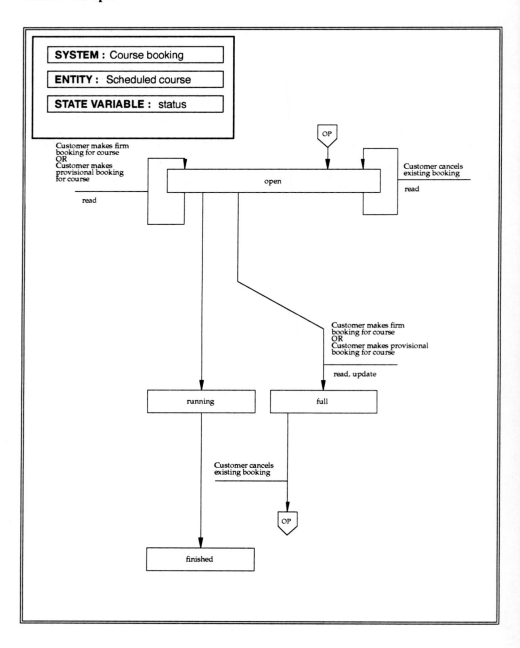

2.3.5.2 Components

These are as for the corresponding enterprise diagram, except for the following:

- if the system is not responsible for creating occurrences of the entity, the initial transition is not shown,
- if the system is not responsible for deleting occurrences of the entity, entity deaths are not shown,
- only events within the system scope are shown as transitions with event and access pairs; other transitions are shown without them.

In effect, all parts of the diagram that the system is not responsible for are suppressed. A simple technique is described above and in the example. However, any other equivalent way of suppressing the events and accesses that the system does not deal with may be used, For example:

- transitions that the system is not responsible for may be suppressed (this leads to 'isolated' states on the system eSTD),
- transitions that the system is responsible for may be shown in one colour and those it is not responsible for in a different colour. This might be useful in an automated environment, where the accesses are initially suppressed. 'Clicking' on a transition that the system is responsible for would then reveal the access.

2.3.6 Rules

The rules should be applied to an enterprise eSTD. System eSTDs are just partial copies of the corresponding enterprise view.

2.3.6.1 Creation and initial transitions and states

1. There must be at least one initial state.
2. A transition with a create access must be an initial transition.
3. An initial transition must always have a create access.
4. Initial transitions may have create and (possibly) update transitions only.

2.3.6.2 Connectivity

1. Every state must have at least one transition leading to it.
2. Every state must be accessible. That is, each state must be capable of being reached by a finite sequence of transitions from an initial state. (This is a stronger requirement than the last requirement.)
3. A transition that does not cause a change of state must have associated read, update or match accesses.

2.3.6.3 Entity deaths

1. A delete access must always correspond to a transition to an entity death.
2. The only accesses allowed on a transition to an entity death are read, match, delete.

3. Any transition to an entity death must have a delete access.

2.3.6.4 Connectors

Definition: each connector links one or more current states to a given new state.

1. All connectors must be named. The name may be an abbreviation

2. Each named connector must be linked to at least one current state.

3. Each named connector must be linked to exactly one state as a new state.

4. Any named connector must appear at least twice. Exactly one of these must be to a new state.

5. Connectors cannot be used on initial transitions.

2.3.7 Guidelines

2.3.7.1 Names of states

1. Names of states should reflect a natural user view of the entity. For example, it is natural to say "the scheduled course is full".

2. Names of states should not duplicate the name of the entity. Remembering that each diagram is for one entity, the state should be called "full", rather than "Scheduled course full".

2.3.7.2 Properties of states

1. The attributes with available values are always the same in a given state, irrespective of the path to that state. (Note: with existing methods, it is likely that this check is too laborious without automated support.)

2.3.7.3 Diagram aesthetics

1. eSTDs should be drawn to minimise the number of crossing transition lines and their length. This can be achieved by use of connectors.

2.4 Entity–Event Table

2.4.1 Purpose

The entity–event table is used to visualise the relationships between events and the stored information used by the enterprise.

It can be used to identify events from entities and relationships, or vice-versa. It can also be use to group entities and relationships so that the influence of an event may be combined to that group. For example, one event may cause deletion of multiple occurrences of several entities and the relationships that depend on them.

In strategic planning, it can be used to help define system boundaries, by choosing entities, relationships and events that are to be within the scope of the proposed system. This is discussed (briefly) in §4.2.4.

2.4.1.1 Enterprise and system use of entity–event tables

An entity–event table used as part of the Enterprise Essential Model is referred to as an enterprise entity–event table.

An entity–event table used as part of the System Essential Model is referred to as a system entity–event table.

The prefix enterprise or system may be dropped if no ambiguity results.

2.4.1.2 Related tools

Section 2.2 on the ERD, gives a more detailed descriptions of associative entities, entities, relationships and subtypes.

If the whole table corresponds to a function (for example, the functions allocated to one system), then the collected accesses along one row correspond to the entry in a function–entity table with that function as a column (see §2.19) for that information component. The entity–event table provides a more detailed view of the information access of that function.

A still more detailed view of the effect of events on information, may be shown for a single information component on an eSTD (§2.3).

2.4.2 Example

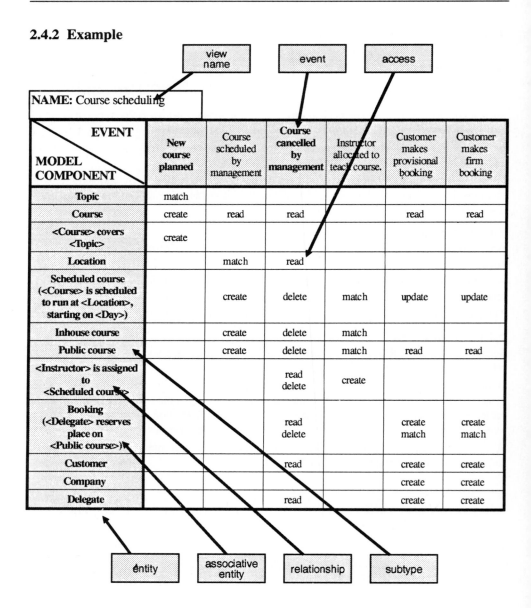

NAME: Course scheduling						
EVENT / **MODEL COMPONENT**	New course planned	Course scheduled by management	Course cancelled by management	Instructor allocated to teach course.	Customer makes provisional booking	Customer makes firm booking
Topic	match					
Course	create	read	read		read	read
<Course> covers <Topic>	create					
Location		match	read			
Scheduled course (<Course> is scheduled to run at <Location>, starting on <Day>)		create	delete	match	update	update
Inhouse course		create	delete	match		
Public course		create	delete	match	read	read
<Instructor> is assigned to <Scheduled course>			read delete	create		
Booking (<Delegate> reserves place on <Public course>)			read delete		create match	create match
Customer			read		create	create
Company					create	create
Delegate			read		create	create

Labels around the diagram: view name, event, access, entity, associative entity, relationship, subtype

2.4.3 Components

2.4.3.1 Access

This gives a view of the access made to the stored attributes and state variables of entities and occurrences of relationships when an event occurs. When the event occurs, this access does not always take place. For example, there might be tests that are carried out when the event occurs; only if the tests are satisfied would the access take place. For example, the event:

> Course cancelled by management

would sometimes cause an occurrence of the relationship "<Instructor> is assigned to <Scheduled course>" to be deleted. In some cases, however, there would be no instructor assigned to that course as yet — there are no occurrences of the relationship to be deleted in those cases. The "delete" access shown in the table indicates a *possible* "delete" access.

The allowed accesses are:

- **create**: create a new occurrence of the entity or relationship;
- **read**: (for entities only) use values of attributes or state variables. These must have been previously assigned in a create or update;
- **update**: (entities only) assign or change of attributes or state variables of an entity;
- **delete**: destroy one or more occurrences of the entity or relationship;
- **match:** check to see if a specific occurrence of the entity or relationship exists;
- **check:** examine a state variable to determine a correct response;
- **change**: change the state of a state variable.

To save space in tables, the following abbreviations are sometimes used:
"C" (create), "?" (match), "R" (read), "U" (update), "D" (delete), "S_c" (state check), "S_a" (state change). (Any other agreed abbreviations could be used.)

2.4.3.2 Associative entity

If an associative entity is included in the table, then its name, together with its relationship frame, is given. It is thus shown in both its roles — as an entity and as a relationship. The row gives potential accesses to this associative entity when each of the events occur.

The relationship frame helps in understanding the dependency of the associative entity on the entities to which it refers. (In the notation shown in the example, the relationship frame is given in parentheses, but any other suitable convention could be adopted.) For example:

> Scheduled course (<Course> is scheduled to run at
> <Location>, starting on <Day>)

indicates that the associative entity "Scheduled course" acts as a relationship and refers to the "Course", "Location" and "Day" entities. An event that causes an access to "Scheduled course" is very likely to cause an access to some (or all) of these entities. It is also true that (for example), deleting an occurrence of "Course" would require one or more occurrences of "Scheduled course" to be deleted.

Showing the associative entity name and its relationship frame help in visualising this inter-dependency.

2.4.3.3 Entity

If an entity is included in the table, it is identified by its name. The row gives potential accesses to stored attributes or state variables of this associative entity when each of the events occur.[1]

2.4.3.4 Event

Each column gives the effects of one event. The name of the event is given at the top of the column. The type of information that may need to be captured by the enterprise or system (depending on what the scope of the table is) includes:

- Which agents (people, machines, organisations, etc.) participated or caused the event. This could correspond to new occurrences of entities or relationships.

- Relevant facts about entities that participated in the event. This will correspond to use of attributes of the entity, shown as "read" or "update" in the table.

2.4.3.5 Relationship

If a relationship is included in the table, then its frame is given here. (The shortened name, as used on an ERD is not a sufficient identification, because several relationships may have the same name, providing they are between different entities.) The row gives the effects of the events on it. For example:

```
<Instructor> is assigned to <Scheduled course>
```

describes a relationship "is assigned to" that is defined to refer to the entity "Instructor" and the entity "Scheduled course" ("Scheduled course" is an associative entity, of course, but in this context it acts as an entity).

2.4.3.6 Subtype

An entity that is a subtype of another entity is not distinguished in any specific way in this table. However, there is an important consistency criterion — any access to the subtype must also be an access to the supertype, if the supertype is also shown in the table.

For example, "Inhouse course" is a subtype of "Scheduled course". Down any column (corresponding to one event) the accesses of "Inhouse course" are always found in the accesses to "Scheduled course".

There may, however, be some accesses to the supertype that are not relevant to this subtype. For example, the event "Customer makes firm booking" does not require any access to "Inhouse course", although an access to "Public course" (which is another subtype of "Scheduled course") may occur.

---

1 Note: not all attributes of an entity need to be stored. The table does not deal with non-stored attributes.

2.4.3.7 View name

Each table restricts attention to a group of entities and relationships and the events that affect them.

No special identification needs to be given to it unless it is retained as a permanent view. A table used as part of the Enterprise Essential Model must have a name that is different from that of any other enterprise entity–event table.

If the table is used to study the relationship between information and the events within the scope of one system, the name should be the name of the system.

Because it is not always convenient to show all the entities and relationships in one table, the table may be split into sections. In this case, an identifying sequence number might be added:

NAME : Course scheduling (part 1)

2.4.4 Organising the table

Listing the entities and relationships in a random order does not lead to a very useful table. For small tables this will not be so important, but for large tables, the interdependencies of the information model components (entities and relationships from either the Enterprise Essential Model or the System Essential Model) will become less easy to identify, because the reader will have to keep switching attention from one portion of the table to another.

Entities and relationships should be shown together (or nearby) in the table if the same event causes an access to them.

For example, an event may occur that establishes an occurrence of a relationship — in fact, it is unlikely that this occurs without some access to the entities that the relationship refers to. Such accesses to the entities will include checking that they exist, adding new occurrences, etc. When entities are deleted, the event that causes their deletion generally requires deletion of occurrences of relationships in which the entity participated.

One of the uses of this tool is to identify groups of relationships and entities, together with the events that affect them. If the events affect this information group and no other components, then it is a coherent grouping and may be considered to be a possible scope of a high-level function. To do this, the relationship and the entities it refers to should be kept close together in the table. Furthermore, for a specific event, the accesses that it causes should be kept as close together as possible.

2.4.4.1 Diagonalisation of entity–event table

For small tables, the optimum arrangement is such that the accesses appear near the diagonal line running from top left to bottom right. This is achieved by grouping some events (group "A", for example) and the information components they access at the top left of the table.[1] Events that do not affect these components are then kept to the right of the table. Any components that are not accessed by events in group "A" are moved to the last few rows of the table. In this way, events in group "A" have their accesses shown in the top left portion of the

---

1 Note: in this context, grouping is used with its everyday meaning. The groups set up in this table may, or may not, correspond to event groups, as defined in event specifications. The grouping here is more for which information components are accessed than anything else.

table. In the bottom right portion appear information components that are not acted on by the events in group "A", but are presumably acted on by some other events (say group "B"), or they would not be accessed at all! These group "B" events are distinct from those in group "A", so the table now has the appearance as shown below:

	GROUP A events	GROUP B events
Information components affected by events in group A		
Information components affected by events in group B		

In practice, things are not usually so simple! If the above organisation cannot be achieved, then consideration should be given to organising it so that "create" and "delete" accesses are near the diagonal. This is the technique used in the example in §2.4.2. Two rough groupings are shown — one to do with 'course scheduling' and one to do with handling bookings from customers.

2.4.4.1.1 Diagonalisation of large tables

If more than two groups of events and information components can be identified, then more rectangular areas appear, but they are still 'on the diagonal'. In practice, this will never be achieved exactly — there will always be some overlap between groups. However, the closer the components are to the diagonal, the better the organisation of the table.

2.4.4.2 Sectioning the entity–event table

Where the table would be too large to show in one section, it may be broken into several sections. Conceptually, if a large table is imagined to exist, with the access entries roughly down the diagonal, then each section should be chosen to show one or two groups, with no components missing from that section. This will often mean that a given row of the table need to be repeated, because many events could affect that row. In such cases, an identifying "∗" tag may be used to identify these rows (particularly if the table is produced in a pencil and paper environment). Similarly, where one event affects many information components, several sections of the table may be used, with that event shown as a column on each section (again, an identifying "*" may be used to indicate this, if required).

2.4.4.3 The entity–event table in automated support environments

The sectioning of the table and use of the "*" notation described above is suitable for pencil and paper methods. In an automated environment, other techniques might be better.

With the correct automated support, it might be possible to provide many different ways of partitioning the table.For example, the following views might be provided:

1. For one entity, show all the events that cause an access to it.

2. For one event, show all entities to which it causes an access of any type.

3. For one event, show all entities that it creates, deletes or updates.

4. For an entity, show all the events that lead to the creation of an occurrence of that entity.

5. For an associative entity, show all events that lead to the creation of that associative entity. All other entities that the associative entity links as a relationship, together with the events that cause their creation, would also appear in this view. This would allow the 'dependency' of one entity upon another to be visualised and help in allocating entity ownership and events to possible systems.

2.4.5 Rules

2.4.5.1 Events (columns)

1. Each event shown in the table must cause access to one of the information aspect components. This means each column must have an access. For tables that are divided into several sections, the event only needs to be shown where it affects any of the information components that appear in that section.

2.4.5.2 Information model components (rows)

1. Each entity, associative entity or relationship shown in the table must have an access. This means that each row must have an access.

2. For tables that have not been divided into sections, each row should be complete. All accesses affecting that entity (or relationship) within the enterprise should appear in the table.

2.4.5.3 Allowed accesses

1. Entities may have create, read, update, delete, match, check, or change accesses.

2. Relationships may have create, delete and match accesses.

2.4.5.4 Table integrity

1. If a relationship appears in a section of a table, then all entities that it relates to must also be shown in section table. (If there is only one section, the rule still applies to that section.)

2. Each section may only show an entity, associative entity or relationship once. In other words, rows may not be duplicated in a single section of the table.

3. If a supertype and subtype of an entity are shown in the table, then for each event:

   the accesses to the supertype must include all accesses to the subtype. There may be additional accesses to the supertype, which do not correspond to accesses of that particular subtype — they may correspond to accesses of the supertype as a whole or accesses to other subtypes.

2.4.6 Guidelines

1. A row or column that is repeated in the table may be tagged with a "∗", particularly using pencil and paper techniques.

2.5 Entity Specification

2.5.1 Purpose

An entity specification is used to define the role of an entity, the way different occurrences of the entity are identified and the attributes of the entity. (Note: associative entities have associative entity specifications and may be regarded as a combination of an entity specification and a relationship specification.)

2.5.1.1 Enterprise and system use of entity specifications

Each entity that is not an associative entity has a corresponding entity specification that is part of the Enterprise Essential Model. This is referred to as an enterprise entity specification.

Each entity used by the system has a specification, describing the system's use of that entity that is part of the System Essential Model. This is referred to as a system entity specification.

Enterprise entity specifications are discussed in §2.5.2 and system entity specifications in §2.5.4.

The prefix enterprise or system may be dropped if no ambiguity results.

2.5.1.2 Stand–alone systems

If the enterprise is identified with the system, then there is exactly one specification for each entity. This is referred to as the entity specification — the Enterprise Essential Model may be regarded as being 'folded over' into the System Essential Model. See also "The Relationship between Systems and the Enterprise".

2.5.2 Enterprise entity specification

2.5.2.1 Purpose
Each entity has an enterprise entity specification that acts as the basic 'repository' of information about the entity.

2.5.2.2 Example

ENTITY : Course

MEANING :
Each <Course> is a standard seminar that may be scheduled to run from time to time on a public basis or on demand for specific customers.

ATTRIBUTES : acronym, name, length, delegate cost

IDENTIFIERS : acronym,
                 name

STATE VARIABLES : status

CONSTRAINTS : —

STORAGE/OCCURRENCE : 120 bytes

OCCURRENCES/TIME :	date	: expected	: minimum	: maximum
	01/10/90	: 16	: 16	: 20
	01/01/91	: 24	: 22	: 30
	01/01/92	: 30	: 24	: 45

2.5.2.3 Components

2.5.2.3.1 Entity
This identifies the entity being specified. For example:

ENTITY : Course

2.5.2.3.2 Meaning
This gives the significance of the entity to the enterprise. For example:

MEANING : Each <Course> is a standard seminar that may be scheduled to run from time to time on a public basis or on demand for specific customers.

In textual specifications an occurrence of an entity is indicated by enclosing the entity name between "<" and ">" Thus "<Course>" should be read as "occurrence of Course".

If the entity is a subtype, then it inherits part of its meaning from the supertypes (see §2.5.3.5).

2.5.2.3.3 Attributes

An attribute associates a value with each occurrence of the entity. For example, the specification of "Course" includes the following:

```
ATTRIBUTES : acronym,
             name,
             length,
             delegate cost
```

In a sense, each Course has a label on it, stating its "acronym", "name", "length" and "delegate cost". All attributes of the entity that are significant to the enterprise are listed here. Many attributes are stored — the enterprise will usually be responsible for assigning or updating a value to that attribute and then using this value at a later time (reading).

Attribute specifications are discussed in §2.8.

Scope of attribute names

The name of each attribute must be unique within the attributes of that entity. It is allowed to be the same as an attribute of another entity. For example, the entity "Course" has an attribute "name", as does the entity "Delegate". In minispecs (using the dialect adopted in this manual) these are referred to by:

```
<Course>.name
```

and

```
<Delegate>.name
```

This is a very important way of controlling complexity in large models. Previously, many authorities advocated an alphabetic list of all data items. For small models, this presented no problems, particularly if the name of the data item gave a good clue to the entity that it was an attribute of. For large information models, however, there were considerable problems and long compound names or arbitrary 'scoping' rules were required. The identification of information items used in YSM largely overcomes these problems.

Attributes of subtypes

If the entity is a subtype, then it inherits attributes from its supertypes (see §2.5.3.2).

2.5.2.3.4 Identifiers

Each entity must have at least one method by which different occurrences of the entity can be distinguished. An identifier is a way of achieving this. Entity identifiers are of two types:

1. **A single attribute:** given the value of this attribute, it is possible to find the single occurrence of the entity that has this value. For example:

```
IDENTIFIERS : part id
```

2. **A combination of attributes:** given the value of each of the attributes that make up the identifier, it is possible to find a single occurrence of the entity that has the same values for these attributes. These compound identifiers are shown as a list of attributes, with a "+" symbol between them (which should be read as "together with"). For example, in a hospital system, the identifier for <Patient> might be:

```
IDENTIFIERS : full name + birth date + town of birth
```

Multiple identifiers
All entities must have at least one identifier. Some entities have more than one alternative identifier. If there is more than one identifier, they are listed in any order. For example, the specification of the entity "Course" includes:

```
IDENTIFIERS :        acronym,
                     name
```

showing that it has two alternative identifiers.

Identifiers of subtypes
If the entity is a subtype, then it inherits identifiers from its supertypes (see §2.5.3.3).

2.5.2.3.5 State variables
An entity may have one or more state variables. Each describes a status type of property of the entity. At any one time, an occurrence of the entity will have one of several distinct values for this status.

Over time, the occurrence of the entity will change its status as events occur. For example, an employee might have two state variables: "employment status" and "marriage status". Each of these could change independently of the other. For example, the "employment status" might be "applicant", "under offer", "on payroll", "left company", "retired"; the "marital status" might be "single", "married", "divorced".

Each state variable of an entity has a characteristic way it may change over time. This is described as an entity life-cycle. Entity state transition diagrams are used to model the way these changes occur (see §2.3).

If the enterprise deals with an entity in a constant fashion over time, then the entity does not have any state variables. The entity is 'always the same', from the moment it is created to the moment it is deleted.

Each state variable has a corresponding eSTD (see §2.3). See §2.5.6.3 for a discussion of the distinction between state variables and attributes.

2.5.2.3.6 Constraints
Where there are specific constraints regarding the entity, these may be entered here. The entity "Month", would have:

```
CONSTRAINTS : number of(Month) = 12
```

This requires that the number of occurrences of this entity is always 12.

Distinction between policy limits and size estimates
The above field should only be filled in if a policy constraint is being discussed. This is fairly unusual. It is much more common to specify size estimates (see §2.5.2.3.8).

2.5.2.3.7 Storage/occurrence

This gives the likely amount of space for stored information referring to one occurrence of the entity.

This estimate is given prior to any implementation. After implementation decisions have been taken, more detailed information of the size of one occurrence will be available. This will be covered in later releases of YSM.

2.5.2.3.8 Occurrences/time

This is used to estimate the number of occurrences of the entity. This estimate is given for one or more dates (planning points).

In most cases, the number of occurrences of a component increases over time. In these cases, the modelling tool shows the growth over time of the storage requirements for this entity.

Date
This is a date for which size estimates are available. This is sometimes called a planning point.

Expected
This gives the 'best' estimate of how many occurrences of the component will be in existence at that date. It should be an unbiased estimate of the most likely value.

Note: if entities are 'retired' and not deleted, this will naturally increase over time.

Minimum
This gives a 'best case' estimate of how many occurrences of the component will be in existence at that date. For practical purposes, the 10th percentile is used, so that it is expected that 'nine times out of ten', this estimate *will* be exceeded by that date.

The minimum number of occurrences expected is usually less significant than the expected number. It is an optional entry.

Maximum
This gives a 'worst case' estimate of how many occurrences of the component will be in existence at that date. For practical purposes, the 90th percentile is used, so that it is expected that 'nine times out of ten', the estimate *will not* be exceeded by that date.

The maximum possible number of occurrences is useful in risk analysis. It is an optional entry.

Alternative presentations
Alternative presentations can be used. In particular, the same information can be presented graphically. For example, the graph below shows the expected growth of a component over time:

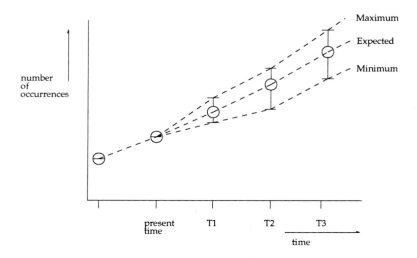

Note: the 10th and 90th percentile estimates are shown as ends of a 'bar'. It is unlikely that the 10th percentile will be of any interest, but it is conventional to show it.

2.5.3 Entities that are subtypes

2.5.3.1 Inheritance
If the entity has been declared to be a subtype of another entity, then all attributes of the supertype are also attributes of this entity (see the discussion of subtypes in §2.2.3.6). In some situations, the entity may 'inherit' attributes from several supertypes (see §2.2.3.6.1).

Although these are attributes of the entity, they are not defined in the same way as the other attributes that are specific to this subtype. The inherited attributes may be modelled by:

- showing them in each subtype, preferably with some indication that they are inherited (this is the convention used in this manual, with inheritance being denoted by italicisation),

- listing attributes only in the supertype, with the subtype only showing its specific attributes.

Other properties such as meaning and identifiers may also be inherited. The general principles will be discussed for attributes and then examples shown for other inherited properties.

2.5.3.2 Inherited attributes

2.5.3.2.1 Showing inherited attributes in subtypes
In this approach, the attributes are displayed in the subtype. For example, "Manufactured item" has a subtype "Finished item" (see §2.2.3.6.1 for ERD). Partial specifications for these two entities are shown below:

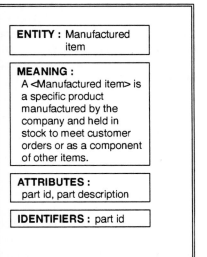

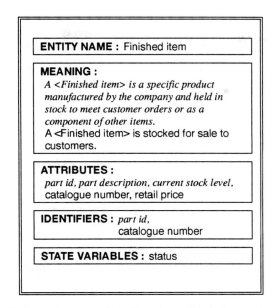

This shows all attributes of "Finished item", including those inherited from "Manufactured item". The inherited attributes have been distinguished by italicising them. The attribute "current stock level" is inherited from "Inventory item", which is another supertype of "Finished item".

The attributes "catalogue number" and "retail price" are not inherited, but specific to "Finished item".

Other conventions may be used to indicate which attributes are inherited, but italicising is suitable for hard-copy.

2.5.3.2.2 Showing inherited attributes in supertype only
This approach does not repeat information "owned" by the supertype in the subtype. Attributes are only shown in one place. For the above example, the specification of "Finished item" would be:

ATTRIBUTES :	catalogue number, retail price

However, these are not the only attributes that each "Manufactured item" has. Only by looking in the specification for all supertypes of "Manufactured item" could the other attributes ("part id", "part description" and "current stock level") be discovered. The check that has to be applied to the EEM is:

If an entity is a subtype of more than one entity, then it must not inherit two (or more) attributes of the same name, except where these are owned by a common 'ancestor'.

2.5.3.2.3 Attribute name clashes

An entity may inherit attributes from several entities that it is a subtype of (see §2.2.3.6.1 for an example). This would lead to ambiguity if the same name were used for attributes of several supertypes of an entity. This is avoided in YSM by requiring that, for each entity, the attribute names of each of its supertypes (if there are any) are distinct.[1] Thus, in the diagram below, B and C must not have any attributes that have the same name:

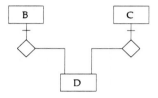

If the supertypes are themselves subtypes of an entity, then attributes of that entity are allowed to be inherited downwards through an 'inheritance net' (see §6.1.3.3.1), an apparent multiple inheritance by a subtype of an attribute belonging to an 'ancestor' is allowed. In the example below, B and C are allowed to have attributes with the same name, providing they are inherited from A:

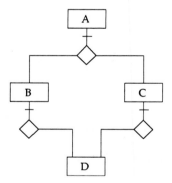

Checking that inheritance obeys this rule is part of the integrity rule checking that must be applied to the Enterprise Essential Model.

2.5.3.3 Identifier inheritance

A subtype always inherits all the identifiers of its supertype. It may also have additional identifiers that are only available for this subtype and not others. Using the notation introduced above, we might have the specification of "Finished item" including:

---

1 More sophisticated inheritance protocols might have been adopted, but this should work and it is a simple solution to the potential ambiguity.

IDENTIFIERS :   *part id,*
                 catalogue number

which indicates that there is an additional identifier, "catalogue number", that is only available to this subtype of "Manufactured item".

2.5.3.4 State variable inheritance
Any state variable of a supertype is automatically a state variable for the subtype. There may be additional state variables that are relevant for the subtype, but not for the supertype. In the example, "status" is used only to keep track of whether a finished item is ready for sale or not (it may be in market research, design, available or obsolete).

2.5.3.5 Inherited meaning
A subtype has the meaning of the supertype. In addition, it may have specific meaning that is not relevant to other subtypes. For example, "Finished item" might include:

MEANING : *A <Finished item> is a specific product manufactured by the company and held in stock to meet customer orders or as a component of other items.*
A <Finished item> is stocked for sale to customers.

Note: because "Finished item" is a subtype of "Manufactured item", the first sentence is obtained automatically from the equivalent statement in the supertype by restricting the statement to the specific subtype concerned. ("Manufactured item" is specified in the example at the beginning of section.)

If an entity is a subtype of several supertypes, then there might be a part of its meaning that corresponds to each one of its several supertypes.

2.5.3.6 Correct way to show inherited properties
The approach used for examples in this manual regards all properties of the subtype as being 'visible' in the subtype. All aspects of the subtype may be seen in one view, without having to cross-refer to possibly many supertypes. With the correct automated support, this is the preferred view of that entity.

Without suitable software support to control the redundancy, showing duplicate information on multiple views is laborious and subject to error. In these environments, inherited properties can either be shown in the supertype only or in both supertype and subtype. If the properties are shown in both supertype and subtype, then careful cross-checking must be carried out.

This situation, where inherited properties are shown in the subtype, although the information is not 'owned' by the subtype, is an example of a more general situation relating to view overlap. See §2.1.8.3.

2.5.4 Specifying system use of entities

2.5.4.1 Purpose
Each entity accessed by the system has a corresponding system entity specification, describing the access to that entity. It allows the System Essential Model to be checked for correct use of enterprise information.

The system may be responsible for creating and deleting occurrences of the entity, or it might need to check whether or not specific occurrences of the entity exist. Any attribute of the entity used by the system is declared in the system entity specification, together with the use that the system makes of it.

Note: if a system uses an associative entity only in its role as entity, then it has an enterprise associative entity specification and a system relationship specification.

2.5.4.2 Example

SYSTEM : Course booking		
ENTITY : Delegate		
MEANING : A <Delegate> is a person who has been booked to attend a standard seminar.		
USE : create, update, read, match, check, change		
ATTRIBUTES :	**attribute**	**: access**
	sequence number	: read, update
	name	: read, update
	job title	: read, update
IDENTIFIERS : sequence number		
STATE VARIABLES : status	: check, change	

2.5.4.3 Components

2.5.4.3.1 System
This gives the name of the system whose use of the entity is being described.

2.5.4.3.2 Entity
This gives the name of the entity whose use by the system is being described.

2.5.4.3.3 Meaning
This gives the significance of the entity to the system. In an automated support environment, the meaning of the entity to the enterprise should also be 'visible' here. In addition, there may be specific information that needs to be recorded about the significance of the entity to this specific system.

In a pencil and paper support environment, only system-specific information should be given here.

See §2.1.8.3 for further discussion of 'visibility'.

2.5.4.3.4 Use
The access to the entity by the system is defined here (access to individual attributes is defined below). The allowed values are:

- **create:** the system is responsible for the creation of occurrences of the entity;

- **read:** the system needs to know the values of attributes of the entity;

- **update:** the system changes the values of some or all of the attributes;

- **delete:** the system may delete occurrences of the entity;

- **match:** the system needs to check whether there are any occurrences of the entity that satisfy a selection criterion;

- **check:** the system needs to examine a state variable to determine a correct response;

- **change**: the system is responsible for changing the state of a state variable.

Any combination of these keywords may be given.

2.5.4.3.5 Attributes

Any attribute that the system reads or changes is visible in this view. Each attribute has a corresponding attribute specification (attribute specifications are covered in §2.8). It is also shown here as being "read" or "updated" by the system. If an attribute is read *and* modified by the system, it is shown as "read, update".

Note: the names of these attributes are not owned by this view — they are inherited from the corresponding enterprise entity specification (see §2.1.8.3 for a discussion of visibility).

2.5.4.4 Identifiers

Each entity must have at least one method by which the system can distinguish one occurrence of the entity from another. An identifier is a way of achieving this. All identifiers used by the system are defined here. There must be at least one, and there may be several. If the system uses more than one identifier they are listed in any order.

Note: any identifier used by the system is inherited by the system — however, not all enterprise identifiers may be used by the system (see §2.1.8.3 for a discussion of visibility).

2.5.4.4.1 State variables

This entry lists all state variables that are checked or changed by the system. If the system needs to know the status of a state variable, then the value entered is "check"; if the system is responsible for changing the status of the state variable, the value entered is "change". If the system does both, then "check, change" is entered.

2.5.5 Rules

2.5.5.1 Attribute names

1. The name of each attribute must be unique within the attributes of that entity. It is allowed to be the same as an attribute of another entity.

2. The name of each state variable must be unique within the entity.

3. No attribute can have the same name as a state variable for the same entity.

2.5.5.2 Identifiers

1. At least one identifier must be given in each entity specification.

2. The identifier must contain only items listed as attributes.

3. For a single-attribute identifier, there cannot be two occurrences of the entity with the same value for that attribute.

4. For identifiers that are composed of more than one attribute, no two occurrences of an entity may have the same set of values for all the components of the identifier.

2.5.5.3 Number of occurrences

1. At any time point, the maximum number of occurrences must be greater than (or equal to) the minimum number of occurrences.

2. Any restriction on the number of occurrences of an entity stated in the constraint entry must be consistent with the maximum (and) minimum number of occurrences at all times.

2.5.6 Guidelines

2.5.6.1 Names of attributes

1. Names of attributes should be as simple as possible and should not duplicate the meaning of the entity. This is because the name of the attribute is never used, except in conjunction with the entity. Thus the attribute "delegate cost" of the entity "Course" is a reasonable name if the attribute corresponds to what a delegate would be charged to attend the course. Using the name "course delegate cost" for an attribute of "Course" just makes the name longer and less understandable.

2. Names of attributes should be a good clue to what they are used for — user review is important here. As a general guideline, the following two sentence structures should make sense to a user:

   The a of that e is v.

   What is the value of a for that e?

   where "*a*" should be replaced by the proposed attribute name, "*e*" by the entity name and "*v*" by a value. For example:

   The name of that course is navigation for master's certificate.

   and:

   What is the name of that course?

   both make sense in ordinary, every-day conversation.

2.5.6.2 Identifiers

1. Identifiers should be naturally used in the activities of the enterprise and not artificially constructed by the information modeller. In the examples above, for the entity "Course", the attribute "acronym" is a little suspicious in this respect, but if it is in the everyday use by sales staff and instructors it would be acceptable. Creating a unique numeric code for each course would not be as acceptable if it did not correspond to the everyday vocabulary of instructors and sales staff.

2. If the only way of distinguishing between occurrences of an entity is by means of a combination of attributes, then it may be convenient to construct an artificial identifier. This should be used with care, as it runs counter to the preceding guideline.

3. Entities with a single attribute (which must be the identifier) are quite rare. The need for the enterprise to keep track of occurrences of such entities should be carefully checked.

2.5.6.3 Distinction between attributes and state variables

Attributes may be given values from abstract data types and then compared with other attributes, possibly of different entities. Attributes may also be defined in terms of a value list, but this is less common (see below). State variables are not defined in terms of an abstract data type, but as a finite list of possible states. The 'value' of a state variable for one occurrence of an entity may be checked against or assigned a literal value, for example:

```
<Course>.status = available
```

but it cannot be used in any other way. They are, in a sense, very restricted types of attributes. Because the value of a state variable necessarily changes over time for one occurrence of an entity, state variables cannot be used as part of an identifier.

If a proposed attribute of an entity is identified with the proposed values being a discrete set of values that are only relevant to that entity, it is a very strong clue that it might be a state variable. If the 'value' can change over time for one occurrence of the entity, then it should be modelled as a state variable; if it only ever has one value for a specific entity occurrence, then it should be modelled as an attribute.

Names of state variables should be given a name that indicates they are of this general nature. If there is only one state variable for a given entity, it should be named "status", or "state". If there are several state variables, then each should be given a name that gives a clue to its 'status' nature. For example, "employment status" and "marriage status".

To avoid confusion, no 'ordinary' attribute should be given a 'status-like' name.

2.6 Relationship Specification

2.6.1 Purpose

A relationship specification is used to define the role of a relationship. (Note: associative entities have associative entity specifications, covered in §2.7. These may be regarded as a combination of an entity specification and a relationship specification.)

2.6.1.1 Enterprise and system use of relationship specifications

Each relationship used by the enterprise has a corresponding relationship specification that is part of the Enterprise Essential Model. This is referred to as an enterprise relationship specification. Enterprise relationship specifications are discussed in §2.6.2.3.

Each relationship used by a system has a corresponding relationship specification that is part of the System Essential Model. This is referred to as a system relationship specification. System relationship specifications are discussed in §2.6.3.

The prefix may be dropped if no confusion results.

2.6.1.2 Stand–alone systems

If the system is truly stand-alone, there is only one relationship specification. This serves as both an enterprise and system specification.

The Enterprise Essential Model may be regarded as being 'folded over' into the System Essential Model. See also "The Relationship between Systems and the Enterprise".

2.6.2 Enterprise relationship specification

2.6.2.1 Purpose

Each relationship has an enterprise relationship specification that acts as the basic 'repository' of information about the relationship.

2.6.2.2 Examples

RELATIONSHIP : covers

PARTICIPATING ENTITIES : <Course> covers <Topic>.

MEANING : A <Course> is said to cover a <Topic> if there is an intention that the course should discuss the topic to some depth.

PARTICIPATION : <Course>   : optional
                 <Topic>   : optional

UPPER LIMITS :   <Course>   : N
                            <Topic>   : N

RULES : None

STORAGE/OCCURRENCE : 20 bytes

OCCURRENCES/TIME :

date	: expected	: minimum	: maximum
01/10/90	: 200	: 180	: 250
01/01/91	: 260	: 235	: 300

RELATIONSHIP : teaches

PARTICIPATING ENTITIES : <Instructor> teaches <Scheduled course>.

MEANING : An <Instructor> teaches a <Scheduled course> means the instructor has been allocated to run the seminar. This is set up at any time after the seminar is scheduled. It may be broken at any time before the Scheduled course starts.

PARTICIPATION : <Instructor>   : optional
                 <Scheduled course> : optional

UPPER LIMITS :   <Instructor>   : N
                     <Scheduled course>   : 1

RULES : The Instructor must be one who can teach the <Course> that the <Scheduled course> refers to.

STORAGE/OCCURRENCE : 20 bytes

OCCURRENCES/TIME :

date	: expected	: minimum	: maximum
01/10/90	: 100	: 80	: 235

2.6.2.3 Components

2.6.2.3.1 Relationship

The name of a relationship need not be unique across the enterprise, but two relationships that involve the same entities must have distinct names.

The relationship name, does *not* act as the identifier for the relationship. This is because there may be several relationships with the same name (as long as they refer to different entities). The participating entities entry acts as the identifier for the relationship.

2.6.2.3.2 Participating Entities

This entry gives the relationship frame (see §2.2.3.4.2). For example, the relationship "covers" seen in the diagram:

has a specification including:

PARTICIPATING ENTITIES : <Course> covers <Topic>.

This should be read as:

> *Each occurrence of the relationship corresponds to the fact that a specific Course covers a specific Topic.*

Without this definition, the entity relationship diagram would be ambiguous — the diagram could be read in either direction.

Higher-order relationships have more than two 'slots' in the relationship frame.

For example:

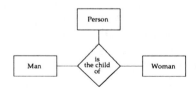

has a specification including:

PARTICIPATING ENTITIES : <Person> is the child of <Woman> and <Man>

would have a specification including:

> **PARTICIPATING ENTITIES :** <Employee> reports to <Employee>(manager)

For symmetric relationships, the position in the relationship frame is of no significance. The relationship "<Person> is friend of <Person>" (see §2.2.3.4.8), for example, does not require the roles to be distinguished.

Choice of relationship frame
There are many ways a frame format can be constructed. Ignoring the other text, there are two orders in which the entity slots can be given for a binary relationship and $N*(N-1)$ possible sequences for a relationship of order N. (There are thus five secondary possible sequences for a relationship of order three.)

The frame should be constructed so that the most natural sentence results when individual occurrences of the entities are entered. The verb should be active (not passive).

2.6.2.3.3 Meaning
This gives a general description of the significance of an occurrence of the relationship to the enterprise. For example, the specification of the relationship "covers" includes:

> **MEANING :** A <Course> is said to cover a <Topic> if there is an intention that the course should discuss the topic to some depth.

This should give the meaning of one ioccurrence of the relationship, not the significance of the whole collection of relationship occurrences that involve one specific occurrence of an entity. For example, in the relationship "<Woman> is mother of <Child>", although it is true that a woman may have several children and a child may only have one mother, this is not stated in the meaning, but in participation and upper limits (see §2.6.2.3.4 and §2.6.2.3.5). For one occurrence of the relationship the meaning of the relationship is clearly stated without any caveats of the form 'from the point of view of'.[1]

---

1 Examining an entity from the point of view is useful in understanding the entity and establishing these limits. YSM regards these techniques as useful heuristics, but there should be a single meaning, describing what one occurrence of the relationship means in real-world terms.

2.6.2.3.4 Participation

For each entity that participates in the relationship, this defines whether the relationship is mandatory or optional.[1] A relationship is mandatory for a specific entity if an occurrence of that entity must participate in the relationship. For each occurrence of the entity there must be at least one occurrence of the relationship. For example, the relationship "<Woman> is mother of <Person>", has a specification which includes:

```
PARTICIPATION :    <Woman>     : optional
                   <Person>    : mandatory (1)
```

The number in parentheses after the mandatory entry denotes the minimum number of occurrences that there must be. If this minimum is "1", then it may be omitted. The parentheses are then not shown.

Examples where there is a minimum number other than 1 include "<Citizen> is member of <Jury>". Another example is the relationship "<Bridge hand> contains <Playing card>", which would have:

```
PARTICIPATION :    <Bridge hand>      : mandatory (13)
                   <Playing card>     : optional
```

This states that a specific bridge hand never has less than thirteen occurrences of the relationship. In other words, each bridge hand contains at least thirteen cards. (In fact, it contains *exactly* 13 cards — see discussion in §2.6.2.3.5.)

Role identification

If the relationship is recursive, then the role is significant and must be identified. For example, the relationship "<Employee> reports to <Employee>(manager)", has a specification which includes:

```
PARTICIPATION :    <Employee>            : mandatory
                   <Employee>(manager)   : optional
```

This states that any employee must always have a manager and that a manager may (or may not) have employees that report to them.

Participation and instance diagrams

If an instance diagram is drawn for the relationship, then there is a simple 'visual' equivalence between the diagram and the participation parameters for the relationship. If the relationship is mandatory for an entity, then there must be at least one line from each entity occurrence; if it is optional for that entity, there may be occurrences of that entity with no lines from them. For the relationship "<Person> is the child of <Woman> and <Man>", for example, the instance diagram would be:

---

1 It is the participating of the entity that is optional, not the relationship as a whole — each occurrence of the relationship corresponds to exactly one <Woman> and one <Person>.

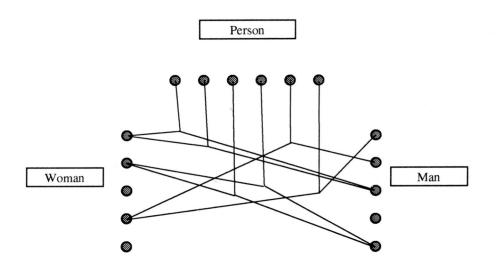

2.6.2.3.5 Upper limits

This entry defines maximum number of times a specific occurrence of one of the entities may participate in the relationship. If there is no upper limit, then this is denoted by "N". For example, the relationship "<Woman> is mother of <Person>", has a specification which includes:

UPPER LIMITS :	<Woman>	: N
	<Person>	: 1

This specifies that a woman may have more than one child and that a person may not have more than one parent.

The most common entries for this parameter are "1" and "N". In some (relatively rare) situations, a fixed limit may be defined. For example, in an enterprise that produces electronic hands for teaching bridge, the relationship "<Bridge hand> contains <Playing card>" would have:

UPPER LIMITS :	<Bridge hand>	: 13
	<Playing card>	: N

This states that a specific bridge hand never has more than thirteen occurrences of the relationship. In other words, each bridge hand contains at most thirteen cards. (In fact, it contains *exactly* 13 cards — see discussion in §2.6.2.3.4.)

Upper limits as a specific type of rule of association

Note: the upper limit entry may be regarded as a specific type of rule of association (see §2.6.2.3.7). For example, if the upper limit is "1", then the rule of association would state that there could not be a pre-existing occurrence of the relationship for that entity occurrence. YSM provides the upper limit entry because this type of restriction is very common. Many relationships can be specified using this, without the need to specify a rule of association. Rules of association are more general, but also more abstract and difficult to use.

2.6.2.3.6 Upper limits and instance diagrams

The upper limit has a simple graphic interpretation for the instance diagram of a relationship. It is the maximum number of lines that link to an occurrence of the entity. In the above example, no <Person> have more than one line linked to them, but a <Woman> or a <Man> may have several.

Cardinality of binary relationships

Binary relationships are so common that a shorthand description is often applied to them. There are three cases:

1. **The upper limit is 1 for both entities.** This is referred to as a '1:1 relationship' (read as 'one to one relationship'). For example, "is married to" is a 1:1 relationship in most societies.

2. **The upper limit is 1 for one entity and N for the other entity.** This is referred to as a '1:N relationship' (read as 'one to many relationship'). For example, "is mother of" is a 1:N relationship. Informally, 'One woman is the mother of many children'. Note: it is the mother that has multiple occurrences of the relationship referring to her. Each child has exactly one occurrence referring to him or her.

3. **The upper limit is N in both cases.** This is referred to as an 'M:N relationship' (read as 'many to many relationship'). For example, "covers" is an M:N relationship.

Annotating ERDs to show cardinality of binary relationships

Entity relationship diagrams are sometimes annotated to show the cardinality of binary relationships as illustrated below:

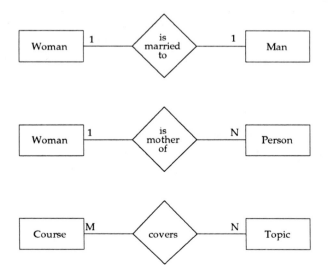

This traditional notation is a little misleading. For example, in the "is mother of" relationship, the multiple occurrences of the relationship relate to the <Woman>. Furthermore, it cannot be extended to higher-order relationships. An alternative notation, shown below, is more general and capable of dealing with all possibilities:

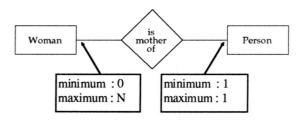

In a suitable support environment, this type of information should be capable of being retrieved (for example 'popped-up') when the graphic view is being examined. It should not be 'hard-coded' on the diagram.

In a 'pencil and paper' environment, there is little choice but to annotate the diagram with 'hard-coded' text. However, diagram annotation should be used with caution — the primary role of diagrams is to declare components and communicate their meaning by suitable choice of names. Information about the components of a diagram is best left as textual support.

Cardinality of higher-order relationships
For higher-order relationships, there is no such convenient terminology as the '1:1', '1:N' and 'M:N' notation. However, the upper limits to the number of occurrences of the relationship are still stated as maximum values for each entity.

For example, the relationship "<Man> was married to <Woman> at <Location>" would require limits to be set for each of the entities. In this case, we might find that each entity might participate in the relationship many times. In the absence of any other constraint, this implies that each man might have been married at any one of many locations. There might be a curious rule stating that each man could only remarry at the first location he got married at — this would not show up as a numeric constraint, but, rather, in the rules of association for that relationship.[1]

2.6.2.3.7 Rules of association
Many relationships and associative entities have rules that specify a condition or conditions that must be satisfied by entities participating in the relationship. These are termed rules of association, or just rules if no ambiguity results. These rules may be stated in:

1. **Unstructured text**: in simple cases, a fairly informal presentation of the rules may be given in text. This is user-friendly, but in complex cases may lead to rules which cannot be understood (and thus cannot be verified).

---

1 See [Len83] for an earlier, rather formal discussion of cardinality. The specification method described here is able to deal with the same aspects of specification as those described in that reference. However, it is (hopefully) less abstract and more suitable for general use. In particular, it will work well in automated support environments.

2. **Rule lists**: a more formal approach may be adopted to defining the rules of association. These techniques are described in more detail in §6.3, but a brief introduction is given below.

Structure of rule list

The simplest case is when the rule only involves the occurrences of the entities that participate in the relationship. These are identified by giving them specific names and then making assertions about them. All of the assertions statements about the occurrences of the entity must be true in order for the relationship to be established.

These assertions, which are called rule clauses, may involve the following:

- statements about values of attributes for that occurrence;

- equality of occurrences;

- negations of either of the preceding types of rule clause.

For example, for the relationship "<Man> is married to <Woman>", we might have:

```
RULES : Man is married to Woman
        satisfies:
    1.      Man.age ≥ 18
    2.      Woman.age ≥ 18
```

This introduces the participants in the relationship and gives them names by which they can referred to ("Man" and "Woman"). The two rule clauses then require that both the man and the woman must be 18 or over before they are allowed to participate in the relationship. Attributes are identified using the "." notation. Each rule clause evaluates to TRUE or FALSE for any specific occurrences used in it. The list of clauses specify whether the specific case, with the named occurrences, is allowed. Negation may be used, so:

```
RULES : Man is married to Woman
        satisfies:
    1.      NOT(Man.age < 18)
    2.      NOT(Woman.age < 18)
```

is completely equivalent to the preceding example.

Choice of name of occurrence

The above rule may also be defined by:

```
RULES : X is married to Y
        satisfies:
    1.      X.age ≥ 18
    2.      Y.age ≥ 18
```

showing that any name may be used to identify the occurrence of the entity.

2.6.2.3.8 Storage/occurrence

This gives the likely amount of space for stored information referring to one occurrence of the relationship.

This estimate is given prior to any implementation. After implementation decisions have been taken, more detailed information of the size of one occurrence will be available. This will be covered in later releases of YSM.

2.6.2.3.9 Occurrences/time

This is used to estimate the number of occurrences of the relationship and is given for one or more dates (planning points). This is as described in §2.5.2.3.8.

2.6.3 System relationship specification

2.6.3.1 Purpose

Each relationship accessed by the system has a corresponding system relationship specification, which describes the use that the system makes of it.

The system may be responsible for creating and deleting occurrences of the relationship, or it might need to check whether or not specific occurrences of the relationship exist.

This specification allows the System Essential Model to be constructed so that system functions use the available enterprise information in the intended way.

Note: if a system only uses an associative entity in its relationship form, then it appears on system ERDs as a relationship. It has an enterprise associative entity specification and a system relationship specification.

2.6.3.2 Example

```
┌─────────────────────────────────────────────────────────────────────────────┐
│ ┌─────────────────────────────────────────────────────────────────────────┐ │
│ │ SYSTEM : Course booking                                                  │ │
│ └─────────────────────────────────────────────────────────────────────────┘ │
│ ┌─────────────────────────────────────────────────────────────────────────┐ │
│ │ RELATIONSHIP : teaches                                                   │ │
│ └─────────────────────────────────────────────────────────────────────────┘ │
│ ┌─────────────────────────────────────────────────────────────────────────┐ │
│ │ PARTICIPATING ENTITIES : <Instructor> teaches <Scheduled course>.        │ │
│ └─────────────────────────────────────────────────────────────────────────┘ │
│ ┌─────────────────────────────────────────────────────────────────────────┐ │
│ │ USE :  create, delete, match                                             │ │
│ └─────────────────────────────────────────────────────────────────────────┘ │
│ ┌─────────────────────────────────────────────────────────────────────────┐ │
│ │ MEANING : An <Instructor> teaches a <Scheduled course> means the instruc-│ │
│ │           tor has been allocated to run the seminar. This is set up at   │ │
│ │           any time after the seminar is scheduled. It may be broken at   │ │
│ │           any time before the Scheduled course starts.                   │ │
│ │           The system must inform the instructor if the course is cancelled.│ │
│ └─────────────────────────────────────────────────────────────────────────┘ │
└─────────────────────────────────────────────────────────────────────────────┘
```

2.6.3.3 Components

2.6.3.3.1 System

This identifies the system which uses the relationship.

2.6.3.3.2 Relationship
This is the name as it appears on system ERDs. Each relationship has a name that is unique to the group of entities that it relates.

2.6.3.3.3 Participating entities
This gives the relationship frame for the relationship. It is the identifier for the relationship.

2.6.3.3.4 Use
The access to the relationship is given here. The allowed values are:

- **create**: the system is responsible for the creation of occurrences of the relationship. Note: this implies that it will enforce all the rules and participation limits defined for that relationship (see §2.6.2.3.7).

- **delete**: the system may delete occurrences of the relationship. If there is a constraint on the minimum number of participations in a relationship for an entity, the system is responsible for checking that the constraint is not violated.

- **match**: the system needs to check whether there are any occurrences of the relationship that satisfy a selection criterion. This is corresponds to finding which occurrences of entities participate in the relationship.

Any combination of these keywords may be given.

2.6.3.3.5 Meaning
This gives the significance of the relationship to the system. Depending on the support environment, enterprise 'meaning' may be visible. Any significance that is specific to the system should be noted.

2.6.4 Rules

2.6.4.1 Identification

1. It is illegal for two occurrences of the relationship to refer to the same occurrences of each of the entities involved in the relationship. Specifically, this is a restriction on M:N relationships — if several occurrences of a relationship may occur between the same occurrences of the entities, then the relationship must be modelled as an associative entity. (For example, "<Manager> gave order to <Worker>" would have to be modelled as an associative entity, with "time the order was given" as a possible additional component of the identifier.)

2.6.4.2 Role identification in recursive relationships:

1. For recursive relationships that are not symmetric, the distinction between the multiple roles that the entity may play in the relationship must be defined for the relationship to be well-defined. At most one of the multiple references to the entity may be left without a role and the others must all be given distinct roles in the relationship. For symmetric relationships, the role for the repeated entities does not have to be distinguished.

2.6.4.3 Participation and upper limits

1. If there is a mandatory lower limit for the number of occurrences of a relationship, then the upper limit for the same entity must be greater than (or equal to) this number.

2.6.4.4 Number of occurrences

1. At any time point, the maximum number of occurrences must be greater than (or equal to) the minimum number of occurrences.

2.6.5 Guidelines

2.6.5.1 Choice of relationship frame

1. The relationship frame should be chosen so that it forms the sentence that is the fact of most importance to the enterprise. For example, "<Course> covers <Topic>" might be regarded as being potentially more important than "<Topic> is discussed in <Course>".

2. Sometimes the existence or not of an occurrence of the relationship involving an occurrence of one of the entities that participate in the relationship is of particular importance. In such cases that entity should be chosen to be the first entry in the relationship frame.

3. The relationship frame should be in the active (rather than passive) voice — in other words "covers" is an acceptable name for a relationship, "is covered by" is not. (In some cases, for example "is mother of" or "is child of", the entities could be given in either order.

2.6.5.2 Conversion to associative entity

1. Trace relationships (those that indicate an interaction at a point in time, rather than a continuing relationship) are more likely to be associative entities than pure relationships. This is because the time at which they occurred is likely to be an attribute.

2.7 Associative Entity Specification

2.7.1 Purpose

An associative entity specification is used to define the role of an associative entity. It defines the role of the associative entity, both as an entity and a relationship.

An associative entity specification may be regarded as a combination of an entity specification and a relationship specification.

2.7.1.1 Enterprise and system use of associative entity specifications

Each associative entity used by the enterprise has a corresponding enterprise associative entity specification that is part of the Enterprise Essential Model. This is referred to as an enterprise associative entity specification. Enterprise associative entity specifications are discussed in §2.7.2.

If a system uses an associative entity as both an entity and a relationship, then there is a corresponding system associative entity specification that is part of the System Essential Model. This is referred to as a system associative entity specification and defines the system's use of that associative entity. (Note: if the system only uses an associative entity as a relationship, then it has a system relationship specification; if the system only uses it an associative entity as an entity, then it has a system associative entity specification.) System associative entity specifications are discussed in §2.7.4.

The prefix enterprise or system may be dropped if no ambiguity results.

2.7.1.2 Stand–alone systems

If no separate enterprise modelling is carried out, then there is exactly one such specification for each associative entity. This is referred to as the associative entity specification — the Enterprise Essential Model may be regarded as being 'folded over' into the System Essential Model.

2.7.2 Enterprise associative entity specification

2.7.2.1 Purpose

Each associative entity has an enterprise associative entity specification that acts as the basic 'repository' of information about the associative entity.

2.7.2.2 Example

ASSOCIATIVE ENTITY : Scheduled Course	

PARTICIPATING ENTITIES : <Course> is scheduled to run at <Location>, starting on <Day>.

MEANING : Each <Scheduled Course> corresponds to a plan to run a standard <Course> at a given <Location> with the first day of the Scheduled Course being <Day>.

ATTRIBUTES : id_no
date scheduled
repeat number

IDENTIFIERS : id_no
<Course> + <Location> + <Day> + repeat number

PARTICIPATION :

<Course>	: optional
<Location>	: optional
<Day>	: optional

UPPER LIMITS :

<Course>	: N
<Location>	: N
<Day>	: N

RULES : The Course must be one whose current state is available.

CONSTRAINTS : —

STORAGE/OCCURRENCE : 130 bytes

OCCURRENCES/TIME :

date	:expected	: minimum	: maximum
01/10/90	: 145	: 130	: 160
01/01/91	: 350	: 330	: 450

2.7.2.3 Components

2.7.2.3.1 *Associative entity*
This is given to identify which associative entity is being specified. For example:

> **ASSOCIATIVE ENTITY :** Scheduled course

2.7.2.3.2 *Participating entities*
The associative entity acts as a relationship linking specific entities. For example:

> **PARTICIPATING ENTITIES :** <Course> is scheduled to run at <Location>, starting on <Day>.

This is referred to as the relationship frame (see §2.2.3.4.2).

2.7.2.3.3 *Meaning*
This gives the significance of the associative entity to the enterprise.

> **MEANING :** Each <Scheduled course>[1] corresponds to a plan to run a standard <Course> at a <Location>, with the first day of the Scheduled course being <Day>.

2.7.2.3.4 *Attributes*
An attribute associates a value with each occurrence of the entity. For example:

> **ATTRIBUTES :**   id_no,
>   date scheduled,
>   repeat number

All attributes of the associative entity that are significant to the enterprise must be defined.

2.7.2.3.5 *Identifiers*
Each associative entity must have at least one method by which one occurrence of the associative entity can be distinguished from another. An identifier is a way of achieving this. The identifier may be a combination of references to entities that participate in the relationship and attributes of the associative entity. Identifiers may consists of attributes, entity references, or a combination of both.

Single attribute used as identifier
For a single–attribute identifier, given the value of this attribute, it is possible to find a single occurrence of the associative entity. For example, the associative entity "Scheduled course" has:

> **IDENTIFIERS :** id_no

---

1 In textual specifications an occurrence of an associative entity is indicated by enclosing the associative entity between "<" and ">". Thus "<Scheduled course>" should be read as "occurrence of Scheduled course". Note: in the relationship frame, the "< >" entry is not an occurrence — it is a 'place holder' for a potential occurrence.

Reference to entity used as identifier

For an entity–reference identifier, the occurrence of the relationship can be uniquely determine from that entity. For example, the associative entity "Birth" (with a relationship frame "<Woman> is mother of <Person>"), could have the following identifier:

> **IDENTIFIERS : <Person>**

Note: a single entity–reference identifier is only allowed if the upper limit of participation for that entity is 1.

Compound identifiers

Compound identifiers are made up of more than one attribute and entity reference. Given the value for each attribute and a single entity occurrence for each entity reference these identify a single occurrence of the associative entity.

These 'compound' identifiers are shown as a list of items, with a "+" symbol between them (which should be read as "together with"). For example, the relationship the associative entity "Birth" (with a relationship frame "<Woman> is mother of <Person>"), could have the following identifier:

> **IDENTIFIERS : <Woman> + child_number**

although there are much better possible identifiers for this example. An identifier that consists of reference to all the entities that participate in the relationship is quite often used as an identifier. Sometimes, however, this is not enough to identify the specific occurrence — there might be several relationships between the same combinations of entities. For example, the associative entity "Meeting" (with a relationship frame of "<Person> met <Person> at a given <Location>") might have:

> **IDENTIFIERS : <Person> + <Person> + time of meeting**

Because the same two people might meet many times (even at the same location), the attribute "time of meeting" is needed to distinguish between difference occurrences of "Meeting". However, the same two people cannot meet at different locations at the same time, so "Location" is not required as part of the identifier.

Multiple identifiers

There must be at least one, but there may be several identifiers and in that case, they are just listed in any order. For example:

> **IDENTIFIERS : id_no,**
> **<Course> + <Location> + <Day>+ repeat number**

declares two identifiers "id_no" and "<Course> + <Location> + <Day> + repeat number".

2.7.2.3.6 Participation

This defines whether the associative entity is optional or mandatory in its role as a relationship.

For example, for the associative entity "Birth", with the relationship frame "<Woman> is mother of <Person>", the following holds:

```
PARTICIPATION :
            <Woman>        : optional
            <Person>       : mandatory
```

This is discussed in more detail in §2.6.2.3.4.

2.7.2.3.7 Upper limits

This defines the upper for the number of occurrences of the associative entity for each occurrence of one of the entities it refers to. For example, for the associative entity "Birth", with the relationship frame "<Woman> is mother of <Person>", the following holds:

```
UPPER LIMITS :
            <Woman>        : N
            <Person>       : 1
```

If there is no upper limit, then this is indicated as "N".

This is discussed in more detail in §2.6.2.3.5.

Cardinality of binary associative entities

A binary associative entity has two entity occurrences for each occurrence of the associative entity. Binary associative entities are so common that a shorthand description is often applied to them. There are three cases:

1. **The upper limit is 1 for both entities**: these are referred to as "1:1 associative entities". In most societies "marriage" is a 1:1 associative entity.

2. **The upper limit is 1 for one entity and N for the other entity**: these are referred to as "1:N associative entities". "Birth", which refers to <Woman> and <Person>, with the time of birth as a dependent attribute, is a 1:N associative entity.

3. **The upper limit is N in both entities:** these are referred to as "M:N associative entities". "Friendship" between one <Person> and another <Person> is an M:N associative entity.

2.7.2.3.8 Annotating ERDs to show cardinality of binary associative entities

In some situations, it is helpful to highlight the cardinality of the associative entity on the diagram. For binary associative entities, the notation below may be used:

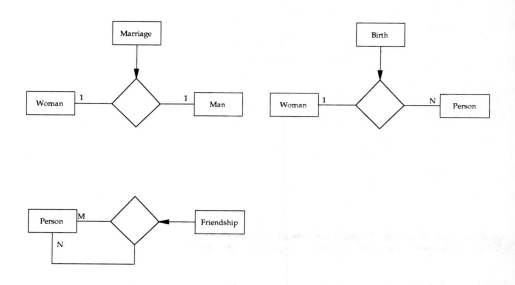

Higher-order associative entities

For higher-order relationship frames, there is no such convenient terminology. However, the upper limits must still be stated in the above way. For example:

```
UPPER LIMITS :
           <Location>      : N
           <Day>           : N
           <Course>        : N
```

2.7.2.3.9 *Rules of association*

Many associative entities have rules that specify a condition (or conditions) that must be satisfied by entities participating in the relationship frame. This is usually referred to as a "rule of association", or just "rule" if no confusion results. These may be stated in unstructured text or a rule list. In the specification of Scheduled course we might have:

```
RULES : The Course must be one whose status is available.
```

In this example, the state variable "status" of the entity "Course" is accessed. (The course might still be in development, or obsolete.)

In a rule list, a more powerful grammar may be used. See §2.6.2.3.7 for further discussion of rules of association.

2.7.2.3.10 Constraints

Where there are specific constraints regarding the associative entity, these may be entered here. See §2.5.2.3.6 for comparable example for an entity.

2.7.2.3.11 Storage/occurrence

This gives the likely amount of space for stored information referring to one occurrence of the associative entity.

This estimate is given prior to any implementation. After implementation decisions have been taken, more detailed information of the size of one occurrence will be available. This will be covered in later releases of YSM.

2.7.2.3.12 Occurrences/time

This is used to estimate the number of occurrences of the associative entity. This is as described in §2.5.2.3.8.

2.7.3 Associative entities that are subtypes

If the associative entity has been declared to be a subtype of another entity, then all attributes of the supertype are also attributes of this associative entity (see the discussion of supertypes in §2.2.3.8). In some situations, the associative entity may "inherit" attributes from several supertypes (see §2.2.3.6.1).

2.7.3.1 Inherited meaning

Note: where the associative entity is a subtype of another entity this description should not repeat information given in either the subtyping specification or the specification of the supertype. This inheritance of meaning by subtypes in general is discussed in §2.5.3.5.

2.7.3.2 Inherited identifiers

A subtype always inherits all the identifiers of its supertype. It may also have additional identifiers that are only available for this subtype and not others. This is discussed in §2.5.3.3.

2.7.3.3 Subtypes of associative entities

A subtype of an associative entity is itself an associative entity and "inherits" the participating entities of its supertype. In this case, the entities that constitute the inherited relationship frame may also usefully be shown.

2.7.3.4 Visibility of inherited properties

See §2.1.8.3 for discussion of 'visibility' in different technical support environments.

2.7.4 System associative entity specification

Each associative entity accessed by the system has a corresponding system associative entity specification which describes the use that the system makes of it.

This specification allows the System Essential Model to be constructed so that system functions use the available enterprise information in the intended way.

If the system does not access any of the attributes or state variables of the associative entity, or deal with any relationships involving the associative entity, then it should have a system relationship specification (see §2.6.3). If the system makes no use of the fact that the associative entity is a relationship, then it should have a system entity specification (see §2.5.4).

2.7.4.1 Example

SYSTEM : Course booking

ASSOCIATIVE ENTITY : Public course

PARTICIPATING ENTITIES : <Course> is scheduled to run at <Location>, starting on <Day>.

USE : match, read

MEANING : A <Public course> is a <Scheduled course> for which bookings may be taken for individual delegates. Each <Scheduled course> corresponds to a plan to run a standard <Course> at a given <Location> with the first day of the Scheduled course being <Day>.

ATTRIBUTES :	**attribute**	**: access**
	sequence number	: read
	date scheduled	: read

IDENTIFIERS : sequence number

STATE VARIABLES : status   : check, change

2.7.4.2 Components

2.7.4.2.1 *System*
This gives the name of the system whose use of the associative entity is being defined.

2.7.4.2.2 *Associative entity*
This gives the name of the associative entity whose use is being described.

2.7.4.2.3 *Meaning*
This gives the significance of the associative entity to the system.

2.7.4.2.4 *Use*
The access to the associative entity is defined here (access to individual attributes is defined below). The allowed values are:

- **create**: the system is responsible for the creation of occurrences of the associative entity. This implies that it will enforce all the rules and participation limits defined for that associative entity.

- **read**: the system uses the entity but does not create or delete occurrences of it. The use may include reading attributes and carrying out match tests. In using the associative entity as a relationship, the system may identify which occurrences of the entities actually do participate in the relationship.

- **update**: the system changes the values of some or all of the attributes

- **delete**: the system may delete occurrences of the associative entity. If there is a constraint on the minimum number of participations in the relationship frame of the associative entity, the system is responsible for checking that constraint is not violated.

- **match**: the system needs to check whether there are any occurrences of the associative entity that satisfy a selection criterion and possibly identify which occurrences of entities they refer to.

- **check:** the system needs to examine a state variable to determine a correct response;

- **change**: the system is responsible for changing the state of a state variable.

2.7.4.2.5 Attributes
Any attribute that the system reads or writes is visible in this view. Each attribute has a corresponding attribute specification (see §2.8). It is also shown here as being "read" or "updated" by the system. If an attribute is read and modified by the system, it is shown as "read/update".

2.7.4.2.6 Identifiers
Each associative entity must have at least one identifier by which the system can distinguish one occurrence of the associative entity from another. All such identifiers that are used by the system must be defined here. There must be at least one, but there may be several and in that case, they are just listed in any order.

2.7.4.2.7 State variables
This entry lists all state variables that are checked or changed by the system. If the system needs to know the status of a state variable, then the value entered is "check"; if the system is responsible for changing the status of the state variable, the value entered is "change". If the system does both, then "check, change" is entered.

2.7.5 Rules
Note: there are guidelines for the associative entity in its role as a relationship and also in its role as an entity. These are listed below, although they duplicate guidelines given for entity specifications and relationship specifications.

2.7.5.1 Names
1. The name of each attribute must be unique within the attributes of that associative entity (but may be the same as an attribute of another entity or associative entity).

2.7.5.2 Identifiers
1. At least one identifier must be given.

2. For each identifier that consists of a single item (either an attribute or a reference to an entity), given values for the item may only have one occurrence of the associative entity with that value.

3. For identifiers that are composed of more than one item, no two occurrences of an associative entity may have the same set of values for all the attributes and the same entity occurrences for the entity references that are part of that identifier.

2.7.5.3 Participation and upper limits

1. If there is a mandatory lower limit for the number of occurrences of the associative entity, then the upper limit for the same entity must be greater than (or equal to) this number.

2.7.5.4 Number of occurrences

1. At any time point, the maximum number of occurrences must be greater than (or equal to) the minimum number of occurrences.

2.7.6 Guidelines

Note: there are guidelines for the associative entity in its role as a relationship and also in its role as an entity. These are listed below, although they duplicate guidelines given for entity specifications and relationship specifications.

2.7.6.1 Attribute names

1. Names of attributes should be as simple as possible and not duplicate meaning of the associative entity (because the name of the attribute is never used, except in conjunction with the associative entity). Thus "date scheduled" is preferred to "date course scheduled".

2. Names of attributes should be a good clue to what they are used for — user review is important here. As a general guideline, the following two sentence structures should make sense to a user:

    The a of that e is v.

    What is the value of a for that e?

    where "a" should be replaced by the proposed name ; "e" replaced by the associative entity name and "v" by a value. For example:

    The date booking made for that booking is 26 August 1990.

    and

    What date was that booking made on?

    (Note that a little re-ordering was carried out to make the language more natural.)

2.7.6.2 Identifiers

1. Identifiers should be naturally used in the enterprise activities and not artificially constructed by the information modeller.

2. If the only way of distinguishing between occurrences of an associative entity is by means of a combination of attributes, then it *may* be convenient to construct an artificial identifier. This should be used with care, as it runs counter to the guideline above.

2.7.6.3 Relationship definition

1. The participating entity field should give a description in a form that can be used as a relationship frame.

2.8 Attribute Specification

2.8.1 Purpose

Each attribute has a specification that is part of the Enterprise Essential Model. This specification describes the meaning and values of that attribute.

If a system uses an attribute of an entity, then this specification is 'visible' to the System Essential Model. To check the System Essential Model, the attribute specification must be used. However, it remains part of the Enterprise Essential Model. See §2.1.8.3 for further discussion of "visibility".

2.8.2 Examples

ENTITY : Course

ATTRIBUTE : acronym

MEANING : A short description used within the organisation for identification purposes.

RETENTION : stored

TYPING : abstract

ADT : alphanumeric

PARAMETERS : minimum length = 3
maximum length = 6

RESTRICTIONS : first character alphabetic

ENTITY : Playing card

ATTRIBUTE : suit

MEANING : A distinguishing characteristic that divides a deck of cards into four equal sets.

RETENTION : stored

TYPING : value list

VALUES : Hearts, Clubs, Diamond, Spades

2.8.3 Components

2.8.3.1 Entity

The name of the entity is entered here. It is used to help identify the attribute being defined. For example:

> **ENTITY :** Course

2.8.3.2 Attribute

The name of the attribute is entered here. It associates a value with each occurrence of the entity. For example:

> **ATTRIBUTE :** acronym

The attribute name, together with the entity name, identifies the attribute being defined. (Two entities may have attribute with the same name — these are two different attributes.)

2.8.3.3 Meaning

The significance of the attribute in terms of what it means for each occurrence of the entity is defined here. Each attribute should correspond to a named quality that has an identifiable value for each occurrence of the entity. The meaning should reflect the meaning of that quality. For example:

> **MEANING :** A short description used within the organisation for identification purposes.

2.8.3.4 Retention

This defines whether the attribute is:

- stored: the attribute will be given a value at one point in time. This value can be read ('retrieved') at a subsequent time.

- temporary: used in intermediate calculation. For example, a data flow may be an input to the system and used in determining response, but its value may not need to be stored.

- derived: recalculated from existing information every time its value is required.

For example:

> **RETENTION :** stored

YSM recommends that all information is attributed (all information is 'about something'). Not all of this information needs to be stored. Consider a reaction vessel, containing a mixture of chemicals that need to be controlled at a certain temperature. This required temperature needs to be stored (perhaps as the attribute "desired temperature" of the entity "reaction"). In controlling the reaction, the "current temperature" and "error" (deviation from desired value) might be used as temporary values. These do not need to be stored. Nevertheless, they are descriptive of the reaction.

See §2.18.3.6.1 for further discussion of non-stored attributes.

Only very few items (such as messages) cannot be attributed. Even in the case of messages, the message is often 'about something' and should therefore be attributed. Attribution is a way of organising and understanding information.[1]

2.8.3.5 Derivation

For attributes that are derived, this gives the rule that is used to calculate the value of the attribute. The minispec does not restate the derivation policy. At any time that this attribute is referenced, the derivation is used to recalculate the value of this attribute. For example, the attribute "number of bookings" of the associative entity "Public course" describes the number of bookings that refer to this course. It is a property of the course, but it can be calculated from other available information. In this case, the definition would be:

RETENTION : derived
DERIVATION : the number of occurrences of booking with status = firm or status = provisional that refer to this public course.

In a System Essential Model, a derived attribute is referred to without explicitly stating the derivation. The derivation is 'hidden inside' the derived attribute. In an implementation, the derivation will need to be explicitly allocated. It *could* be allocated to an application process that used this attribute. However, it might also be hidden inside an object-oriented data base. The approach used in YSM allows this decision to be deferred. The policy of how the derivation is carried out is stated only once — in the specification of the attribute.

2.8.3.6 Typing

The values for the attribute may either correspond to an abstract data type (ADT) or be explicitly listed. Attributes that use ADTs may be used with any operations defined for that ADT. Attributes that are defined as a list of values may only be tested for equality with one of the values.

Allowed values in this entry are "abstract" or "value list", for example:

TYPING : abstract

2.8.3.7 Values

If the entry for "typing" is "value list", then the individual allowed values are listed here. For example:

Attributes that are defined in this way may only be:

- checked for equality with one of these values (which are treated as constants);
- compared with the value for the same attribute of a different occurrence of the same entity.

Note: if the entry for "typing" is "abstract" then this entry is not present.

---

[1] The approach used by Yourdon in the past (the DeMarco approach) that there are many data items, each with a unique name is no longer supported. YSM is now very clearly committed to the use of entity–relationship–attribute modelling.

2.8.3.8 ADT

If the entry for "typing" is "abstract", then name of the abstract data type is entered here. For example:

```
ADT : date
```

An attribute defined in terms of an ADT may be:

- compared with another attribute of the same abstract data type;
- compared with a value contained in a data flow of the same abstract data type;
- compared with a fixed value, using a constant in units that are supported by that abstract data type (see §2.10.4.3 for a further discussion of supported units).

Abstract data types may be existing 'standard' ADTs, or they may be user-defined (see §2.10).

2.8.3.8.1 Examples of use of abstract data types

Dates

In commercial systems, many attributes are of type "date". In order to ensure that they are all in a consistent format, it is important to define how they will be stored in one place only. This is achieved by using the standard ADT "date" for all of these items.

Heights

For a system that tracks flights over mountains and also helps 'talk pilots in' when they are landing at different airports, we might identify the entities "mountain", "aeroplane", "runway", "pilot" ....

The aeroplane would have an attribute, its "current altitude" that would need to be compared with the "elevation" of the mountain and the "height above sea level" of the runway. It is not enough to say that all these attributes are all numeric, or even to say that they are all integers — they are all *heights*.

These attributes should all therefore be declared to be of type "height". What a "height" represents is defined in one place only — in an abstract data type specification.

2.8.3.8.2 Distinction between ADTs and 'values'

An abstract data type is a set of values that are interpreted as representing a certain class of real-world values. For example, the ADT "height", is not merely a set of possible values (real numbers) but is also an intention to use these values for measuring things. It is clearly different from "weight".

2.8.3.9 Parameters

This field is only applicable to attributes that are defined in terms of an ADT that has associated parameters. For such parameters, the value of the parameter is entered here.

Not all ADTs have such parameters. If there are no parameters associated with that ADT, then this list is not present. Some of the parameters are optional and then they do not have to have a value entered for them.

2.8.3.9.1 Example of use of associated parameters
The attribute "lean body weight" is typed as a "weight":

ENTITY : Patient

ATTRIBUTE : lean body weight

MEANING : Estimated weight of body, after subtracting an estimate for subcutaneous fat.

TYPING : abstract

ADT : weight

PARAMETERS :

parameter	: value
accuracy	: 0.5 kg
noise	:
resolution	: 0.1 kg

RESTRICTIONS : lean body weight < 200 kg

The "weight" ADT has three associated optional parameters — "accuracy", "noise" and "resolution". The accuracy is defined to be 0.5 kg, whereas the resolution is defined to be 0.1 kg. No specific value is allocated to the "noise" parameter, so the default value is assumed.

Note: other attributes could be defined to be a "weight" — for example, the attribute "amount of drug", that is an attribute of the associative entity "actual treatment". Different values would be associated with this attribute. For example, the resolution could be "1 mg" — presumably the drug is weighed more carefully than the patient!

Note: each of these parameters has a type, which is often the same as the type of the attribute. The value entered in such a parameter field must be a legal constant for that type (so where units are used, they must be ones supported by that type). In the examples used here, the supported units are "kg" and "mg".

See §2.10.3.11 for further discussion of the definition of such parameters.

2.8.3.10 Restriction
If the entry for "typing" is "abstract", then the possible values for the attribute are 'inherited' from that data type. It is possible to define a subset of the ADT as the allowed values for the attribute. This is termed a restriction.

Restrictions may be written down fairly informally, or they may use a more formal grammar. An example of an informal restriction is:

RESTRICTION : First character is alphabetic.

which could be used to demand that the first character of an alphanumeric attribute was not numeric.

Each restriction evaluates to TRUE or FALSE for each value inherited from the parent ADT. The allowed values for the attribute are those that give TRUE for the restriction. A more formal definition of this restriction grammar is given §6.5.

2.8.4 Rules

1. For a given values list, the values must be distinct from one another. (The value may be repeated in another attribute value list. It may also be the same as a value in an ADT. However, they must not be compared or assigned in any of these cases.)

2. For an attribute defined using an ADT with associated parameters, the value of each non-optional parameter must be given.

3. Restrictions may only be used with attributes that have an abstract typing.

4. Restrictions may only use functions and operations defined on that ADT.

2.8.5 Guidelines

1. If a set of values are commonly used for different attributes, then it is a good idea to create an ADT with these values (provided the values are used with the same meaning).

2. If an ADT is restricted in the same way for many different attributes, then it is advisable to create a new ADT with these restrictions. The attributes can then be directly defined in terms of this new ADT.

3. If a set of values can be made easier to understand by giving it a name, then an ADT should be created with those values and that name.

2.9 Subtyping Specification

2.9.1 Purpose

The subtyping specification defines the significance of a subtyping in terms of its meaning. There must be a subtyping specification for each subtyping that is used in the Enterprise Essential Model.

The individual subtypes are listed, with a definition of what it is that distinguishes that subtype from others in the subtyping.

See §2.2.3.6 for further discussion of the concept of subtyping.

Note: properties that are common to all subtypes of an entity are not defined here, but in the entity specification for the supertype. If a supertype is defined as having an attribute, then all its subtypes inherit that attribute (see §2.5.3.1).

2.9.2 Examples

SUPERTYPE : Scheduled course

SUBTYPING : availability

MEANING : Shows the distinction between Scheduled courses on which individual delegates can reserve a place from those which they cannot.

COVERAGE : Complete

SUBTYPES :
Public Course : *A course that has been scheduled and advertised by the management in the hope that individual places will be sold for it. Any person can be booked on the course as an individual delegate.*
Inhouse Course : *Individual delegates cannot be booked on the course. Inhouse courses are provided on a contract basis for customers, who then have the responsibility for ensuring that their own employees or contractors attend as required.*

SUPERTYPE : Employee

SUBTYPING : Responsibility

MEANING : Shows the distinction between those employees who can have other employees reporting to them and those employees who cannot.

COVERAGE : Partial

SUBTYPES : Manager : *An employee who may have other employees report to him/her.*

2.9.3 Components

2.9.3.1 Supertype
This gives the name of the supertype is entered. This is an entity having occurrences that fall into several distinct groups. Each categorisation into groups is termed a subtyping.

An entity may have more than one subtyping and the supertype name, together with the subtyping name is used to identify which subtyping is being defined.

2.9.3.2 Subtyping
This gives the name of the subtyping and identifies which of the subtypings of a specific entity is being defined. For example:

```
SUPERTYPE : Scheduled course

SUBTYPING : availability
```

indicates that the subtyping "availability" of the entity "Scheduled course" is being specified here.

For a specific supertype, at most one subtyping may be unnamed — all other subtypings must be named to identify them. For the unnamed subtyping, this entry may be left blank. The subtyping still requires a subtyping.

2.9.3.3 Meaning
The significance of the distinction that the subtyping highlights should be described here. Possible reasons for the distinction include:

- Some of the attributes are only relevant to certain well-defined groups of individual occurrences of the supertype. Other occurrences of this entity do not and could not have values for these attributes, which are not relevant for them.

- Some relationships that the supertype participates in may be restricted to a well-defined group of occurrences of this entity and there are policy reasons why other occurrences of the entity could not take part in the relationship.

For example:

```
MEANING : Shows the distinction between Scheduled courses on which individual
          delegates can reserve places from those which they cannot.
```

involves a distinction on the basis of participation in a relationship. This may be made more explicit by using the relationship frame in the meaning, for example:

```
MEANING : Subtyping on basis of participation in the relationship
          "<Delegate> reserves place on <Scheduled course>".
```

2.9.3.4 Coverage
Subtypings may be either complete or partial as described in §2.2.3.7.2. Allowed values for coverage are "complete" or "partial". For example:

> **COVERAGE** : complete

2.9.3.5 Subtypes

Each subtype of the subtyping has an entry which consists of its name and a description of what distinguishes it from other subtypes. These description should be in terms of how the subtypes are used and what makes them distinct from the other subtypes in terms of meaning. Where this information duplicates the information in the subtype meaning specification, it may be regarded as 'inherited'. (This manual uses a convention of italicisation for inherited information.) For example:

> **SUBTYPES : Public Course** : *A course that has been scheduled and advertised by the management in the hope that individual places will be sold for it. Any person can be booked on the course as an individual delegate.*
> **Inhouse Course** : *Individual delegates cannot be booked on the course. Inhouse courses are provided on a contract basis for customers, who then have the responsibility for ensuring that their own employees or contractors attend as required.*

A subtyping may have a single subtype. For example:

> **SUBTYPES :Manager:** *An employee who may have other employees report to them.*

In this case, there will be occurrences of "Employee" that are not managers, but there is no requirement that another subtype has to be created to include them. They are occurrences of "Employee". Note: a subtyping with one subtype must always be partial.

The inheritance of meaning by subtypes is further discussed in §2.5.3.5.

2.9.4 Rules

1. All subtypings must have at least one subtype.
2. A complete subtyping must have at least two subtypes.
3. The names of the subtypes must be distinct.
4. The name of a subtype must not be the same as the name of the supertype.

2.9.5 Guidelines

1. If several entities have similar relationships or attributes of the same abstract data type, then it may be helpful to create a supertype entity of which they are subtypes. If there is no distinction made between the subtypes, the subtypes should be replaced by the supertype.
2. If an entity is used in different ways by different people and each group uses different criteria for using that entity, then it may be helpful to subtype the entity. If there are no operations, relationships or attributes that are common to all the subtypes, replace the supertype by its subtypes.

2.10 Abstract Data Type Specification

2.10.1 Purpose

Abstract data type specifications are used to define an abstract data type (ADT). An ADT is a set of values, together with the allowed operations on these values. The 'abstract' in the name signifies that there is no commitment to any physical representation.

The use of ADTs in defining in attributes is covered in §2.8.3.8.1.

2.10.1.1 YSM use of typing

In YSM all data items (attributes, non-attributed elemental data flow, temporary items) have an abstract data type. The value of this item may only be used in a way that is appropriate for this data type. Enterprise operations are defined to operate on certain types of data and only those types of data. It is a misuse of the model and illegal in YSM to try to use them with an incorrect type of data. For example, a real number cannot be 'added to a string', the square root of an integer is illegal, etc.

This is the basics of 'strong data typing'. It allows data assignments and comparisons to be checked for semantic correctness.[1]

2.10.1.2 Avoiding ADT specifications

ADT specifications are a key part of the textual support for the Enterprise Essential Model and the System Essential Model. However, they will be invisible to the model builder in many situations, as YSM provides standard ADTs. These standard ADTs will be sufficient to build enterprise and system models in most situations. They are listed in §6.4.

Note: this section need only be read if the standard ADTs are insufficient and "user-defined" ADTs are required.

2.10.2 Examples

ABSTRACT DATA TYPE : day of week	
MEANING : Usual seven days.	
STRUCTURE : Simple	
NUMBER OF VALUES : Finite	
VALUE DEFINITION : value list	
VALUES : Sunday, Monday, Tuesday, Wednesday, Thursday, Friday, Saturday	
ORDERING : Cyclic	

---

1 There are also semantic rules of a similar nature that are applied in mapping onto implementation technology. These are also semantic constraints in type of unit and the way it is used.

ABSTRACT DATA TYPE : row
MEANING : Used in rectangular board games to describe position.
STRUCTURE : Simple
NUMBER OF VALUES : finite
VALUE DEFINITION : inheritance
PARENT ADT : cardinal
RESTRICTIONS : row > 0 AND row <= number of rows
ORDERING : linear

PARAMETERS :	parameter	: type	: default
	number of rows	: cardinal	: —

2.10.3 Components

2.10.3.1 Abstract data type

Each ADT has a name that must be distinct from all other ADT names. This name identifies which ADT is being defined.

The name should be one that communicates the meaning of the quality or property (see "meaning") that the ADT represents. It should not contain any implication that it will be used for specific attribute. For example, "customer address" is unlikely to be an appropriate name for an ADT, but "address" might be.

2.10.3.2 Meaning

The meaning of the ADT is its intended use. It should correspond to a 'quality' that could be associated with occurrences of entities by defining a named attribute. For example, "length" is a property that could be associated with many different kinds of entities. In a sense, the ADT provides the 'ultimate' detail — an entity may have several attributes; each attribute has several possible values. These values are the ADT.

What is a good ADT for one enterprise is not always relevant or correct for another enterprise. This is an example of the 'relative' philosophy of information modelling. For example, a "colour" might be considered as a possible ADT, but in organisations that make paint (for example) there might be a requirement to store information about different "colours". If this is the case, "colour" cannot be defined as an ADT, but would have to be defined as an entity.

With this proviso (i.e. what is generic in one organisation not being so in another), an ADT should be a generic property or quality that is capable of being used for many different attributes.

2.10.3.3 Structure

YSM allows both simple and compound ADTs to be defined. This is indicated by the value in the structure field, which may be:

- **simple:** one which has a single, indivisible value. Examples include "day of week" and "row".

- **compound**: one which has several component parts, each of which may itself be compound or simple. For a compound ADT, the composition is then given (see §2.10.5).

2.10.3.4 Number of values

For simple ADTs, this is used to describe 'how many' values there are for this ADT. The allowed entries are:

- **finite**: a finite set contains a finite number of values. For such a finite set, the values may either be listed (see §2.10.3.6) or defined as a restriction of another finite ADT. Provided only a finite number of values satisfy the restriction, finite ADTs may also be defined as a restriction of a discrete or continuous ADT (for example, all the integers between 10 and 100).

- **discrete**: a discrete ADT is one with an infinite number of values with 'gaps' between any two values. The "integer" ADT, with the values ..., $-2, -1, 0, 1, 2, 3, ...$ is an example of a discrete ADT.

- **continuous**: a continuous ADT is one where there is a continuum of values. Between any two values there is always at least one other value. Continuous ADTs necessarily have an infinite number of values. The "real" ADT is an example of a continuous ADT. Continuous ADTs often are often used for 'dimensioned physical quantities' (see §2.10.4).

2.10.3.5 Value definition

For each simple ADT, there are three possible methods by which the allowed values may be defined:

- **value list**: the values will be explicitly listed (see §2.10.3.6).

- **inheritance**: in this case, the values are those of a pre-existing ADT, which is referred to as the parent ADT. The parent ADT will be given, together with any restrictions that apply (see §2.10.3.9). The values are 'inherited' from the parent ADT.

- **defined**: in this case a rule will be used to define the values. This rule is any unambiguous definition of what constitutes membership of this set of values. These rules will be based on standard set theory methods. ADTs such as "real", "integer" are examples of defined ADTs.

Example:

```
VALUE DEFINITION : inheritance
```

2.10.3.6 Values

For ADTs whose value definition is by value list, the list of values is given. For example, "day of week" is given values by:

> **VALUES** : Sunday, Monday, Tuesday, Wednesday, Thursday, Friday, Saturday

The same value may appear in more than one value set definition. For example, for an ADT that would be used to specify attributes that were to do with results of examinations, we might specify the ADT "exam grading" by:

> **VALUES** : distinction, credit, passed, failed

For an ADT that would be used to specify attributes in control systems, we might define the ADT "plant status" by:

> **VALUES** : not required, ongoing, succeeded, delayed, failed

Both of these value lists contain the value "failed", but they are unrelated, as they are part of different ADTs.[1]

2.10.3.7 Parent ADT

For ADTs whose value definition is by inheritance, the name of the parent ADT is given. The values of the ADT that is being defined will be the same as those of the parent ADT, except that a restriction may be imposed (see §2.10.3.9). For example, the "row" ADT is defined by:

> **PARENT ADT :** cardinal

which should be read as "takes values from the cardinals".

2.10.3.7.1 Parameter settings

Where the parent ADT has one or more parameters (see §2.10.3.11), the defined ADT may associate values with these parameters. If the parameter is mandatory, then a value *must* be given to this parameter. The assignment of values to the parameters is given after the parent ADT name. For example:

---

[1] This is sometimes referred to as overloading.

ABSTRACT DATA TYPE : rank
MEANING : A 'row' on a chess board.
STRUCTURE : simple
NUMBER OF VALUES : finite
VALUE DEFINITION : inheritance
PARENT ADT : row, number of rows = 8
RESTRICTIONS : —

defines a new ADT "rank" from the parent ADT "row" (defined in §2.10.3.11.2) by giving a value of 8 to the parameter "number of rows".

Similarly, the ADT "character" is defined in terms of its parent ADT "string" by giving the value "1" to the two parameters "minimum length" and "maximum length" (see §6.4.6).

If the parent ADT has parameters that are optional and this entry does not associate values with the parameters, then the parameters (and their defaults) are inherited by the 'child' ADT. For example, the ADT "alphabetic" is defined from the ADT "string" by restricting all characters to be from a set of characters (an alphabet). The "minimum length" and "maximum length" parameters are inherited by "alphabetic". Any attribute defined as an "alphabetic" can use (or override) these parameter settings.

2.10.3.7.2 Operation inheritance

An ADT defined in this way inherits the operations of its parent, insofar as they are defined on the restricted set of values. For example, "positive reals" may be defined as a restriction of "real". All the operations of reals are allowed on positive reals, except that "subtract" is only defined when the result is not negative.

2.10.3.7.3 Type conversion between inherited ADTs

A value in an inherited ADT may always be converted back to its parent type. This may be done explicitly, by using the name of the parent ADT. In the above example, if "X" is a positive real, then:

```
real(X)
```

is the equivalent real number. This is referred to as explicit type conversion. This is probably too formal for most people, who would prefer implicit type conversion, where the inherited ADT is used in a place where the parent ADT is required. If this is done, an implicit conversion is assumed to take place.

The conversion may take place in the other direction too. For example:

```
positive real(Y)
```

converts a real to a positive real. In this case, however, the value must be checked to be an allowed value before the conversion is carried out — positive real(−1.6) is meaningless!

Inheritance of operations is discussed in more detail in §2.17.3.4.3.

2.10.3.8 Definition

For ADTs whose value definition is by "definition", the definition of what constitutes membership of that set of values is given. This could be in terms of standard set notation, or it could be rather less formal. For example, the "real" ADT has the definition:

> **DEFINITION :** A decimal number, (optionally) preceded by a "+" or a "−" sign.

2.10.3.9 Restrictions

For ADTs whose value definition is by "inheritance", there are often some restrictions on the values that are allowed. This is defined in the "restriction" field. A restriction must evaluate to true or false and the allowed values for the ADT are those taken from the parent ADT for which the restriction is true (if no restriction is given, the allowed values are all those of the parent ADT). For example, definition of an ADT "working day" as a restriction of the ADT "day of week" by:

> **ABSTRACT DATA TYPE :** working day
>
> **MEANING :** A normal working day, excluding weekends.
>
> **STRUCTURE :** Simple
>
> **NUMBER OF VALUES :** finite
>
> **VALUE DEFINITION :** inheritance
>
> **PARENT ADT :** day of week
>
> **RESTRICTIONS :** NOT(working day = Saturday OR working day = Sunday)
>
> **ORDERING :** linear
>
> **PARAMETERS :** —

which defines the set of values to be all those "working day"s that substitute in this expression to give the value TRUE, in other words "Monday", "Tuesday", "Wednesday", "Thursday", "Friday". A full definition of allowable restrictions is given in §6.5.

2.10.3.10 Ordering

An ordered ADT is one where the values have a significant order. Not all ADTs have an ordering, but many do and then it is an important property of the ADT. The allowed values in this field are:

- **None**: in this case there is no significance in terms of position and operators such as "<", etc., are not available.

- **Linear**: for a linear ADT, any element may be compared in size with any other. The operations "<", ">", "≤", "≥" are always available, with their usual meanings. For discrete ADTs with a linear order, the operations "First" and "Last" return a boolean value, with obvious meaning. The operations "Next" (except for the last value) and "Prior" (except for the first value) give obvious results.

- **Cyclic**: a cyclic ordering means that the values 'wrap round in a circle'. Examples include "day of week", "time of day". For discrete ADTs with a cyclic order, the "Next" and "Prior" operations are always available. Thus "Next(Saturday)" has the value "Sunday", irrespective of whether the calendar being used indicates they are in different "weeks".

The availability of the order operators is:

Operation	Order Number	Linear			Cyclic	
		Finite	Discrete	Continuous	Finite	Continuous
First		Y	Y	Y		
Last		Y	Y	Y		
Next		Y	Y		Y	
Prior		Y	Y		Y	
<		Y	Y	Y		
≤		Y	Y	Y		
=		Y	Y	Y	Y	Y
≠		Y	Y	Y	Y	Y
≥		Y	Y	Y		
>		Y	Y	Y		

2.10.3.11 Parameters
For some ADTs, there may be an associated parameter that gives a degree of flexibility in the use of that ADT to define attributes (see §2.8.3.8) and other ADTs (see §2.10.3.7.1). These parameters are defined in the "parameters" entry.

2.10.3.11.1 Optional parameters
Some parameters are optional and some are mandatory. An optional parameter associated with an ADT must always be given a default in the definition of that ADT. This default is used if an item declared to be that ADT does not associate a value with this parameter. For example, the standard YSM ADT "string" has two optional parameters associated with it, as follows:

PARAMETERS :		
parameter	**: type**	**: default**
minimum length	: cardinal	: 0
maximum length	: cardinal	: infinity

These parameters may be used in a specification of an attribute of type "string". Values that are allocated to these items must be "cardinal" (in other words 0, 1, 2, ...). The default values are "0" and "infinity", so that if no values are entered for these parameters, the data item may be of any length. An attribute may also be declared to have a minimum length greater than zero, or a finite maximum length by over-riding the defaults (see §2.8.3.9 for an example).

2.10.3.11.2 Mandatory parameters

If a parameter associated with an ADT is not given a default value in the definition of that ADT, then any data item declared to be of that type must always allocate values to that parameter. This is carried out in the attribute specification. As an example of a mandatory parameter, the ADT "row" is defined to have one such parameter:

PARAMETERS :		
parameter	**: type**	**: default**
number of rows	: cardinal	: —

As no default value (indicated by "—") is given, any data item that is typed as a "row" *must* fill in a value (a cardinal) for this parameter.

Thus in a game of chess, we might declare (possibly several) data items as being attributes of type "row". Each of these (because of the nature of chess boards), would have to use the value "8" for this parameter in the attribute specification for each of the items). Alternatively, an ADT "rank" could be defined once, using the same restriction (see §2.10.3.7.1). The ADT "rank" could then be used to define all these attributes. In practice, it would be better to define a compound ADT "square" (see §2.10.5).

In another well-known game (TIC-TAC-TOE), attributes would be specified as "row" with the parameter value of 3.

Note: in both these cases, the constant ("8" or "3") is not a value for the attribute — it defines how many different values there may be.

See §2.8.3.9 for examples of the use of mandatory parameters in defining attributes.

In the above examples, the parameter was a different ADT from the one being defined. For dimensioned qualities, the parameter is often the same ADT as the one being defined (see §2.10.4).

2.10.4 Dimensioned ADTs

Many physical quantities are associated with dimensioned measurements that may be made on physical objects. The ADTs associated with such qualities ("length", "velocity", "mass", etc.) have all the properties of any ADT, as discussed above. However, they also have some specific entries that only refer to them. These are described below.

2.10.4.1 Example of dimensioned ADT

ABSTRACT DATA TYPE : height

MEANING : Used in air traffic control for airplanes, runways, obstacles, flight paths, etc.

STRUCTURE : Simple

NUMBER OF VALUES : continuous

VALUE DEFINITION : inheritance

PARENT ADT : real

RESTRICTIONS : height <= maximum height

ORDERING : linear

DIMENSIONS : L

UNITS :
ft
miles
metres

PARAMETERS :	**parameter**	**: type**	**: default**
	accuracy	: height	: 0
	repeatability	: height	: 0
	resolution	: height	: 0

2.10.4.2 Dimensions

Each physical quality must have a well-defined dimensionality. For example, "velocity" is always obtained by dividing a "length" by a "time", "force" is defined to be "mass" * "acceleration" etc. Although YSM does not use this information in any formal way, it is a useful guideline to how such a quality can or should be used. This information can either be entered into the "meaning" field, or a separate entry can be created, as shown in the example.

One obvious rule is that qualities corresponding to different dimensionality can never be compared (a "length" cannot be compared with an "area"). This dimensionality of the quality can be expressed using the mass, length, time, current and temperature dimensions, which are conventionally abbreviated: "M", "L", "T", "I" and "K".[1] For example, "length", "area" and "velocity" would have the dimensions:

---

1 There are two others, but they are unlikely to be of use in information systems.

```
DIMENSIONS : L
```

```
DIMENSIONS : L²
```

and

```
DIMENSIONS : LT⁻¹
```

respectively. A superscript has been used to show the re-use of a dimension — thus an area is always obtained by multiplying two lengths ; a velocity is always obtained by dividing a length by a time. (If a printer that cannot produce superscripts is used then any other convenient alternative may be chosen.)

The dimensions of a derived quality can be calculated from its definition. For example, a "force" is 'that quantity that gives a given acceleration to a known mass' and "acceleration" is defined as 'change in velocity in unit time'. The dimension of "acceleration" is therefore LT^{-2} and the dimension of "force" MLT^{-2}.

Note: this part of the specification is of limited use at present. However, there are some heuristics (not covered by this release of YSM) that relate to scaling of quantities and choice of units, for which this entry is very relevant.

2.10.4.3 Units

Fundamental to the concept of ADTs is the concept that there is no commitment to the internal way a "height" is stored (for example). However, there are situations where the models built, particularly the System Essential Model, may need to make some reference to constant values (sometimes called literal values).

For example, in a system that is controlling air traffic, an ADT "height" would be used to declare some obvious attributes for the entities "Mountain", "Airplane", "Runway". The System Essential Model might include the following within a minispec:

```
"If <Airplane>.altitude less than  <Mountain>.height  +  200 ft  then  ... "
```

The intention is to indicate a fixed height difference which is a minimum clearance for overflying the mountain. The constant "200 ft" makes use of a supported unit for the ADT "height". Supported units are given in the units field. For example:

```
UNITS :
          ft,
          miles,
          metres
```

defines three possible units that may be used to specify any "height".

Some data flows to or from terminators will also have required units — in other words, there is a policy constraint that the output or input value must be converted to/from these specific units. These data elements may be defined as having that type. (Warning — many transducers and actuators use values that are, at best analog or digital values. These should not be typed with units, if the system has the responsibility of carrying out conversion to/from this format to a typed value.)

2.10.4.4 Measurement parameters

For data items that are declared to be of a dimensioned ADT, there will be some additional parameters that relate to the measurement. Any data item declared to be of that type may have values defined for those parameters. For example, each data item declared to be of type "height" will have its own inherent "measurement resolution" and "accuracy". The following parameters are often appropriate for dimensioned ADTs:

- **Accuracy**: this is used to give an indication of the absolute mean error between the real-world quality and system value obtained from measuring it.

- **Noise**: this is used to give an indication of the difference between one measurement and a repeat of that measurement carried out immediately afterwards (before the real-world quantity being measured has changed). This parameter is particularly significant in control system specification.

- **Resolution**: this is used to give an indication of the smallest difference that can be detected in the value. It is an indication of the 'fineness' of the measurement. It is particularly important in control systems where analog to digital converters have truncation errors associated with a finite number of bits being used.

The above may be supposed to be absolute values. However, there is nothing to stop an ADT being defined with associated parameters which correspond to relative errors, percentage errors, etc.

As an example of a declaration of a data item that uses this type of parameter, consider "runway elevation", which would have the following parameter values entered in its attribute specification:

```
PARAMETERS :
     parameter     : value
     accuracy      :
     noise         :
     resolution    : 1 metre
```

This attribute is now assumed to be measured as a whole number of metres (whether 'rounding' or 'truncation' is used is not defined here, but either in the specification of the data flow from the terminator, or a minispec). As no values for the other two parameters are given, this attribute uses the default values for these parameters. These defaults are given in the ADT specification.

Note: the attribute specification used a constant height value, which is given in terms of metres. Metres is a supported unit for the ADT for the height ADT, which is the required type for the parameter.

2.10.5 Compound ADTs

Compound ADTs have two or more components, each of which is also an ADT.

2.10.5.1 Examples of compound ADTs

ABSTRACT DATA TYPE : vector
MEANING : A displacement in three dimensional space.
STRUCTURE : Compound
COMPONENTS :   <u>**component**</u>   <u>**: type**</u> dx   : length dy   : length dz   : length

ABSTRACT DATA TYPE : address
MEANING : Used for any entity which needs to be mailed.
STRUCTURE : Compound
COMPONENTS :   <u>**component**</u>   <u>**: type**</u> house   : residence street   : alphabetic city   : alphabetic post-code   : post-code

ABSTRACT DATA TYPE : square
MEANING : Position on chess board.
STRUCTURE : Compound
COMPONENTS :   <u>**component**</u>   <u>**: type**</u> rank   : rank file   : file

2.10.5.2 Components

For each component the specification gives:

- **The name of the component**: each component has a name which is unique within that ADT (but may be the same as components of other ADTs). For example, "dx", "dy", "dz", "house", "street", "city", "post-code".

- **Type**: each component must correspond to a named ADT. This item gives the name of that ADT. (There must be a corresponding ADT specification for that ADT.) The examples shown use the ADTs "length", "residence", "alphabetic" and "post-code".

An ADT may be re-used for several components, as in the first example.

The name of the component may be the same as the name of the ADT that it is typed as, for example, "post-code". This is very common and causes no problems.

When an attribute is declared to be of a compound ADT, the whole structure is defined. For example, if the entity <Customer> was defined to have an attribute "mailing address" which had an attribute specification, which showed it to be of type "address", then:

1. **The whole attribute may be referenced**: in an assignment from a data item that is part of a data flow to an attribute, for example:

   ```
   <Customer>.mailing address := new address
   ```

   is allowed, providing that the data item "new address" has also been defined to be of type "address".

2. **A component of the attribute may be referenced**: for example:

   ```
   If <Customer>.mailing address.city = Paris
   ```

   would give the value "true" for those customers who live in Paris.

If a component of the ADT is itself compound, then not only does that component have to be defined (in a similar way), but that inner component may also be referenced. Thus, supposing that "residence" was defined to be "house number", together with "apartment number" (which could have a null value in some cases). In that case:

```
<Customer>.mailing address.residence.house number
```

would refer to the value of this lowest level component for that occurrence of <Customer>.

Note: the use of compound ADTs is only recommended when the set of qualities form a natural grouping, which is often treated as if it were a single unit of information (even though it has components). Using compound ADTs in other situations will decrease, rather than increase clarity.

2.10.6 Rules

1. The name of the ADT must be unique within the enterprise. Note: the standard YSM abstract data types are: alphabetic, alphanumeric, boolean, cardinal, character, currency, date, integer, real, string, text, time, time interval, time of day.

2. The name of the ADT must not be "generic". (This is, in effect, a reserved term.)

3. No value may be repeated in a value list.

4. The parent ADT must be different from the ADT being defined. Further, no circular dependency is allowed; thus it is illegal to define A in terms of B ; B in terms of C ; C in terms of A etc.

5. A restriction of a finite ADT is itself finite.

6. A restriction of a discrete ADT is itself discrete or finite.

7. A restriction of a continuous ADT is a continuous ADT (usually) or (rarely) discrete or finite).

8. A continuous ADT must have a "linear" or "cyclic" order.

9. Any named element in an existence list must be used in the corresponding existence test. (See §6.5 for a definition of "existence test" and "existence list".)

10. A named element in an existence list must only be used once in that list.

11. A named element in an existence list must have a different name from the ADT being defined.

12. The ADT being defined must be used in an existence test if there is one.

13. The names of the components of a compound ADT must be distinct from each other.

14. The name of a component of a compound ADT must be distinct from the ADT that it is a component of. (This rule must be checked for cyclic dependency. Thus it is illegal to have A a component of B and B a component of A. See §6.1.3.3.1 for a reference on acyclic graphs.)

2.10.7 Guidelines

1. The name should be one that communicates the meaning of the quality or property (see "meaning") that the ADT represents. It should correspond to a 'quality' that could be associated with occurrences of entities by defining a named attribute. This name should reflect a property that is suitably generic for the enterprise. For example, "date", "name", "height", "temperature", "dollars", "sterling", "counter" would be suitably generic, but, in most enterprises, "customer name", "reaction time" would not. What is a relevant level of generality depends on the enterprise — in a company that deals with word processing "string" is suitably generic, but "name" too specific ; in many companies "string"s would not be relevant to their activities whereas "name" would be suitably generic.

2. The name of the ADT should not contain any implication that it will be used for a specific attribute. For example, "customer address" is unlikely to be a good name for an ADT, but "address" might be.

3. The ADT is not just a value set with associated operations and two ADTs should be regarded as distinct if they are used for a different type of real-world information. Examples of such a distinction are "height" and "weight" — both of which are reals restricted to be non-negative. However, it is not meaningful to compare a "weight" with a "height" even though they have the same value set. The key point is that the ADT conveys a meaning, as well as a set of values.

4. The names given to the values in a value list should reflect their intended meaning.

5. If an attribute (for a specific entity) does not inherit all of the defined operations for an ADT then a new ADT should be defined for that attribute.

6. ADTs are mostly useful when a large number of data items are of the same type. If only one data item takes values from a discrete set, then an abstract data type should not be declared (the value set should be defined under that data item).

7. Names of ADTs should be lower case.

8. A simple ADT that is defined in terms of another should generally have some restrictions. If not then the two ADTs are just aliases. These aliases should only be used if a useful distinction is being made between their intended use.

9. If it is anticipated that the same kind of restriction will be applied to many (or all) of the data items that will be specified as being that type, then a parameter should be created in the ADT and the restriction worded in terms of that parameter and placed in the ADT specification. For example, it seems likely that many data items declared to be of type "height" will have a maximum value. Rather than having to build this restriction into all such attribute specifications, the parameter was defined and the restriction placed in the ADT specification for "height". Generally, speaking, avoid the use of restrictions in attribute specifications as much as possible — they are more difficult to understand than giving parameter values.

10. If the meaning of a component of a compound ADT is not clear from its name and "parent" type, a new ADT should be created so that the meaning of the component is clarified.

2.11 Statement Of Purpose

2.11.1 Purpose

The statement of purpose gives a very high-level overview of the responsibilities of the system Its informal structure allows easy review by all project members and users. It serves as a vehicle for early agreement on the system's overall functions.

This tool is not meant to be detailed or predictive. It is meant to ensure that all parties agree upon the highest-level view of what the system is to accomplish. If a statement of purpose cannot be written down, then the probability is that the analysis team and users do not really know why the system exists.

Producing and refining a statement of purpose in parallel with the system scoping tools (context diagram, system ERD and event list) is a great help in clarifying concepts.

2.11.2 Example

SYSTEM: Satellite Camera Control System

GENERAL DESCRIPTION :
The purpose of the system is to take and display pictures of land resources from a satellite. The satellite will contain an automatic electronic camera that will (when activated) collect light until the exposure is complete. The image may then be read out before sending a signal to start another exposure.

RESPONSIBILITIES :
1. Accepting requests for photographs from accredited researchers and scheduling a suitable time for these photographs to be taken.
2. Controlling operation of the camera and storing pictures.
3. Using the satellite attitude control instrumentation to point the camera at the target.
4. Displaying pictures on request.
5. Keeping an up to date record of the satellite's orbit so scheduling can be achieved. To do this an on-board accelerometer is used, together with ground tracking information.

SPECIFIC EXCLUSIONS :
1. Changing the satellite orbit.
2. Tracking the satellite. This function is carried out by ground tracking stations.

2.11.3 Components

2.11.3.1 System

This gives the name of the system whose purpose is being specified

2.11.3.2 General description

The general description gives an overall indication of why the system exists or will exist.

2.11.3.3 Responsibility

Responsibilities are high-level descriptions of duties that are within the system scope. Each responsibility should be a general functional statement or define a group of entities and relationships that must be 'tracked'.

2.11.3.4 Specific exclusion

This specifically excludes an area of possible responsibility from the scope of this system. It is a record of past negotiations on scope. Any responsibility that might be assumed to be within the system's scope but has been excluded should be listed here. This serves as a clarification for all involved in developing the system on why these potential responsibilities are not within the scope of the system.

2.11.4 Rules

Because there is no formal grammar or structure for the statement of purpose, there are no rules for it.

2.11.5 Guidelines

1. The statement of purpose should be small enough to review in less than an hour. A statement of purpose that is very long is counter-productive to its intention. Very large (in terms of development resource required) systems can still have quite concise statement of purpose — it's just that each functional responsibility covers many smaller details.
   Large statements of purpose should be re-organised into higher-level groupings of responsibilities (and possibly exclusions). If this is not possible, it may be a clue that the 'raison d'être' for the system is not understood, or it is a poor choice of system boundary (in the sense that the system is not a coherent collection of responsibilities).

2. Each specific responsibility should not be longer than two or three sentences. If it is difficult to describe the responsibility more concisely, then either the responsibility is not well-understood, or it is a miscellaneous collection of responsibilities. To achieve this criterion, the statement of purpose may have to be re-organised into more coherent areas of responsibility.

3. The statement of purpose should give a good clue as to whether a specific real-world event is within the system's scope.

4. The statement of purpose should give a good clue as to whether a specific real-world entity or relationship is within the system's scope.

2.12 Context Diagram

2.12.1 Purpose

The context diagram shows the interface between the system and its environment. It highlights the required outputs of the system and the inputs it requires to accomplish its purpose. The sources for inputs and destinations for outputs are also shown. These may be other systems, hardware devices, people, or organisations

By implication, the context diagram defines the scope of a system — i.e. what is necessary to transform the given inputs into the required outputs.

The interface between the system and the environment cannot be changed unilaterally by the system analyst or designer — any changes in the content or structure of this interface may only be carried out after negotiation with other parties, outside the system's scope.

The context diagram documents the required inputs and outputs regardless of anticipated or previous implementations of the system.

2.12.1.1 Supporting specifications

The components of the context diagram have corresponding textual specifications. These are:

- terminator specifications — covered in §2.13;

- data flow specifications — covered in §2.18;

- data store and access flow specifications — covered in §2.20;

- event flow specifications — covered in §2.25.

2.12.1.2 Context diagram as a data flow diagram

The context diagram is a data flow diagram which shows the system as a single process group. This manual has provided separate chapters for these modelling tools. Logically, they should be combined as a single chapter.[1]

See §2.14.1.2 for a discussion of showing terminators on data flow diagrams.

---

1 We kept them as two, largely for 'public relations', rather than any logical reason.

2.12.2 Example

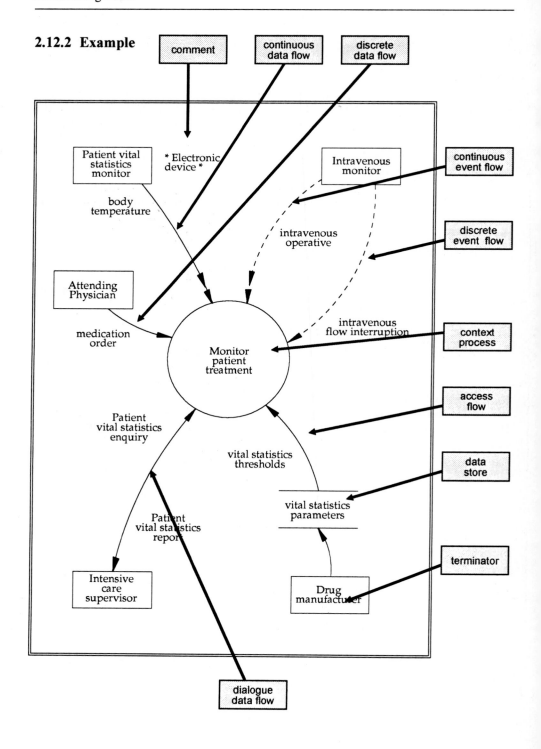

2.12.3 Components

The context diagram has many of the same components as the data flow diagram. In fact, the context diagram is a particular type of data flow diagram in which there is a single process and terminators are shown.

Not all components will be used in a given system — event flows are only likely to occur if the behaviour of the system changes over time; continuous flows are only found when the system monitors or controls the environment.

2.12.3.1 Access flow

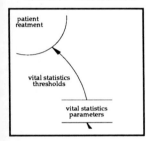

An access flow indicates that the system uses stored data that is shared between it and its environment. Any store shown on the context diagram must be accessed by the system and at least one terminator.

Access flows are discussed in more detail in §2.14.3.1.

Access flows may be continuous or discrete. The use of continuous access flows on context diagrams is discussed in §2.12.3.4. Discrete access flows on the context diagram are discussed in §2.12.3.11.

2.12.3.2 Comment

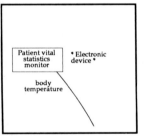

As with all diagrams, free-format comments may be added to the context diagram to elucidate or stress any important points. It may also be used to highlight any aliases, ties to preceding or anticipated implementations, characteristics of terminators, etc. It is not specified with additional textual support.

2.12.3.2.1 Annotating context diagram to check against event list
The context diagram is a useful medium for checking events against flows into and out of the system. This can be carried out by annotating the flows on the context diagram with the numbers of the events that they are associated with. This is discussed in §5.6.5.1.

2.12.3.3 Context process

The context process is a process group which represents the system. All requirements for

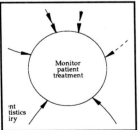

behaviour or data storage by the system is considered to be within this process. On the context diagram, the system is regarded as a 'black box'.

The context process is named to describe the main function or 'purpose' of the system.

2.12.3.4 Continuous access flow

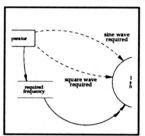

If the system or a terminator continually updates or monitors the value in a store that appears on the context diagram, this is shown as a continuous access flow. This is relatively unusual, but an example is shown below.

Continuous access flows are discussed in more detail in §2.14.3.3.

2.12.3.4.1 Function generator

This example relates to a system that produces output electrical waveforms. An operator can set the desired frequency of the generated waveform and select "sine wave" or "square wave" as the desired output. The system needs to continuously access the required frequency. At any time, the output frequency is given by this value. The interface is better modelled as a store, rather than a continuous input data flow because the operator can set the required frequency at a point in time and leave it; the frequency may also be varied continuously. (A possible implementation of this interface would use a dial that could be rotated.) The context diagram below illustrates how the system needs to continually access the "required frequency":

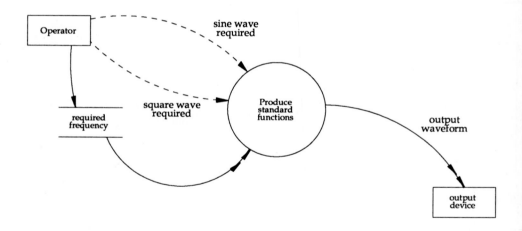

2.12.3.5 Continuous data flow

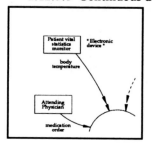

A continuous data flow represents a value (or values) that is available over a period of time. An input flow would be continuously available over time to the system; an output data flow would be continuously available to a terminator. Continuous data flows are discussed in more detail in §2.14.3.4.

2.12.3.5.1 As system input

System inputs are generated by terminators — they will only be available to the system when the terminator is active; when the terminator is not active, the flow will not be available. Generally speaking, the system cannot control the periods of time at which the terminator is active (although it may output event flows to the terminator that are intended to 'turn it on' or 'turn it off'). The presence of a continuous flow may be checked by means of the standard operation "available(id)", where "id" is replaced by the name of the flow — see §6.9.3.

In the example, the terminator "Patient vital statistics monitor" is a device that is always on, and always reads and outputs the value "body temperature". If the device is not working, "body temperature" will not be available.

If the method of 'taking the patient's temperature' was not yet agreed, then it is better to model "temperature" as a continuous input data flow from the terminator "Patient".

2.12.3.5.2 As system output

Continuous data flows output by the system are values that are required by the terminator on a continuous basis. The terminator must be continuously active to use these. At times when the terminator is not active, it effectively 'ignores' these system outputs. For example, a lift system might continuously output the position of each lift cabin to a passenger. This is shown as below:

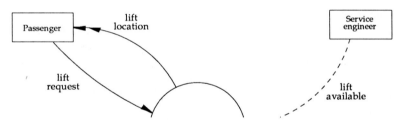

At any instant, there may be no passengers, or potential passengers may not be 'looking' at this output. None the less, the output is continuously generated by the system and is available for use at all times by the terminator.

2.12.3.6 Continuous event flow

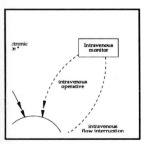

A continuous event flow represents a status that is indicated over a period of time. Several such statuses may be grouped together onto one such flow.

Continuous event flows are discussed in more detail in §2.14.3.5.

2.12.3.6.1 As system inputs

A continuous event flow generated by a terminator will only be available to the system when the terminator is active; when the terminator is not active, the flow will not be available. Generally speaking, the system cannot control the periods of time at which the terminator is active (although it may output event flows to the terminator that are intended to 'turn it on or off'). The presence of a continuous flow may be checked by means of the standard operation "available(id)", where "id" is replaced by the name of the flow — see §6.9.3.

In the example, "intravenous operative" is an event flow output by the device "Intravenous monitor". This level is raised if the Intravenous monitor is working normally, and lowered when it is not. If the monitor is not active, the flow is not present.

2.12.3.6.2 As system outputs

A continuous event flow that is output to a terminator is often used to indicate that a specific mode of behaviour should continue to occur in the terminator. Sometimes continuous output event flows indicate an ongoing behaviour of the system (or monitored device) that the terminator needs to know about.

2.12.3.7 Data flow

A data flow on the context diagram either represents information produced outside the system that the system uses (system input), or information produced by the system that is needed by the outside world (system output).

Data flows are discussed in more detail in §2.14.3.7.

Data flows may be continuous or discrete. The use of continuous data flows on context diagrams is discussed in §2.12.3.5. Discrete data flows on the context diagram are discussed in §2.12.3.12.

2.12.3.8 Data store

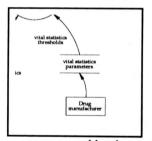

A data store on the context diagram shows an interface between the system and its environment. The store indicates that the system and the terminator are 'decoupled', in the sense that one can change values in the store independently of the other. Unlike a data flow, there is no synchronisation (this distinction is discussed in more detail in §2.12.6.5).

YSM demands that information in data stores is organised around entities, relationships and attributes. Showing a store on the context diagram indicates that entities and relationships included in the store are accessed by the system and also some device, person or organisation outside the system. The store holds information that is created by the system and used by the environment, or created by the environment and used by the system, or both.

See also the corresponding entry for the data flow diagram (§2.14.3.11).

2.12.3.8.1 Systems that collect information for ad-hoc enquiries

If a system has the responsibility for collecting information for ad-hoc enquiries, then there will be a large number of stores on the context diagram. The system will store information in them, ready for use by the terminators. These terminators are people who will access this information at unpredictable times. It is not the responsibility of the system to anticipate the events that cause these accesses to be made, no the policy used by the terminators. It is a general principle of YSM that ad-hoc processes are excluded from the system boundary.

The requirement that the system collects this information is best modelled using other tools such as the ERD (see §2.2) or function–entity tables (see §2.19). This is particularly true if there are many terminators that access the same information.

2.12.3.9 Dialogue data flow

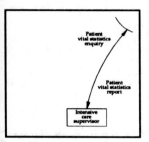

A data flow generated by a terminator for use by the system may be combined with a data flow that is created by the system for use by the terminator and shown as a single dialogue data flow. This is very helpful if the data flow generated by the system is in response to the incoming data flow. Dialogue data flows may also be used when the system outputs a data flow to a terminator, causing it to generate a data flow, used by the system.

A dialogue data flow may also be used to show a group of system inputs, together with corresponding response data flows. Groups of output flows to a terminator, together with the responses back from the terminator may also be shown as a dialogue flow.

Dialogue data flows are discussed in more detail in §2.14.3.12.

2.12.3.10 Dialogue event flow(not shown on example)

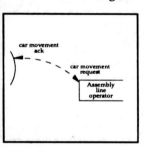

A pair of event flows that occur at the same time may be combined and shown as a dialogue event flow, provided they are to/from the same terminator. This is a 'request/response handshake'. (Note: the example on the left is expanded in §2.12.4.1.)

A dialogue event flow may also be used to show a group of input flows, together with corresponding response flows, either initiated by the terminator or the system. This provides an overview of more complex interfaces.

See the corresponding entry in §2.14.3.13.

2.12.3.11 Discrete access flow

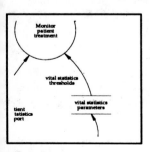

A discrete access flow is an access to a store that takes place at a particular instant of time. Once the data is in the store, it may be accessed at any time by the system (or by the terminator).

See also data store — §.2.12.3.8.

Discrete access flows are discussed in more detail in §2.14.3.14.

The example shows the system accessing the store "vital statistics parameters"; using a named access flow "vital statistics thresholds". This store is also accessed by the terminator "Drug manufacturer".

2.12.3.12 Discrete data flow

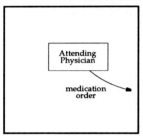

A discrete data flow on the context diagram represents a data flow that is only present for an instant of time, as opposed to a continuous data flow (the value of a continuous data flow is present over a period of time).

Discrete data flows are discussed in more detail in §2.14.3.15.

2.12.3.12.1 As system input

An incoming discrete data flow may be detected by the system. The value (or values) of the flow are available to the system at that point in time. See §2.22.5.8 for a discussion of the 'stimulus' concept.

In the example, the "Attending Physician" produces a "medication order" which the system receives and immediately processes. Note: if the "medication order" was a data store, the system would not immediately receive the data; it would have to initiate an access to the store.

2.12.3.12.2 As system output

A discrete data flow generated by the system is only present at that point in time. The terminator must be assumed to be in a position to act on the data when it occurs (sometimes, it may choose to ignore it, depending on its internal dynamics).

2.12.3.13 Discrete event flow

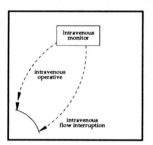

A discrete event flow on the context diagram represents an event flow that is only present for an instant of time, as opposed to a continuous event flow (which is present over a period of time).

Discrete event flows are discussed in more detail in §2.14.3.16.

2.12.3.13.1 As system input

An incoming discrete event flow may be detected by the system. It is used to identify events that have occurred in the environment that require changes in the system's behaviour. See §2.22.5.8 for a discussion of the 'stimulus' concept.

In the example, the "Intravenous monitor" generates a discrete event flow "intravenous flow interruption". This signals that the event "Intravenous flow is interrupted" has occurred. This signal is generated once, and must be detected by the system at that time or it will be missed.

2.12.3.13.2 As system output

A discrete event flow generated by the system is only present at that point in time. The terminator must be assumed to be in a position to act on the event flow when it occurs (sometimes, it may choose to ignore it, depending on its internal dynamics).

2.12.3.14 Event flow

An event flow shown on the context diagram indicates a required synchronisation between the system and its environment.

See continuous event flow (§2.12.3.6) and discrete event flow (§2.12.3.13).

See also the corresponding entry for the data flow diagram in §2.14.3.18.

2.12.3.15 Terminator

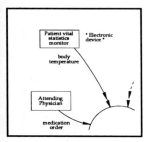

A terminator is a producer or user of data that interfaces with the system. A terminator may be a person, device, department, organisation, or another system.

Terminators show a potential role, rather than individual instances. Thus the terminator "Attending Physician" is used, rather than "Dr Smith". Terminators are often entities (see §2.2.3.3). Events (see §2.22 and §2.23) occur 'in' or are detected by terminators. A terminator is the last producer of data that added essential content to an input flow, or the first user of data that uses the essential content of an output flow. (§2.12.6.4 discusses the distinction between terminators and 'intermediate handlers')

2.12.3.15.1 Repeating terminators

For some systems there may be many flows to or from the same terminator. Trying to show them connected to the same terminator may lead to an unreadable diagram. In these cases, several copies of the same terminator may be shown on the context diagram. Each copy is tagged by a "*" after its name to indicate that it appears elsewhere on the same diagram.

2.12.3.15.2 Showing multiple instance terminators on the context diagram

Sometimes the system needs to interface to several instances of a terminator. This is indicated by showing a "ghost" behind the terminator. Multiple attending physicians might be shown:

This notation is optional, but, if used, gives insight into the fact that the system needs to distinguish between the different instances of the terminator. It is not clear that this would be a

requirement — the system might accept "medication order" from any physician and deal with it in the same way. However, if the system produced a summary or report that was specific to a particular physician (and possibly sent them a copy), it *would* need to distinguish between the different instances.[1]

---

1 The system is more than the automated system. As a consequence, it is incorrect to say "The sorting of the reports according to which physician they are destined for is a manual activity; the system does not need to deal with this" is incorrect. This sorting activity is part of the system requirements and is therefore within the system boundary.

2.12.4 Partial context diagram

2.12.4.1 Example

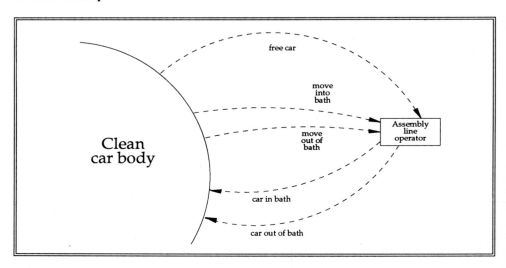

2.12.4.2 Purpose

A partial context diagram is used to view a smaller portion of the entire system's interfaces at one time. Each diagram presents a portion of the interface. This allows more specification of detail than a single context diagram and provides a more concentrated focus to the reviewer. For example, the diagram above is a detailed view of the interface seen in §2.12.3.10.

In a partial context diagram, the system may be shown as a partial arc of a circle, as in the example. As an alternative, a complete circle may be used, providing the existence of multiple context diagrams is clearly understood.

A partial context diagram has the same components as the context diagram. The only difference between the two is the amount of information on one page. All rules and guidelines for context diagrams apply to partial context diagrams.

2.12.4.3 Effect of technical support environment on partial context diagrams

A system will have several partial context diagrams if they are used. The entire set of these is equivalent to the context diagram. These may be used instead of, or in addition to, the single context diagram, depending on technical support.

In pencil and paper environments, the use of multiple partial context diagrams should be regarded as an alternative to a single context diagram. One may be produced from the other as a temporary, working document, but they are not both required. (This is because of the redundancy.)

In automated support environments, the single context diagram and the multiple partial context diagrams are just different selections of the same information. The support tool acts as a 'filter', showing only the required information on a specific partial view. This filtering may be chosen to highlight one event (or a group of events), one terminator, some of the flows to one terminator, etc. This sophistication is only really appropriate to automated support environments — in pencil and paper environments, the work involved in checking completeness and consistency would be counter-productive.

2.12.5 Rules

2.12.5.1 Structure

1. The context diagram shows one process. (Note: showing terminators on DFDs is allowed, so a diagram with terminators and more than one process is a DFD and not, specifically, a context diagram.)

2. A context diagram must contain at least one terminator.

3. The context process must either (or both):
   - have at least one output event flow or data flow.
   - have a create, update or delete access to a store (the system collects information for use by other systems).

4. A data flow on the context diagram must flow between the context process and a terminator. Flows between terminators are not shown on the diagram — they may occur, but the system does not care.

5. A data store on the context diagram must be accessed both from a terminator and from the system. A store must show both input and output access flows.

6. An event flow on the context diagram must be between a terminator and the context process. No event flows should be shown between terminators, or into or out of a data store.

2.12.5.2 Names

1. The context process must be named.

2. Each data flow must be named.

3. Each data store must be named.

4. Each event flow must be named.

5. Each terminator must be named.

2.12.6 Guidelines

2.12.6.1 Structure

1. It is very rare for a system to be able to carry out its purpose without any inputs. If the system uses information that has been collected by another system, corresponding stores should be shown on the context diagram. Rare examples of systems that do not have any inputs include 'clocks' and those producing random outputs.

2. Comments should be used sparingly as they may clutter the diagram. Large amounts of commentary should be specified in the dictionary under the item.

2.12.6.2 Complexity

1. A context diagram should be able to be reviewed and commented upon by a reviewer within one hour. If it takes longer the complexity should be reduced.

2.12.6.3 Names

2.12.6.3.1 For all components

1. A name should show no trace of a current or imagined implementation. If specific technology is already known or mandated, it should not generally compromise the essential nature of the context diagram. Implementation tags (comments) may be added to the diagram, or (better) added as comments to the meaning entry of the specification of the corresponding component. For example, if a system is being designed to control a camera, then the specific make of camera should be suppressed from the diagram. However, the general type of camera, (electronic, photographic film, etc.) should be indicated, because it will affect the generic type of interface the system must support.

2. A name should be specific enough such that a subject-matter expert can state its meaning without consulting the textual specification.

2.12.6.3.2 For data flows

1. An output data flow should usually have different content from any other output data flow. It may have the same content, but different meaning. If two output data flows have the same meaning *and* content, then they are the same flow and should be given the same name (this data flow may still be shown twice on the diagram — for example, as an output to two different terminators). If two output data flows have the same content, but different meaning, it may be helpful to use qualified names (see §2.14.3.7.2).

2. An input data flow should usually have different content from any other input data flow. It may have the same content, but different meaning. If two input data flows have the same meaning *and* content, then they are the same flow and should be given the same name (it is allowed, though rare, to show the same input twice on the diagram — for example, as an input from two different terminators). If two input data flows have the same content, but different meaning, it may be helpful to use qualified names (see §2.14.3.7.2).

3. If an output data flow has the same content and meaning as an input data flow, then a single name should be used for both. If there is any difference in meaning, then a qualifier should be used for the name (e.g. the system is responsible for checking, or 'filtering' the flow). Sometimes, the system is responsible for ensuring that a data flow is routed from one terminator to another — in this case, the name should be the same.

4. The name of a data flow should describe a single occurrence of the data flow. Although there may be many "Deposits" in the course of a day, the flow associated with the event "Customer makes a deposit" should be named "customer deposit".

5. Qualified data flow names should only be used when the difference in meaning is small. This is discussed in more detail in §2.14.3.7.2.

6. If there are several input data flows that are associated with one event, it is helpful to group them together. For example, "medication order" contains a group of data items, all of which are associated with the event "Attending Physician orders medication".

7. If the system produces several output data flows in response to an event, it is helpful to group those going to a particular terminator together.

8. If there are flows associated with different events to/from a terminator, it may be helpful to group them together; for example, an input flow "patient vital statistics" containing the flows "body temperature" and "blood pressure", etc. This should only be done if a good, honest name can be found for the grouping.

2.12.6.3.3 For data stores and accesses

1. Guidelines for data store names are given in §2.20.5.2.

2. Guidelines for access flows are given in §2.14.5.2.1.

3. Qualified attribute accesses should only be used when the difference in meaning is small. This is discussed in more detail in §2.14.3.1.7.

2.12.6.3.4 For event flows

1. An event flow should be named so that a user with subject matter knowledge is able to state its purpose without consulting the textual specification.

2. A discrete event flow should be named such that the event causing the flow can be described without having to look it up in an event list.

3. To name an event flow that is output from the system, the following should be applied:

   - if the system detects a condition, but is not responsible for the action to take based on that condition, the flow should be named as a condition, not the decision or action. For example, a system detecting an emergency condition and reporting to an operator who is responsible for taking action, should generate the flow "plant is on fire" rather than "shut down the system".

   - if the system is the decision maker, and the receiver of the flow neither makes decisions nor needs to know why they were made, the flow may be named for a command. A flow from a system responsible for camera control generates the event flow "open shutter" to a camera, rather than the flow "aimed at the target".

2.12.6.3.5 For terminators

1. A terminator should be named according to the role they play in interactions with the system. An individual's proper name should not be used. For example, "Attending Physician" is preferred to "Dr Smith" even if there were only one doctor.

2. Terminator names should not be abbreviated, except where it is normal business practice to do so (e.g. HQ, IRS, etc.). Any abbreviation must be one used in the business, not just for the convenience of the modeller.

2.12.6.4 Distinguishing terminators from intermediate 'handlers'

Terminators must affect the essential content of the data flow to be valid. It would be incorrect, on the example given at the beginning of this section, to use "Terminal" as the terminator originating the flow "Patient vital statistics inquiry". A terminal might provide the data to the system in a chosen implementation, but it does not affect the essential data content.

Some devices do affect data content and are properly shown as terminators. "Intravenous monitor" serves as a detector of the state of the Intravenous system, and produces the essential event flows to inform the system of events that occurred in that system. The intravenous hook-up itself could not do that, although it is the original source of the data. If a device will be used to interpret, calculate, validate, sense, or otherwise affect essential data, rather than just provide the data given to it by some other source, it is properly shown as a terminator.

Although "Patient" is the original source for body temperature, the patient is not the sensor of the body temperature. This sensor, "Patient vital statistics monitor", provides the actual data value for temperature, and is the terminator. This device is the also the source of data on blood pressure, heart rate, etc., although this is not shown in this 'cut-down' example. (If the only value sensed by the device was the temperature, it would be better to name the terminator "Thermometer".)

2.12.6.5 Distinction between data store and data flow

Both data flows and data stores may be used to model the interface between a system and its environment. Data flows should generally be used, except when the stored information is accessed by the terminator and the system. Enterprise information shared between more than one system thus appears on context diagrams for each system as a store; registers (for example) in a controlled device also are shown as stores.

In other cases, the interface is shown as a data flow.

A store on the context diagram may be read whenever required, rather than only when the information is created. There must be an event that causes this information to be read from the store.

An input data flow on the context diagram indicates that the system must use the information when it occurs. A discrete data flow input may be detected by the system (see §2.22.5.8 for a discussion of the 'stimulus' concept). Even if the system cannot respond immediately to a data flow input, there will be an internal store, together with a process whose role is to store this input when it occurs. These are not visible on the context diagram, but are hidden 'inside' the process.

Continuous data flow inputs may be used at any time they are available

2.12.6.6 Choosing what to show on each of several partial context diagrams

Partial context diagrams are meant to focus on the interfaces between the system and an individual terminator. A sufficiently small portion of the systems interfaces should be chosen for any one view so that all flows, and terminators shown can be easily read and understood by the reviewer. No crowding should occur.

This presentation can be used to show the system's interfaces that deals with specific events. When used in this way, it is recommended that data flows and stores be grouped such that all input from one terminator associated with one event is in one flow (of the same type), and all output associated with one event and one terminator is in one flow. No flows should group inputs or outputs associated with different events. Flows should have the exact same names seen in the event table stimulus, system inputs and system outputs columns.

If practical (that is, sufficiently small), all inputs and outputs associated with one event should be seen on one partial context diagram.

In pencil and paper environments, to aid in detailed study of the interfaces, a terminator should appear on only one partial context diagram wherever this is possible. If there are many flows to the same terminator, this may not be possible and then flows associated with one event or group of events should be shown on each partial context diagram. This aids in reviewing. More than one terminator may appear on the diagram, but they should support the same functional area of the enterprise. (See §2.12.4.3 for a discussion of automated support.)

2.12.6.7 Aesthetics

The context diagram serves two purposes

1. It is a high-level overview of the entire system.

2. It is a detailed view of the system interfaces.

Unfortunately these can contradict each other. The choices made when grouping flows and terminators can dramatically affect the ability of the diagram to do its job.

2.12.6.7.1 Use as an overview

If the context diagram is used to present a very high-level picture of the system, it should be easy for a non-technical reviewer to absorb. To do this, the reviewer must be able to read the entire diagram without moving his head such that a part of it is obscured. This means the terminators must be near the process. This may require the use of a very large process circle to allow the connection of many flows and to provide a larger space between flows.

As the reviewer is interested in 'the big picture' he may not require that the data flows be very specific on content in this presentation. To avoid crowding on the diagram, and to improve comprehension of the whole system, flows should be grouped sufficiently to keep the number small. Flows that are to or from the same terminator may be grouped as long as the name is still meaningful and will communicate their content to the reviewer.

Data stores may also need to be grouped. If a store is used by more than one terminator, it could be repeated on the diagram, using an "$*$" to show the reviewer that other terminators use the store.

The reviewer must be able to spot the names of the flows easily. The name should be in normal reading position, situated such that there is no doubt which flows they apply to. Either of the following approaches may be adopted:

1. Place the name close beside, rather than overlapping, the flow.

2. Hide the line by the label, so that the label appears where the flow would be, except that the line is suppressed for a short distance on either side of the label.

The choice of which approach to use will depend on the technical support environment.

If there is no way to group the system's interfaces such that a sufficiently simple yet meaningful presentation can be made, use partial context diagrams.

2.12.6.7.2 Use for detailed review of system's interface

When a reviewer's charge is to make certain that all components of a given interface are present, the context diagram must be presented to emphasise each of the interfaces. The reviewer will generally study the system's interfaces between the context process and a single terminator. This means that he must be able to clearly 'isolate' this view from the rest of the diagram. Sufficient space should be left between sets of flows going to different terminators. Flows should be grouped meaningfully such that no more than around five (a very rough rule of thumb) appear for any one terminator. If necessary, multiple copies of the terminator, tagged with a "$\star$" may be used.

To examine the interface, the reviewer must be able to see the flow as a link between the terminator and the system. No flow lines should cross.

If there is no way of achieving this presentation without grouping so much that all meaning is lost, partial context diagrams should be used.

2.13 Terminator Specification

2.13.1 Purpose

The terminator specification describes a terminator in detail. Each terminator on a context diagram should have a corresponding specification.

2.13.2 Examples

Name : IRS

Meaning : The "Internal Revenue Service" is the government authority charged with collecting taxes. (In countries other than the U.S., it would have a different name.)

Number of instances : 1

Name: Lift

Meaning: The system controls the movement of lifts.
Each lift may be moved up or down by a motor.

Number of instances: 4

Identifier : <Lift>.number

Name : Attending Physician

Meaning : A fully licensed MD employed by or practicing at this hospital.

Number of instances : many

Identifier : <Physician>.id

2.13.3 Components

2.13.3.1 Name
Each terminator has a unique name.

2.13.3.2 Meaning
A full description of the terminator, and its role in this system. Generally speaking, subtle points about the way the terminator should be handled are entered here, rather than as comments on the context diagram.

2.13.3.3 Number of instances
Some terminators have a single instance. An example would be the terminator "Pilot", in a system to control an airplane.

Other terminators have multiple instances. An example would be "Instructor" in a system that was responsible for running scheduled courses, including allocating instructors to them. This entry may have the value:

1. **constant**: where the number of instances is fixed and known in advance. For example "4" if the system is known to be dealing with exactly 4 lifts. If there is only one instance "1" is entered.

2. **variable**: where the number of instances is greater than 1, but the number varies and cannot be known in advance.

2.13.3.4 Identifier
Where the number of instances is greater than 1, this gives the way one instance of the terminator is distinguished from another.

2.13.4 Rules

1. If the number of instances is any value other than "1", then there must be an identifier.

2.13.5 Guidelines

1. If the terminator has a corresponding entity, the terminator is likely to have multiple instances, because single-occurrence entities are rare. In these cases, a suitable terminator identifier is the identifier for the entity. Any system that accesses multiple instances of the entity will have to distinguish between them using this identifier. It is therefore a convenient and natural identifier for the terminator.

2.14 Data Flow Diagram

2.14.1 Purpose

The data flow diagram highlights the functions of the system and how they use stored information and transfer information between each other.

2.14.1.1 Supporting specifications

The components of the data flow diagram have corresponding specifications. These are:

- process specifications — covered in §2.15;

- minispecs — covered in §2.16;

- data flow specifications — covered in §2.18;

- data store — covered in §2.20; and access flow specifications — covered in §2.21;

- behavioural state transition diagrams — covered in §2.24;

- event flow specifications — covered in §2.25.

2.14.1.2 Showing terminators on data flow diagrams

Showing terminators on DFDs is optional. This chapter does not show them in the example, but it is allowed. To show terminators on a DFD, any flow shown on a DFD is shown attached to that terminator. If the DFD is drawn showing terminators in this way, then any flow that is not connected to a process or terminator must be connected to another process within the system (but not shown on this diagram).

If terminators are shown, then they may be repeated to reduce diagram complexity. If terminators are repeated, then it is conventional to indicate this in some way. For pencil and paper support environments, a "$\star$" tag on each of the repeat is conventional.

The context diagram is therefore a type of data flow diagram. Traditionally, what is referred to as the context diagram shows the whole system as a single process group. Equivalently, multiple context diagrams may be used (see §2.12.4.3) or terminators shown on the event–response DFD (see §5.7.1.2).

2.14.1.3 Example explanation

The example is taken from a process control system that controls the cleaning of car bodies. When a car body has been welded it is too dirty to paint properly and is thus immersed in an oxidising solution to clean it. The car body is moved from the body shop to a holding area, known as a mix bank, to await the availability of the tank that holds the oxidising solution. It is then immersed for the necessary period of time, before being moved to another mix bank to await painting. The solution must be the correct one for the type of car body, and of the correct strength. A monitor will read the solution strength to allow the control system, based on preset thresholds, to know when it needs to be changed to be effective. A schedule is used to give pre-planned times when the solution needs to be changed. Note: this example is also used for describing events (see §2.22 and §2.23) and bSTDs (see §2.24).

2.14.2 Example

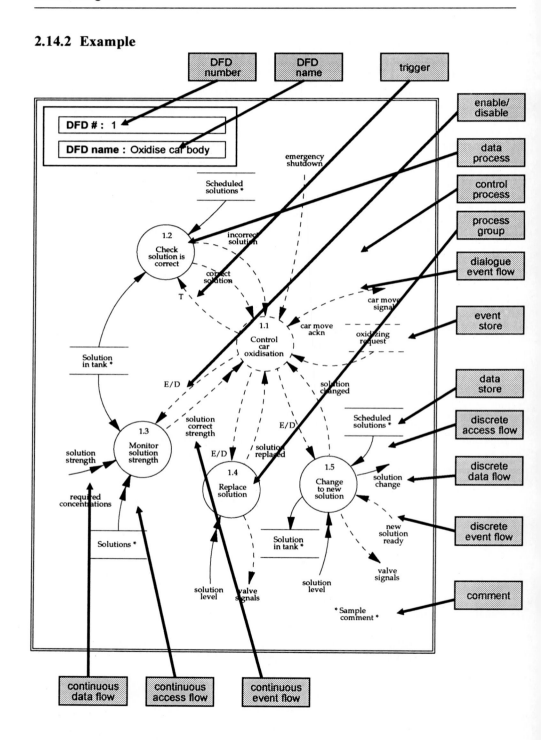

2.14.3 Components

Not all the possible components of the data flow diagram will be required for a specific system. Control processes and event flows will only be required if the behaviour of the system changes over time; continuous flows will only be required if the system monitors or controls the state of the devices in its environment; stores are only used if information available at one time is needed subsequently (most systems use stores).

2.14.3.1 Access flow

An access flow is used to show that a process needs to use or change stored information to accomplish its purpose. The information in the store corresponds to entities, and/or relationships. On the DFD, this is indicated in general terms only. The access flow specification (see §2.21) gives more detail of this interface.

2.14.3.1.1 As process input

An access flow from a store shows that the process uses information held in the store. This may correspond to a:

- **match access**: the process checks whether an entity or relationship occurrence matching a particular criterion exists;
- **read access**: the process uses the values of one (or more) attributes for a selected occurrence of the entity;
- **check access**: the value of a state variable needs to be checked.

2.14.3.1.2 As process output

An access flow going to a store shows that the process will alter information held in the store. This may correspond to a:

- **create access**: to create a new occurrence of an entity or relationship;
- **delete access**: to delete one (or more) occurrences of an entity or relationship;
- **update access**: to modify the value of an attribute;
- **change access**: to change the state of a state variable.

2.14.3.1.3 Access flow notation

The standard accesses described above have the following icons:

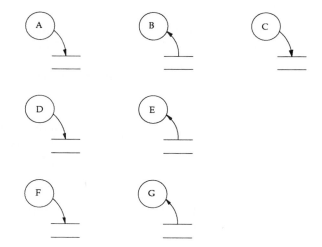

These show a "create" (A), "read" (B), "update" (C), "delete" (D), "match" (E), "change" (F) and "check" (F).

If a process has more than one access to a store, then the icons are combined. This means that a flow from a process to a store is very rarely seen — very nearly all write accesses require some match or read access first.

2.14.3.1.4 Persistence of access flows

Continuous access flows persist over a period of time (see §2.14.3.3); discrete access flows (see §2.14.3.14) take place at a specific instant of time and are transient.

The only accesses which are allowed to be continuous are match, update and read accesses. Only continuous processes can have continuous access to stores.

Discrete accesses are more usual than continuous accesses, but continuous accesses are allowed.

2.14.3.1.5 Multiple accesses to the same store

Only one access flow is required for any one use of a store by a process. It may be obtained by a logical combination of accesses to the store. More than one access can be shown, if separation of accesses is helpful. This is only relevant if the access flow is named.

2.14.3.1.6 Named access flows

An access flow may be named sometimes makes the DFD more understandable. Naming the access flow sometimes makes the DFD more understandable.

Access flows may be labelled in one of two ways:

- **attribute access**: if there is a single entity in the store, then the name of the attribute that is accessed may be used to label the flow. This necessarily requires that only a single attribute is accessed. This type of access may be qualified (see §2.14.3.1.7).

- **as a named access flow**: in this case, the access may include access to several attributes (and possibly other access operations, too). The name is used to describe the access flow. Note: the name may be the same as other access flows to other stores. Access flow names must be chosen to be distinct from any attributes of an entity in the store, as otherwise the name would be interpreted as an attribute access (see above).

The access flow between "Monitor solution strength" and "Solutions" is named "required concentrations" to suggest which information is needed by the process. The specification for this flow will detail the exact attributes used by the process.

Optionality of access flow naming
The use of named access flows is optional. Each access flow between a process and a store has an access flow specification, containing all the details about the access. The access flow does not have to be named to identify this specification. Naming of access flows is only provided for 'upwards-compatability' with older, paper-based techniques. Named access flows are an optional part of YSM and would not have to be provided by an automated support environment.

Note: the same information about access to stored information in the store is given in the input-output list of the minispec (see §2.16.3.7). Either this input-output list or the access flow specification, or both, may be present, depending on the support environment.

2.14.3.1.7 Qualified attribute access flow
An attribute access flow can be qualified. A qualifier is an adjective adding further description to the name of the attribute. It is placed in parentheses before the attribute name.

The qualifier is used to distinguish the difference due to the processing of an attribute. Two accesses with the same names, but different qualifiers, would represent use of the same attribute, but in a different context. It is a semantic distinction.

Possible qualifiers include "updated, new, adjusted, cleared". Particularly when the attribute is updated, it may be helpful to distinguish between the 'old' and 'new' values. For example, in a stock control system, there might be a process that handled deliveries. One of its functions would be to update the "on-hand balance" of the entity "Inventory item", as shown below:

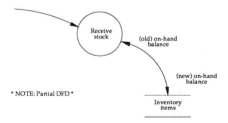

Within a procedural minispec for "Receive stock", we might find:

```
(new) <Inventory item>.on-hand balance :=
        (old) <Inventory item>.on-hand balance + number delivered
```

depending on the style adopted. (See §2.16.3.7.4 for discussion of use of qualified access names in minispecs.) Qualifiers should not be used to allow multiple use of an attribute for different meanings. For example, "(current) on-hand balance" and "(last month's) on-hand balance" would describe attributes (possibly the same attribute) of different occurrences of an associative entity "Stock history". These attributes would have different names — they are different attributes. A qualifier would be an incorrect means of making this distinction.

Note: the use of qualified attribute access flows is redundant with the information contained in the access flow specification. Qualifiers are not allowed with named access flows — it would make no sense to do so.

2.14.3.2 Comment

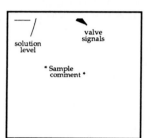

As with all diagrams, free-format comments may be added to the DFD to elucidate or stress any important points. It may also be used to highlight any aliases, ties to previous or anticipated implementations, etc. It is not specified with additional textual support.

2.14.3.3 Continuous access flow

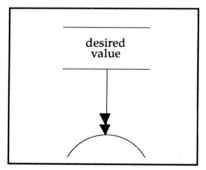

A continuous access flow is a specific type of access flow — see §2.14.3.1.

In some situations, it may be necessary to continuously monitor the value in a store. This is carried out using a continuous data process. On the DFD, the link between this data process and the store is shown as a continuous access flow.

A store acts as a buffer between a continuous process and a discrete process, as shown in the examples given in §2.14.3.3.1 and §2.14.3.3.2. See also §2.12.3.4.1 for a continuous access flow on a context diagram.

2.14.3.3.1 Example: set point control logic

In a system with a set point, there is a continuous process with the task of maintaining a measured value to some set point value. This set point may be varied by the operator on a discrete basis. To model this, a discrete process updates the store with the desired value. A continuous process carries out control, using the currently available required values from the store. This is sketched below:

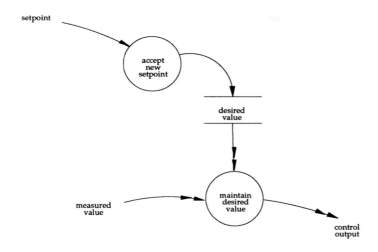

2.14.3.3.2 Example: continuous 'tracking' of satellite orbit

The orbit of a satellite continuously changes due to various effects. These may be divided into non-gravitational forces that act on the satellite and the fact that it is not moving in a constant, spherical gravitational field. The equations governing this change are differential equations for the rates of change of the orbit parameters. (These are discussed, for example, in [Roy88].)

A system that needed to predict what the orbit is over a period of time needs to solve these equations. With perfect technology, this would be achieved by integrating the equations. To do this, a continuous data process is required.

The current orbit parameters would be attributes of the satellite and would be shown as a store "orbit" (say) on a DFD. The continuous process would continuously access this store, which could also be used by a discrete data process that needed the current orbit parameters. (A process to schedule pictures that were to be taken from the satellite would be such a discrete process.) This is shown below:

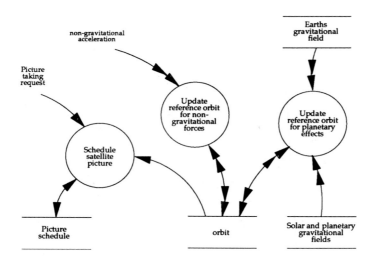

Note: the two stores "Earths gravitational field" and "Solar and planetary gravitational fields" contain mathematical models of the gravitational fields that the satellite is moving in.

2.14.3.3.3 Implementation issues

As soon as the system is mapped onto a specific technology, the only data access operations are those supported by the architecture.

For analog processors, both discrete and continuous operations may be possible. For digital processors, all current architectures restrict physical store accesses to be discrete and any continuous access flows must be replaced by discrete access flows, carried out at some sampling frequency.

2.14.3.4 Continuous data flow

A continuous data flow is a data flow (see §2.14.3.7) that persists over time.

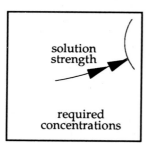

Continuous data flow can be thought of in a 'passive' or 'active' sense. Used in the passive sense, it describes the value of a currently changing quantity (or values of several quantities) that the receiver needs to know about. Used in the active sense, it informs the receiver of a *desired* value that continuously changes.

2.14.3.4.1 Time graph of a continuous data flow

The general form of a continuous data flow is sketched below:

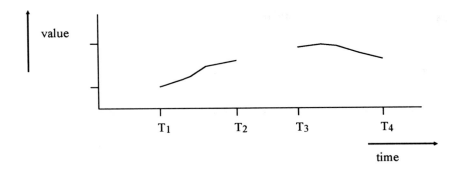

The value is present over a period of time and may change during that time. Its value is 'time–continuous'. From time "T$_1$" to "T$_2$", its value is available (and changing) ; from "T$_2$" to "T$_3$", its value is *not* available; from time "T$_3$" to "T$_4$", its value is again available.

2.14.3.4.2 As output from process or terminator
A continuous data flow represents data that is generated by its source, whenever that source is active. This source may be a terminator, or a continuous data process.

2.14.3.4.3 As input to process or terminator
The continuous data flow may be received by a terminator or data process. Both discrete and continuous data processes may receive continuous data flows.

2.14.3.4.4 Presence/absence of continuous data flow
A continuous data flow is only available when its source is active and generating this data flow as an output. If the source is a process, it must be enabled. When the source stops, the flow ceases to exist. This occurs when the source process is disabled or a terminator changes to a mode of behaviour that does not generate the output.

The data flow receiver does not have to be continuous. At any time when its source is operating, however, this flow is available to the receiver and may be used.

The presence of a continuous data flow may be checked by means of the standard operation "available(id)", where "id" is replaced by the name of the flow. (The standard operation "available" is defined in §6.9.3.)

2.14.3.4.5 Example
The example given in §2.14.2 shows a continuous data flow "solution strength" used by the process "Monitor solution strength". This is the current concentration of the solution. The value comes from an analog device (seen as a terminator on the context diagram) and is continuously present.

2.14.3.4.6 Implementation issues
As soon as the system is mapped onto a specific technology, continuous data flows have to be mapped onto the available mechanisms. For analog processors, continuous data flows are possible. For digital processors, continuous data flows must be replaced by discrete data flows, or data stores, generated at some sample frequency.

2.14.3.5 Continuous event flow

A continuous event flow is one that is available over a period of time, such as the status of a device. Continuous event flows represent a situation or status that is (or should be) either true or not.[1]

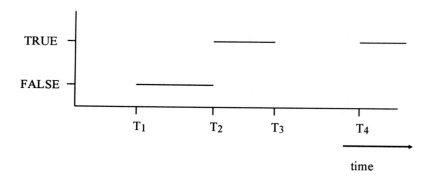

A continuous event flow can be thought of in a 'passive' or 'active' sense.

Used in the passive sense, it describes the status (or statuses) of some type. This could be the status of a real-world device, or it could be an indication of ongoing behaviour, output be a control process. The receiver needs to know this status to determine correct behaviour.

Used in the active sense, it informs the receiver of a *desired* status that can change over time.

2.14.3.5.1 Time graph of a continuous event flow

The general form of a continuous data flow is sketched below:

The value is present over a period of time and may be either TRUE or FALSE during that time. Its value is 'time–continuous'.

From time "T_1" to "T_2", its value is available (and FALSE) ; from "T_2" to "T_3", its value is TRUE; from time "T_3" to "T_4", its value is *not* available; from time "T_4", its value is again available (and TRUE).

2.14.3.5.2 As output from process or terminator

At any time when the source is active, the event flow may be asserted to be TRUE by that source. At other times, although the source is active, it asserts that the situation does not hold by generating the continuous event flow, with the value FALSE.

---

1 Note: a continuous event flow is sometimes referred to as a 'flag' — however, it should not be confused with actual device mechanisms of this name provided in many run-time software architectures — it is a conceptual mechanism.

If the source is not active, the event flow does not exist, and can therefore not be regarded as either TRUE *or* FALSE. It just does not exist. (Some modellers refer to this as the value being undefined, or *bad*.)

Note: continuous event flows may be generated by any type of continuous process.

2.14.3.5.3 As input to a process
The event flow user does not have to be continuous. At any time when its source is operating, however, this flow is available to the receiver. The presence of a continuous event flow may be checked by means of the standard operation "available(id)" (see §6.9.3).

Within the process, the event flow acts like a boolean data item, although there are some restrictions on its use in state logic (see §2.24 for a discussion of the use of continuous event flows in bSTDs). Other standard YSM operations may also be used: rising(id) generates a discrete event flow at the moment in time when the value goes from FALSE to TRUE; falling(id) generates a discrete event flow at that moment in time when the value goes from TRUE to FALSE. (In all these functions, "id" is replaced by the name of the flow.)

Note: continuous event flows may be used by any type of process.

2.14.3.5.4 Example
In the example, the continuous event flow "solution correct strength" shows that the car bathing solution is either appropriately strong for use (if it has the value TRUE) or not (FALSE). This allows the control process to determine what the state of car bath should be.

2.14.3.5.5 Implementation issues
As soon as the system is mapped onto a specific technology, continuous event flows have to be mapped onto the available mechanisms. For analog processors, continuous event flows are possible (logic signals). For digital processors, continuous event flows must be replaced by discrete event flows, semaphores, flags, or some other mechanism.

2.14.3.6 Control process

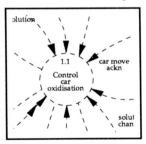

A control process can:

- coordinate the activation of processes;

- 'signal' to other components (terminators, control processes, or event stores) of the System Essential Model.

The coordination is carried out by interpreting and generating event flows, and controlling processes by triggering, or enabling or disabling them.

2.14.3.6.1 Continuous control processes
A control process must be continuous if either (or both) of the following are true:

- any process that it control is continuous;

- it tracks a persistent state of a terminator.

Most control processes are continuous as they must remember the state of the entities that they are controlling between events. In the example given in §2.14.1.3, "Control car oxidation" must remember the state of any car body in the oxidation area. It is therefore a continuous control process.

Example
The control process in the example, "Control car oxidation", shows that control logic must be present to coordinate the monitoring, changing and replacing of solutions. It also drives the movement of the car body by interpreting the state of the oxidation bath and generating required signals such as "move car signal" to move the car body.

2.14.3.6.2 Discrete control processes
Discrete control processes, while rare, are allowed. A discrete control process can only:

- controls discrete processes;
- generates discrete signals;
- does not need to track the state of anything over time.

At the time a discrete control process is activated, it may use continuous event flows and discrete event flows that are part of its stimulus to determine required actions. These actions can include triggering discrete data processes and generating discrete event flows. All of the actions of this discrete control process are performed in 'zero time'. The control process does not persist, and needs no memory of the state of an external entity between events.

Discrete control processes are usually only used when the result of a data process may produce one of several different possible event flows, which, when compared to several different possible combinations of continuous event flows, will lead to complicated decision logic.

2.14.3.7 Data flow

A data flow represents the information generated by a source (data process or terminator) and used by another system function or terminator. On a data flow diagram, the source of the data flow may be:

- **a data process**: the function represented by the data process is the one that generates the information
- **a process group**: the function that generates the information is within this group. See §2.14.3.21.1 for a discussion of what it means to state that 'a data process is within a process group'.

If the source of the data flow is not on this diagram, then it must be a terminator (but see comments in §2.14.1.2) or a process not shown in this diagram. In either case, the data flow will appear in the parent DFD. On the parent DFD, the parent process group is shown as a user of the information; the source may, or may not, be visible on this diagram.

The user of the information contained in the data flow may also be a data group, process group in the diagram. If the user of the data flow is not on this diagram, then it must be a terminator (but see comments in §2.14.1.2) or a process not shown in this diagram. In either case, the data flow will appear in the parent DFD.

2.14.3.7.1 Content and identification of data flow

The information in a data flow may be one indivisible item of information (an element) or a group of such items. This packaging is described in the corresponding data flow specification (see §2.18). All data flows have a name which is used to identify them for the purpose of this specification.

2.14.3.7.2 Qualified data flow

A data flow name may have a qualifier, which is an adjective that adds insight to the meaning and use of the data flow. Qualifiers are used to distinguish flows that have identical content, but slightly different meaning in this specific case.

For example, in a hospital system, there might be a data flow "actual treatment", representing the treatment given to a patient. If a nurse notified the system of the actual treatment that had been given, the system would probably need to validate the information, to try to 'filter out' incorrect values. This would be carried out by a data process — "Validate actual treatment"

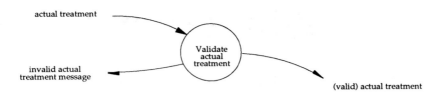

(say). After this check, the information has the same structure and values as before, but the semantics are rather different in this context — the information is now valid. This can be indicated by using a qualified data flow "(valid)actual treatment". This is shown below.

Note: the qualifier is shown in parentheses as a prefix to the data flow name. Qualified data flows do not have a separate specification, but may be described under the corresponding entry for the unqualified flow name (see §2.18.3.2.1).

Qualified data flows may also appear on context diagrams.

2.14.3.7.3 Persistence of data flows

Data flows may be continuous or discrete. The use of continuous data flows on data flow diagrams is discussed in §2.14.3.4. Discrete data flows on data flow diagrams are discussed in §2.14.3.15.

2.14.3.8 Data flow diagram name (DFD name)

A data flow diagram has the same name as the corresponding process group shown on the parent DFD. The context diagram is treated as a DFD with one process group.

Note: the name specifies the function. This DFD is used to specify the function. The same function may appear on different DFDs (with a different process number). The data flow diagram can, therefore, act as the child of process groups on several different DFDs. The child diagram serves to specify each of these processes.

2.14.3.9 Data flow diagram number (DFD number)

A data flow diagram usually (see below) has the same number as the process group it describes in more detail. The child diagram of the context diagram is numbered "0".

The diagram may describe more than one identical process group. This may be achieved in one of three ways:

1.By having multiple copies of the process group. See §2.14.3.20.4 and §2.15.

2.By using the same named DFD to specify two processes with the same name, but different process number.

3. By having one process group that is a copy of another, with some of the input and output names changed.

In the latter two cases, the lower-level DFD specifies different process groups with different numbers. This is also discussed in §2.15.

2.14.3.10 Data process

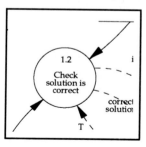

A data process is a process that solely transforms data. It is not responsible for the coordination of process activation. Data processes may use continuous event flows as boolean items. Data processes may be continuous or discrete. A continuous data process may generate continuous or discrete event flows. A discrete data process cannot generate continuous event flows.

See also §2.14.3.20, which defines the term "process".

2.14.3.11 Data store

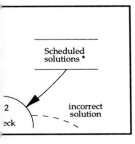

Stores act as 'buffers' between processes that are active at different times. A data store between two processes 'decouples' them; a data flow between processes 'synchronises' them. A store holds data 'at rest'.

If two discrete processes need to share information and they are active at different times, they may use a data store. One process writes the data into the store when it is activated. Subsequently, when the second process is activated, the information may be read from the store. (If two processes need to share data and do not use a store, then they may communicate by means of data flows, *providing both processes are active at the same time*.)

Data stores represent the collection of information about entities or relationships. The content of the stores is specified using a data store specification (see §2.20).

2.14.3.11.1 Use of stores by process groups

A store is said to be used by a process group if any process in that group uses the store. On any DFD, any store used by a process and another process (or terminator) must be shown, even if the other process (or terminator) is not visible on that diagram. Each process group that uses the store is shown with an access to the store.

If the process group includes any process that carries out a continuous access to the store, the process group is shown linked by a continuous access to the store, even if it also includes discrete accesses to the store. In other words, 'a continuous access + a discrete access = a continuous access'. However, as an alternative, the process group may be shown connected to the store by a continuous access flow *and* discrete access flow.

2.14.3.11.2 Showing a store several times on the same diagram.

If a data store is accessed by more than one process such that it is impossible to avoid crossing access flows on the diagram, a store may be shown more than once on a single diagram. When this is done, a "*" is used after the store name for each of the copies of the store. This indicates that it is used elsewhere on the diagram. See §2.14.5.5 for a discussion for how this is used.

2.14.3.11.3 Example

The data store "Solutions" seen in the example (see §2.14.2) is a collection of entity occurrences of all possible solutions used in the bathing of car bodies. One "Solution" has certain attribute values, such as the thresholds for its strength, the solution name, etc.

The data store "Scheduled solutions" contains the occurrences of an associative entity.

2.14.3.12 Dialogue data flow

A dialogue data flow contains several data flows that act as an 'interface'.

At the lowest level, there can be two data flows in direct relation to each other: one causes the other to occur, such as a question and its answer, or a stimulus and its response. This is a dialogue pair. Dialogue flows can also be used to summarise complex interfaces (see §2.14.3.12.2).

2.14.3.12.1 Dialogue pairs

A dialogue pair can be used to emphasise the close relationship between the two flows. For example, a request for a report and the ensuing report might be modelled by a dialogue pair.

The output of a dialogue pair must be part of the response to the input. For example, if a customer complaint got an immediate response, the complaint and its reply could be modelled as a dialogue pair. The process "Deal with customer complaint" would have both the input and the output coming from it, as shown below:

If, however, the complaint could not be dealt with immediately (perhaps because the system might need to interface to a terminator "Customer service"), the flows "customer complaint" and "complaint response" would not occur at the same time. It would not be appropriate to model this as a dialogue pair.

2.14.3.12.2 Levelled dialogues

If there are many flows between a pair of processes, they do not all have to be shown on higher-level DFDs. A single dialogue data flow can be shown. The specification for this flow would define it as multiple, containing the lower-level flows. These lower-level flows can be 'ordinary' data flows, or they can themselves be dialogue data flows.

2.14.3.12.3 Identification and specification of dialogue data flows

A dialogue data flow is identified using the pair of names. The specification of this flow defines wheter it is a dialogue pair or a multiple flow.

2.14.3.13 Dialogue event flow

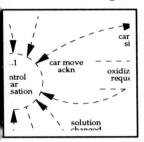

A dialogue event flow is a packaging of several event flows between two processes (or between a process and a terminator).

Except that the flows are event flows, rather than data flows, the description is exactly as for dialogue data flows, as described above.

2.14.3.14 Discrete access flow

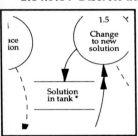

A discrete access flow is a specific type of access flow — see §2.14.3.1. A discrete access flow is an access to a store that takes place under the control of the system at a particular instant of time.

In the example, the access flow from "Change to new solution" to the store "Solution in tank" shows that the store is being updated by the process, although the diagram does not explicitly distinguish between updates, creates and deletes. If the modeller wished this distinction, the flow could be named and qualified, or tagged with a comment.

The process "Check solution is correct" accesses the store "Solution in tank" to insure that the solution is correct for the car that is about to be bathed. This is a match access.

2.14.3.15 Discrete data flow

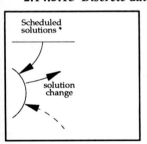

A discrete data flow is a data flow that is only available at a particular instant of time. A discrete data flow is transient and must be processed at the moment it occurs. If a process using a discrete data flow is not active at the time it occurs, the data flow will be lost. Information that must persist over time must either be a continuous data flow, such as a temperature reading, or in a data store, such as "last temperature read". (See also the entry for data flow in §2.14.3.7.)

2.14.3.15.1 Time graph of a discrete data flow

The general form of an elemental, discrete data flow is sketched below:

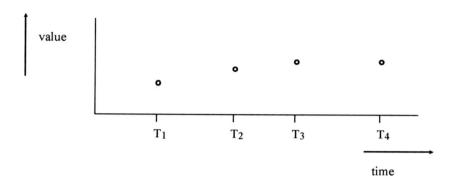

The value is only present at specific instants of time. It has a value at those times and no value at any other time.

2.14.3.15.2 Example of discrete data flow

In the example (see §2.14.2), "solution change", generated by "Change to new solution" represents the need to tell the operators that they must change to a certain solution. For each change this flow is sent once, and is acted upon by the operators when they receive it. If the operators are otherwise occupied, they may miss the flow. "Change to new solution" will wait a certain amount of time, and, if it does not receive "new solution ready" from the operator, it will re-generate the flow.

If the policy were to continually alert the operators until they responded, the flow should be modelled as a continuous data flow.

2.14.3.15.3 Discrete data flows as 'packets of data'

A discrete data flow can be thought of as a 'packet of data'. This is generated at an instant in time. All required components of the flow are generated at that time, as part of the 'packet'. Sometimes there may be optional components of the data flow — these are not always given values in a particular 'packet'.

Data flows may be grouped together. This is not visible on the diagram, but can only be seen by examining the data flow specification. For example, the data flow "credit card payment" might be defined to have:

COMPOSITION : credit card number + amount

This might be seen on a DFD as the data flow "credit card payment". Each time payment occurs, there are two values available. One is "credit card number"; the other is "amount". Both of these are components of "credit card payment". Choosing whether to show the two components as separate flows, or as one higher-level flow is a matter of controlling detail. However, in reality, they both occur together. On the other hand, if several types of payment occurred, they might be declared as a multiple flow. For example, "payment" might contain "credit card payment" and "cash payment". These do not occur together. This is a multiple flow.

For further discussion of data flows, see §2.18.3.3.

2.14.3.16 Discrete event flow

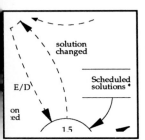

A discrete event flow is an event flow that is only available for an instant of time.[1]

A discrete event flow must be processed at the moment it occurs. If a control process detecting a discrete event flow is not active at the time it occurs, the event flow will be lost. If the event flow is used by a control process, it must be in a state to detect this flow when the flow occurs, otherwise the event will be missed.

2.14.3.16.1 Time graph of a discrete event flow

The general form of an elemental, discrete event flow is sketched below:

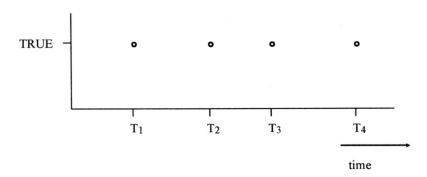

The value TRUE is only present at specific instants of time. It is meaningless to refer to a discrete event flow having the value 'false' — it is asserted TRUE at certain instants of time. It has *no* value at any other time. The absence of this TRUE value could be because:

- the source is not active;
- the source is active, but has determined that the event flow does not need to be generated.

The receiver of a discrete event flow has no way of determining which of these is true at any time it has not received the discrete event flow. (In control systems, this usually necessitates an additional continuous event flow, indicating whether the source is active and *capable* of generating the discrete event flow, when required.)

---

1 Discrete event flows are sometimes referred to as 'signals' — however, discrete event flows should not be confused with actual device mechanisms of this name provided in many run-time software architectures — it is a conceptual mechanism.

2.14.3.16.2 *Example of use of discrete event flow*

In the example, the discrete event flow "solution changed" shows that the process "Change to new solution" tells the control process when the solution has been switched to the new solution.

If the knowledge that the event occurred must persist over time it should be represented as a continuous event flow, such as "solution correct strength", or in an event store, such as "oxidising request".

A discrete event flow may represent a grouping of event flows; any one of these flows can occur independently of the others. This is discussed in §2.25.

Event flows in general are discussed in §2.14.3.18.

2.14.3.17 Enable/Disable

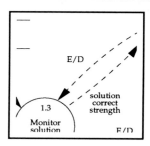

This represents a process being enabled and disabled from a control process. An "E/D" represents the fact that the control process enables the process at one time and disables it subsequently. In certain circumstances, an "E" in isolation is used. This represents a process that is 'turned on' by a control process which does not 'turn it off'. This occurs when the process being enabled and the control process are both part of a group that is enabled and disabled by a higher-level process (see §2.14.3.17.4).

A process which is enabled will run whenever its stimulus occurs (if discrete) or run continuously (if continuous) until disabled.

Note: Both "Disable" (see §6.9.7) and "Enable" (see §6.9.8) are standard YSM operations.

2.14.3.17.1 *Enabling/disabling continuous data processes*

When a continuous data process is enabled, it will start transforming its inputs to produce its outputs. It continues to do this until it is disabled.

In the example, the continuous process "Monitor solution strength" is enabled, and will do its job until told by the control process to stop by a disable (which occurs when the solution is being changed or replaced).

2.14.3.17.2 *Enabling/disabling discrete data processes*

Some discrete data processes are triggered when required and some are enabled, so that they are 'ready' when their stimulus occurs. When a discrete data process is enabled it is ready to transform its inputs to produce its outputs. However, it does not actually do this until a discrete flow (called its stimulus) occurs. This must be an input to the data process. At that instant of time the process runs. It may use any information in the stimulus data flow, together with stored data (seen as an access flow into the process) or continuous data flows.

When the process is disabled, it is no longer able to transform its inputs, even when the stimulus occurs.

See §2.15.3.6.1 for further discussion of the relationship between enabling/disabling discrete processes and the occurrence of their stimulus.

Example of enabling/disabling discrete processes
In a bank, deposits might only be accepted between certain hours. This is essential policy (say).

Deposits would be handled by the discrete data process "Accept deposit". A control process would enable the "Accept deposit" process at the beginning of those hours, and disable it at the end of those hours. However, "Accept deposit" is stimulated by the arrival of a discrete data flow, customer deposit. At the time that flow arrives, "Accept deposit" can validate and record the deposit in zero time.

If "Accept deposit" were always active, it would run at any time that a deposit was submitted — this is not what is required, as the deposit should *not* be accepted outside banking hours.

If "Accept deposit" were triggered, it would run at that instant, and the deposit would have to be present at the moment of triggering. If another deposit arrived later, the control process would have to trigger the process again.

2.14.3.17.3 Enabling/disabling process groups
There are two situations:

1. **The process group does not contain a control process**: when the parent process group is enabled, all processes in the group are enabled; all are disabled when the parent process group is disabled.

2. **The process group contains a control process**: when the parent process group is enabled, the child control process and all child processes not specifically enabled/disabled or triggered by the control process are enabled. The other child processes are considered to be disabled at this point in time and are enabled (or triggered) by the control process at specific points in time. When the parent process group is disabled, all processes in the group are disabled.

Example of enabling/disabling process group
In the example given in §2.14.2, the process group "Replace solution" is enabled and disabled by "Control car oxidisation". "Replace solution" is a process group, containing a control process and several data processes controlled by that control process.

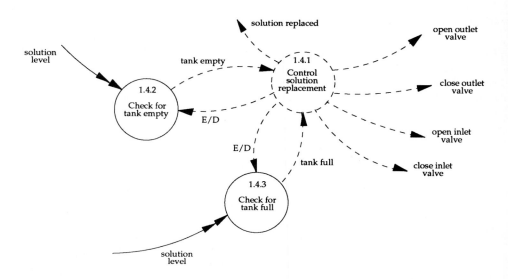

When "Replace solution" is enabled, there might be processes within it that are not controlled by "Control solution replacement" — these are enabled at the point when "Control car oxidisation" enables "Replace solution". Other processes within "Replace solution" will be explicitly enabled/disabled (some might also be triggered) by "Control solution replacement", as and when required. The behaviour of "Replace solution" will vary over time.

Although it reaches a final state, "Control solution replacement" persists over time while the tank is filling. Should there be an intervening event during this time, such as "emergency shutdown", "Control car oxidisation" may need to disable "Replace solution", even though it has not completed.

If "Control solution replacement" completes, it will remain active in a final state until "Control car oxidisation" disables it. Usually, "Control car oxidisation" disables "Replace solution" after it receives the "solution replaced" event flow; this indicates that "Control solution replacement" has reached its final state.

2.14.3.17.4 Unpaired enables

If a process is enabled by a control process, it is usually disabled by the same control process. The only exceptions to this are processes that are enabled by the control process and continue to remain active until the parent process group is disabled on the parent DFD.

2.14.3.18 Event flow

An event flow represents the occurrence of an event (discrete), or the state of something (continuous). Event flows may include other event flows.

See continuous event flow (§2.14.3.5) and discrete event flow (§2.14.3.16).

2.14.3.19 Event store

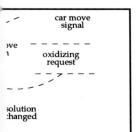

An event store is a mechanism for storing events relating to resources until they can be used by a control process. When an event detector signals an event store (rather than use a discrete event flow), it allows the event to persist until its receiver is in a state where it can respond, or until the store is re-initialised.

There are three operations provided for use with an event store:

- **initialise**: this creates the event store and sets it up to contain zero events. If the event store already exists, it clears the event store of all outstanding events that have already been signalled. Any waits remain active. This is shown as a flow into the event store from the process that initialises the store.

- **signal**: if there are no processes waiting on the event, this adds one more occurrence of the event to those in the store; if there are processes waiting on the event store, *exactly one* of them is released and the number of events in the store remains zero. This is shown as a flow into the store from the process.

- **wait**: if there are events held in the store, the number present is decremented by one (the event is 'consumed'); if there are no events in the store, the process is suspended and put into a queue for the event store. In general, there will be more than one process in this queue; when the event store is signalled, *exactly one* of them will be resumed (and the event is 'consumed'). On the DFD, a wait is shown as a flow from the store to the process; in effect, it is a 'destructive read'. Note: YSM assumes a simple (first come, first served) queue; implementation architectures may provide other queueing strategies.

None of the flows to or from an event store are named. An event store is shown on a DFD if more than one process on the diagram uses it, or a process on the diagram uses it and a process not shown (or included within a process group on the diagram) also uses it.

2.14.3.19.1 Example

In the example, "oxidising request" will collect any events of the type "Car waiting for the bath" and signal them by "signal(oxidising request)". Each time it does this, another occurrence of the event is put in the store (depending on how many events the event store has been defined to hold — see §2.26.2.2). When "Control car oxidisation" is ready to move a car into the bath, it will treat any event in this store as if it were a discrete event flow sent at that precise moment. The event, when consumed by the control process, is no longer in the store.

This allows the control process to ignore the event until it is ready to respond to it without fear of missing the event.

If "oxidising request" were modelled as a discrete event flow, the control process would have to detect it at the moment it was sent. If it missed this event flow, it would be unaware that a car was waiting for the bath.

If "oxidising request" were modelled as a continuous event flow, the process generating it would have to remain active and continually monitor the bank. The control process would have to detect the moment that the event flow was raised to know that a car was waiting. If it missed that moment, it would have no stimulus to cause it to take action. As this event flow would really represent the state of the mix bank, it would be better named "cars present in mix bank".

2.14.3.20 Process
A process is a system function. It is either a data process, a control process, or a process group. It may be either continuous or discrete. All processes have a corresponding process specification (see §2.15).

2.14.3.20.1 Process naming and numbering
All processes have a name that describes the function that they carry out. There is a lower-level DFD (for a process group), minispec (for data process and some discrete control processes), or bSTD (for a control processes) with this name.

Each process has a number that is obtained from the number of the diagram by adding a decimal point, followed by a digit. This number uniquely identifies the process — there is a process specification with this number. The diagram number may be omitted if the number becomes too long. For example, on diagram 3.2.4.5, rather than "3.2.4.5.6", a process could be numbered ".6".

2.14.3.20.2 Continuous process
A continuous process takes time to carry out its function. It 'persists'. Only continuous processes can generate continuous outputs.

2.14.3.20.3 Discrete process
A discrete process, although not distinguished graphically from a continuous process, is a process that can start and finish immediately when it runs. A discrete process may be triggered, enabled (and disabled) explicitly by a control process, or be permanently enabled. As the conceptual processor runs at ultimate speed, these processes are said to take 'zero time'. When it runs all of its inputs must be present.

A discrete process does not persist over time and is not able to produce continuous outputs.

Discrete processes may be data, control or process groups.

2.14.3.20.4 Showing multiple instance processes on the DFD
Some systems require identical copies of the same process to carry out parallel actions relating to different real-world entities. For example, in a system designed to control four lifts, there might need to be four instances of the process logic — one instance for each lift. As an optional notation, this may be shown on a DFD as a process with a 'ghost' behind it. For example:

This notation may be used with any of the three types of process. A single 'ghost' is used, with the number of instances being defined in the process specification (see §2.15).

2.14.3.21 Process group

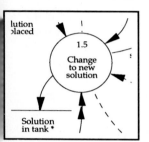

A process group represents a group of processes as a single icon on the diagram. This group may also contains internal stores and flows to accomplish its purpose — these are 'hidden' within the process group. This process group is usually specified by another data flow diagram.

Process groups are used to reduce the complexity of any one data flow diagram by combining related functions and naming the combination for the general function that this group carries out. For example "Change to new solution" is a process group comprising processes "Control solution change", "Determine new solution", "Detect full tank", "Detect empty tank", etc.

A process group is drawn on the diagram as a solid circle; it cannot be visually distinguished from a data process without looking at its specification.

2.14.3.21.1 The 'levelling' convention

The diagram on which the process group appears is referred to as the higher-level or parent diagram. The process group itself is referred to as the parent process.

The data flow diagram that is the expansion of the process group is referred to as the child diagram and is a lower-level DFD. Any process that appears on this diagram is referred to as a child process with respect to the process group. The process is also described as being within the process group.

The number and name of the child DFD are the same as the number and name of the parent process group. Thus, the diagram that expands the process group "Change to new solution" would be numbered "1.5" and named "Change to new solution".

A child process of a process group may itself be a process group. It will then have a child DFD that defines it. It will have child processes on that DFD. This may go on indefinitely (at least in principle).

See §5.7.1.4 for an example of a levelled set of DFDs.

Continuous process groups

Process groups are considered continuous if any of the processes they group are continuous.

Discrete process groups

A process group is discrete if all the processes in the group are discrete, including any control process. An example of a discrete process group might be "Validate and Update Solution Threshold" which might contain two processes, one, to do the validation, and the other, to do the updating. All information required for these processes is there, and the processes will run in 'zero time', that is, they have no need to persist.

2.14.3.21.2 Activation of process groups

A process group may be enabled/disabled or triggered explicitly by a control process. When this occurs the effect on the processes within the group depends on the structure of the group. The way this interpreted depends on whether there is a control process within the group:

1. If there is no control process in the group, the activation is interpreted as being of all the processes in the group in the same way. For an enable, all the processes in the group are enabled at that time; for a trigger, all the processes are triggered. Note: some of the processes in the group may themselves be process groups. If so, the same rules apply to them.

2. If there is a control process in the group being activated, any processes within the group are of two types:

   - **Those activated by the child control process**: these are activated when (and only when) activated by the child control process.

   - **Those not activated by the child control process**: these are activated when the process group is activated.

See §2.14.3.17.3 for further discussion of enabling/disabling process groups.

See §2.14.3.23.1 for further discussion of triggered process groups.

2.14.3.22 Prompt

A prompt is an activation mechanism. There are three types of prompt:

   - **trigger**: the control process determines that a process must run immediately;

   - **enable**: the control process determines that a continuous process should be running or a discrete process should run when its stimulus occurs;

   - **disable**: the control process determines that a continuous process should not be running or a discrete process should not run even when its stimulus occurs;

A prompt represents no data or reason for activation other than that the process should be activated or stopped. Prompts are labelled by the type of activation mechanism — see trigger (§2.14.3.23) and enable/disable (§2.14.3.17).

2.14.3.22.1 Controlled and permanent processes

The concept of 'controlled' and 'permanent' processes is useful. A process is controlled if either of the following are true:

   - it is shown on a DFD as being enabled/disabled or triggered;

   - its parent process is controlled.

This can be continued until the parent process is the context process. Any processes not defined as controlled are permanent. A permanent process is always running (for a continuous process), or ready to run (for a discrete process). A permanent discrete process does not actually run until its stimulus arrives.

Any process is either controlled or permanent.

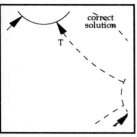

(§2.14.3.20.3).

2.14.3.23 Trigger

A prompt labelled "T" represent a trigger. The trigger will activate a discrete process, which will then 'run to completion' and 'stop'. The process takes no time to complete in an Essential Model.

In the example, "Check solution is correct" is triggered, as it will run in "zero time" and immediately reply with one of the two event flows "correct solution" or "incorrect solution". There is no possible time for another event to occur when "Check solution is correct" is running. Only discrete processes may be triggered. See discrete process

2.14.3.23.1 Triggered process groups

There are two situations:

1. **The group contains no control process**: when the group is triggered, all processes within the group are considered to be triggered at that time and run immediately.
   Data flows may be used within these processes. One process may selectively read attributes and output a discrete data flow to a second process which performs calculations on those values and outputs a report. The intermediate data flow, if output from the first, will be immediately present as the discrete process that produces it runs in zero time. The second triggered process can always depend on it being there. This intermediate data flow is not the stimulus for the second process. The second process is triggered, which is its stimulus.

   In order for this inter-relationship between the processes to be 'well-defined', the triggered processes should be considered to execute in the sequence implied by such dependencies. 'Circular dependencies' ("A" outputs a flow, which is an input to "B" and "B" outputs a flow, which is an input to "A", or extensions of this dependency) are not allowed.

2. **The group contains a control process**: all processes in the group must be triggered by this control process.

2.14.4 Rules

2.14.4.1 General rules

1. It is illegal to show two processes with the same name on the same data flow diagram. Processes with the same name may appear on several DFDs; each will have a process specification (see §2.15); there is one minispec (see §2.16) that describes the function that they carry out. There is a single function, which has a single minispec . Each instance of this function is a separate process with a process specification .

2. It is illegal to show two processes on the same DFD with the same number.

3. Every process must have at least one output. The output may be an access, data, or event flow, subject to the restrictions given by other rules.

2.14.4.2 Separation of data and control

1. A data process is not allowed to receive an incoming discrete event flow. A data process may use continuous event flows as boolean data items. [1] Data processes may sometimes only use continuous event flows and no data flows (see §2.16.3.13.12 for an example).

2. A discrete control process is not allowed to produce any outputs, other than triggers and discrete event flows.

3. A control process must produce at least one prompt or event flow. (Note: it is possible to have a control process that has no event flow inputs — it could rely solely on clocks, to prompt required processing at given times.)

4. The source of a continuous flow must be a continuous process (or terminator). On the diagram, these flows 'come out' of these continuous processes (or terminators).

5. Data processes are not allowed to trigger or enable/disable other processes.

6. Process groups are not allowed to trigger or enable/disable other processes.

7. A control process is not allowed to generate or use data flows.

2.14.4.3 Use of event stores

1. Only a continuous control process or continuous process group may wait for an event flow from an event store. In the case of a control process, the occurrence of an event in this store acts as a condition for the control process. In the case of a process group, one or more control processes 'inside' the group use the event as a condition.

2. Only one process may initialise an event store. (This is a 'global' constraint, not just a restriction on a single DFD.)

2.14.4.4 Miscellaneous

1. A trigger must activate a discrete process. (Note: This cannot be checked without looking at the process specification — see §2.15.)

---

1 A process group is allowed to use a discrete event flow and data processes and process groups are shown with the same icon. Any process that is not a control process with an incoming discrete event flow must therefore be a process group.

2. A specific continuous event flow may be produced from only one data or control process. This process is the one responsible for monitoring the status that this flow represents. It may appear on several DFDs, as an output from a process group containing the source of the event flow.

2.14.4.5 Names

Most of the components of the DFD must be named. Specifically, all:

- data flows;

- data stores;

- event flows;

- event stores;

- processes

must be named. Naming access flows is optional — it is not usually worth the effort to do as the access to the entities contained in the stores is better seen in the minispecs.

2.14.5 Guidelines

2.14.5.1 General

1. A DFD should not generally have more than one control process. The event–response DFD is allowed to have more than one control process (it shows all control processes), but when grouping processes, it is strongly advised that no group has more than one control process. If two control processes need to communicate, they should be placed in separate groups, each with data processes or process groups that they control. In effect, a DFD with two control processes should be 'split' into two groups. It is a deliberate restriction, which will usually simplify the interpretation of behaviour; occasionally, it may result in a process group that only contains a control process.

2. Every process must have at least one input. This may be a data flow, access flow, event flow or prompt. (This is not an absolute rule, because there are a few examples of processes with no inputs.)

3. A data flow diagram should be able to be reviewed and commented upon by a reviewer within one hour. If it takes longer the complexity should be reduced.

4. Comments should be used sparingly as they may clutter the diagram. Large amounts of commentary should be specified in the description of the item. The need for large amounts of commentary may indicate poor partitioning of the model, or an unhelpful choice of names.

2.14.5.2 Names

1. A name should show no trace of a current or imagined implementation for the system. (The specific terminator may be 'fixed', particularly in embedded systems, but that is outside the system boundary. Even then, the role of the device should be described, rather than its specific technical type. Thus "Lift" is correct; "Model a12/b lift" would be incorrect.)

2. Discrete event flow names should be named to allow the reviewer to understand what the presence of the flow means in terms of real-world events. The full specification is given in an event flow specification (see §2.25).

2.14.5.2.1 Access flows

The following is a 'rule of thumb' for labelling access flows:

1. Access flows on the lowest-level diagram should be labelled. If the access flow includes all the attributes of the entity, the flow need not be named (often the case for a create or delete).

2. Access flows on high-level diagrams are not be named. They will suggest general store access. At lower-levels, access flows are named as described in the preceding guideline.

A more pedantic approach would demand that all access flows are named — this is not usually cost-effective in pencil and paper support environments. A radical approach is not to label any access flows — in fact all information about the access flows can be derived from the minispecs.

The correct strategy with automated support environments is more subtle (see §2.14.6).

2.14.5.2.2 Data flows

1. A data flow should be given a name that describes the meaning of a single occurrence of the information it contains. Although there may be many "Deposits" during a day, the flow associated with the event "Customer makes a deposit" should be named "Customer deposit". (See §2.14.3.15 for further discussion of this.)

2. Data flow names should allow a subject-matter expert to understand what its value represents (for elemental data flows) or what its likely components are (for compound data flows) from the name alone. The full specification is given in a data flow specification (see §2.18).

3. Discrete data flow names should be named to help a reviewer understand what the presence of the flow means in terms of real-world events.

4. Qualified data flow names should only be used when the difference in meaning is small. A major difference in the meaning or purpose of the flow merits a different data flow name and full specification. As a general rule, if the data content of the flow has changed, it requires a new, unqualified name. Restrict qualifiers to clarify the information learned about a flow as it is checked by a process. Qualifiers are used to make a a semantic, rather than structural, distinction between data flows.

2.14.5.2.3 Store names

See §2.20.5.2 for guidelines on data store names.

See §2.25.5 for guidelines on event store names.

2.14.5.2.4 Event flows

1. A discrete event flow should be named such that the event causing the flow can be described without having to look it up in an event list.

2. If the process which outputs a flow detects a condition, but is not responsible for the action to take based on that condition, the flow should be named for the condition, not the action. For example, a process detecting an emergency condition and reporting to an operator who is responsible for taking action, should generate the flow "plant is on fire" rather than "shut down the system".

If, however, the process is the decision maker, and the receiver of the flow neither makes decisions nor needs to know why they were made, the flow may be named for a command. A flow from a process "Take picture" generates the event flow "open shutter" to a camera, rather than the flow "aimed at the target".

2.14.5.2.5 Processes

1. A process should be named such that a subject-matter expert can describe, in general terms, its internal functions without referring to other documentation. The full specification is given in a minispec (see §2.16).

2. A process name should be made up of a specific imperative verb followed by a specific direct object.

2.14.5.3 Distinction between data flows and data stores

A data store between two processes 'decouples' them; a data flow between processes 'synchronises' them.

If two processes respond to different events, they can only communicate data values by means of stores; if they deal with the same event, they may communicate by means of data flows.

2.14.5.4 Distinction between event flows and data flows

In a sense, event flows and data flows may be distinguished by the number of bits they carry when they occur. This may be:

- **0 bits**: this is a discrete event flow;

- **1 bit**: this is a continuous event flow;

- **many bits**: this is a data flow.

In addition, they each carry information that they are 'present' or 'available'. Thus a discrete event flow can only convey the information that something has happened (or should happen) at that point in time; a continuous event flow can only convey the information that something is true (or should be) at certain times and false at others.

The concept of event flows has proved useful because of their different semantics from data flow diagrams. Data flows convey information that tends to be transformed, stored, used in calculations etc.; event flows are not used in this way, but modify system behaviour over time.

2.14.5.4.1 Defining process groups

This is discussed in more detail in §2.14.5.4.1, but the main ideas are repeated below:

1. It must be possible to give the group a name which is an honest description of the group's responsibilities. This name should be of the general 'verb phrase + object phrase' format. It should not constitute an essay — if a short meaningful name cannot be found, the group is not a good, natural grouping. If all functions contribute to the same, clearly defined function, the group is said to be 'cohesive'.

2. Control processes should always be placed on the same diagram as the processes they control. This has already been forced on the designer by the rules given previously.

3. The interfaces between process groups should be as simple as possible. If this is not the case, it may be possible to find better groupings that reduce the complexity of the interfaces. This is referred to as the 'coupling' of the data flow diagram.

4. A data flow diagram should either deal with part of the system's response to a single real-world event, or it should deal with all of the system's response to one or more events. It should never deal with parts of more than one event. This ensures that the 'levelled set of DFDs' is event-partitioned — the responses to an event can be localised to one process and its child diagrams.

2.14.5.4.2 Correct use of event stores

Distinction between event and data stores
Event stores are very different from data stores:

- **data stores**: are totally passive, waiting for data processes to access it. Data may be put in the store — it remains there until it is retrieved, changed, or deleted.

- **event stores**: are much more 'active'. They work together with processes in a cooperative way, determining 'what happens next'. A process, in effect, transfers control to the event store when it executes a 'wait'. The event store transfer control back to the (chosen) process at some subsequent time. In the meantime, the process is 'suspended', waiting for the event store to 'resume it'.

It multiple event flows occur and the system cannot respond when they occur, then event stores should be used. If the system can always respond when they occur, then discrete event flow should be used. If only one occurrence of an event needs to be recognised, then a continuous event flow can be used.

Requests and 'availability' events
An event store should only contain event flows that represent an external resource, or requests for the use of an external resource. Its proper use is to delay recognition of events until sufficient external resources are available to respond to that event.

Example, showing 'queuing' of requests
In the example (see §2.14.1.3), the event store represents requests for use of the oxidation tank. These requests cannot be honoured until the tank is available. If the tank is available the moment the request is stored, the request is used by the control process immediately, and the event response occurs immediately.

Example, showing 'queuing' of resource availability
The store can reflect the resource availability, rather than the request. If a bank lobby were modelled, the store may be "bank clerk available". Each time a clerk was free to deliver service, an event would be posted to the store. When a customer used the service, the event would be 'consumed', so that it was no longer in the store.

2.14.5.5 Aesthetics
For a data flow diagram to accomplish its purpose, it must be able to be understood easily by its viewers. This means care must be taken to ensure the diagram is legible and uncomplicated. If the viewer's first reaction is shock (usually suggested by a deep breath, slowly shaking head, and blinking eyes) this has not been accomplished. The diagram may be correct, but the analyst cannot know this until the work has been validated by at least one human.

To aid in good presentation, the quantity of information on the diagram must be sufficiently small to be taken in one careful glance. The reviewer should not have to study for hours to see what is on the diagram; if the reviewer cannot absorb the material rapidly, too much time will be spent just comprehending the contents of the diagram and not enough on whether that content is correct.

2.14.5.5.1 Easing individual process understanding

Legibility
In reviewing data flow diagrams, the reviewer will spend some time concentrating on each individual process. He will determine if it has all required inputs to produce the outputs. To do this, the reviewer must be able to see the names easily (without the aid of a magnifying glass). This may require the circles to be enlarged. If that cannot be done because of crowding, the diagram should be rearranged. If no uncrowded arrangement is possible, the number of items on the diagram should be reduced. This can be achieved by grouping data flows, data stores, or processes.

Inputs and outputs to processes
The reviewer must be easily able to spot which are inputs and outputs to one process. This requires a prominent arrowhead. It may also help to place the inputs to the left of the process, and the outputs to the right to give a standard input–process–output grouping.

The reviewer must be able to consider the inputs and outputs of one process all at once; this means that their number should not exceed his capacity to remember them. A very general rule of thumb is about seven flows for any one process. If there are too many, try grouping flows into multiple flows), or grouping stores. If there is no good way of grouping then try reducing the size of the process by placing some of its subfunctions in other process groups.

Naming flows
The reviewer must be easily able to spot the names of the flows. The name should be in the normal reading position, situated such that there is no doubt which flows it applies to. (Asking the reviewer to read names written in a curve over the flow will give him a pain in the neck. Asking him to read upside down will tempt him to inflict pain.) Generally placing the name close beside, rather than overlapping, the flow will increase legibility. (This depends on the technical support environment. For automated support, labels over the flows may be recom-

mended. However, in this case, the flow line should be suppressed, rather than 'going through the name.)

The reviewer should be easily able to locate all flow names while concentrating on one process. The flows around one process should be sufficiently short that the reviewer can see all the names at once, without having to hunt for them. He should not be expected to have to use many fingers to keep track of names and flows.

Separation of diagram components
Flows and stores not accessed by a process should not be situated so close to it that the reviewer may mistakenly think that the process does use them. Some white space around the process makes it easier to focus on it without error.

2.14.5.5.2 *Easing general diagram understanding*
The reviewer will also have to take in the diagram as a whole to determine that the sources and receivers of flows are correct, and to determine that the processes needed to accomplish the general process are all present.

This requires that the diagram be sufficiently small to be looked at without having to move the head such that any one part will be excluded from vision. The print must be large enough to be seen from the distance at which the reviewer can review the entire diagram in one glance.

It also requires that individual pieces of the diagram be easily distinguishable from each other. Names must be easily associated with what they identify. The flows must be quickly traced from source to receiver. If they cross each other, this will become very difficult. The diagram should be rearranged to avoid most crossed lines. A general rule of thumb is no more than three crossed lines on a diagram. If lines do cross, breaking the flow with a small gap around the crossing flow is preferred to placing a curved line over the flow as a way of showing the cross.

If the diagram cannot be rearranged without causing more problems, another way of avoiding crossed lines is to duplicate the data stores. If a data store is accessed by many processes, it should be repeated. To help the reviewer, place an asterisk on the store (in all places it is used) to show the reviewer that the store exists elsewhere on the diagram. Only duplicate stores to avoid crossed lines; the addition of another item also adds to general diagram complexity.

2.14.5.5.3 *House style*
It is worth while cultivating a 'house style' for data flow diagrams. Reviewers react poorly to 'messy' diagrams, with different size icons, labels and data flows that are tangential to process icons. If stores are used, they can be 'spread around' the process in a visually appealing way, or they can be 'all over the place'. Access flows look better if they enter store icons at right angles, rather than acutely.

While, strictly speaking, the definition of such house style and visual guidelines is outside YSM, it is suggested that some effort to prepare visually appealing diagrams is worthwhile.

2.14.6 Dependency of the data flow diagram on the support environment
It is important to avoid being 'locked into' a pencil and paper way of thinking of the DFD. It is one view into the system. With pencil and paper it has become rather 'over-loaded' with roles. It tries to show functions, data, some dynamics, use of data by functions.

With a third generation approach (properly supported by software), this is an outmoded approach — the DFD should be mainly regarded as a view of how system functions 'fit together'. There are 'windows' into the dynamics and information viewpoints (store and control process icons), but the diagram should not regard them as any more than that. The DFD does not specify data (or data accesses). Nor does it show dynamics.

The ideal solution would be provided in an automated support environment, with the following approach:

- show system functions as data processes on a DFD. (The functions visible could be selected dynamically, of course).

- system dynamics could be shown (or not) in various ways — one rather traditional way would be to show control processes. These could be suppressed, if wished.

- show store icons for any entities or relationships used by each function. These would be given a name that reflects its likely content — however, they are only 'clues' that data is modified.

- 'clicking' on a store would cause the entities accessed by that process to be displayed (probably in frame format).

- 'clicking' on the access flows would cause a summary of the access to entities by that process (probably in a similar format to that shown for the minispec frame format).

- 'clicking' on the data process would cause the minispec and/or process specifications to be displayed.

- 'clicking' of a control process would cause a bSTD to be displayed.

2.15 Process Specification

2.15.1 Purpose

This specification describes the type, number of instances and activation mechanisms for any type of process. There is one process specification for each process in the System Essential Model.

The functionality of the process is described by other modelling tools:

- data flow diagrams (see §2.14);

- minispecs (see §2.16);

- bSTDs (see §2.24) or behavioural state transition and action tables (see §2.24.4).

Each of these can describe more than one process (see §2.15.3.9).

2.15.1.1 Effect of support environment

Whether this modelling tool is used as a separate specification tool depends, to some extent, on the support environment.

In a pencil and paper environment, much of the information included in this specification could be included as annotation of the DFD or combined with that in the minispec, or bSTD. This is not ideal, because the same minispec or bSTD may specify several processes, but it is possible.

In an automated support environment, this specification tool acts as direct support for the DFD. Every process on a DFD has a process specification. Each process specification refers to a minispec, bSTD, or lower-level DFD that is used to specify the function that this process is an instance of.

2.15.2 Examples

PROCESS : Update instructors' salary with pay award

MEANING : Deals with pay increases

PROCESS TYPE : data

PERSISTENCE : discrete

ACTIVATION MECHANISM : stimulus

STIMULUS : pay award

NUMBER OF INSTANCES : 1

SPECIFICATION : minispec

MEAN TIME BETWEEN FAILURES : 6 months

MEAN TIME TO REPAIR : 1 day

RESPONSE TIME : 1 day

PROCESSING/ACTIVATION : 8.5 M instructions

PROCESS : Control lift

MEANING : Controls 1 lift

PROCESS TYPE : group

PERSISTENCE : continuous

NUMBER OF INSTANCES : 4

IDENTIFIER : lift number

SPECIFICATION : bSTD

MEAN TIME BETWEEN FAILURES : 5 years

MEAN TIME TO RECOVERY : 1 day

MINIMUM SAMPLE FREQUENCY : 1 second

PROCESSING/ACTIVATION : 300 k instructions

2.15.3 Components

2.15.3.1 Process
This gives the name of the process that is being specified. For example:

> **PROCESS** : update instructors' salary with pay award

2.15.3.2 Meaning
This is used to give a general description, or comments about the process. Care should be taken to avoid specifying the process in this field — it is meant for any significant information that does not 'live' anywhere else. For example:

> **MEANING** : Deals with pay increases

2.15.3.3 Process type
This defines the type of process. A process may be a:

- control process (see §2.14.3.6);
- data process (see §2.14.3.10);
- process group (see §2.14.3.21).

The allowed values are "control", "data" or "group", for example:

> **PROCESS TYPE** : data

2.15.3.4 Persistence
This indicates whether the process is:

- continuous: the process will continuously produce its outputs. Examples would include data processes that continually monitor and control external equipment using analog signals. Control processes are usually continuous, but discrete control processes are allowed (see §2.14.3.6.2).

- **discrete**: at specific instants of time, it carries out its function. It will be explicitly triggered or require a stimulus (see §2.14.3.20.3).

The allowed values are "continuous" or "discrete". For example:

> **PERSISTENCE** : discrete

If the process is continuous, then it is specifically enabled/disabled or is always active; if the process is discrete, then there is an activation mechanism that causes its function to be carried out.

2.15.3.5 Activation mechanism
For discrete processes only, the type of activation mechanism used to cause the process to run is entered. The allowed values are:

- **stimulus**: there is a discrete data or event flow that acts as a stimulus for the process. When this flow occurs, the process runs.

- **trigger**: the process is triggered when a specific control process identifies the need for it. The process itself has no control over this, nor does it 'know about' anything before or after it runs; it just runs.[1]

- **time**: the process is activated when a particular temporal event occurs (temporal events are discussed in §2.22). The process 'wakes itself up' when this event occurs and carries out its function. The event causing the activation is documented in the event specification (see §2.22.5.10).

The activation mechanism is entered in this field. For example:

ACTIVATION MECHANISM : stimulus

2.15.3.6 Stimulus

If the entry for "activation mechanism" is "stimulus", then this entry gives the name of the flow that acts as the stimulus.

This entry is only present for discrete processes only. When this stimulus occurs, the process carries out its function, taking no time. Having carried out its function, the process then waits for it to occur again. For example:

STIMULUS : pay award

Note: a discrete data process can have a discrete data flow as stimulus; a discrete control process can have a discrete event flow as a stimulus. Other combinations are not allowed.

2.15.3.6.1 The interaction of enable/disable and the stimulus

For discrete processes, the distinction between "enable" and the activation of the process when the stimulus occurs is important. Consider a process that is enabled and disabled as required by a control process. When enabled, it 'waits for' a stimulus data flow to occur. When it does occur, the process runs to completion (taking no time) and then 'waits for' the stimulus to occur again. This may be shown in a state transition diagram:

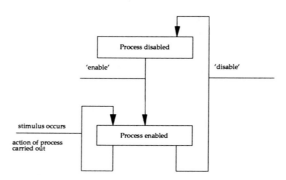

---

1 This is an example of a general modelling principle that no unit knows *why* or *when* it is invoked, but only *what* it should do.

Note: this diagram describes the state of the process from an 'external' point of view (as seen by a hypothetical 'operating system'). The process itself is a discrete data process and is modelled using a minispec. It is not modelled as having 'states'. The diagram above describes how any discrete process with a stimulus behaves.

The diagram below shows the relationship between persistence, activation mechanism and stimulus:

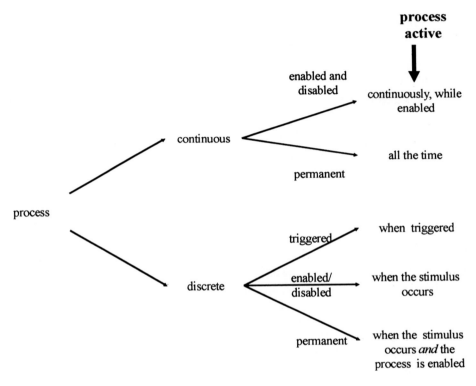

1. Any process is either continuous or discrete.

2. Continuous processes are either enabled and disabled or permanent.

3. A discrete process can be enabled and disabled, or triggered, as shown in a DFD. Other discrete processes are permanent.

4. For both enabled/disabled and permanent processes, the process can only run at a certain time. This is given as the stimulus entry in the process specification (if a process is triggered, this is also shown here, redundantly with the DFD).

On the right hand side of the diagram, the actual time that the process runs is given.

2.15.3.7 Number of instances

Some processes have more than one instance. The number of instances of a process may be:

- **constant**: the number of instances are 1, 2, 3, ... , but this number does not vary over the life-time of the system. The usual situation is that there is one instance, but there may be more. For example, in the case of a system that controlled four lifts, there would be four instances of a process group "Control lift".

- **variable**: the number of instances changes over a time period that is less than the life of a system. In an air tracking system, we might allocate a copy of a process to deal with each aircraft as it entered the controlled airspace.[1]

This entry gives the actual numbers of instances. The entry is the actual number of instances, or "variable". For example:

> **NUMBER OF INSTANCES : 1**

Process that have more than one instance are referred to as multiple instance processes. Multiple instance processes may be indicated graphically on the DFD if desired (see §2.14.3.20.4).

2.15.3.8 Identification

If number of instances is not 1, then the identifier is entered here. This identifier is used by other processes that need to deal with this multiple instance process. For example:

> **IDENTIFIER : lift number**

Note: the fact that a process is multiple instance is transparent to the process itself — it does not know, or need to know, about the other instances of itself.

Only data processes can use instance information — they may send and receive flows to multiple instance processes, including process groups.

Process groups may contain a control process that enables and disables processes within this group — this control process does not know about multiple instances or need to identify individual instances. It only controls processes within that group. It may generate signals, to which instance identification is added, but that can only be accessed by a data process outside the group.

See §2.16.3.1.2 for a discussion of dealing with flows to and from multi-instance processes.

See also terminator identification (§2.13.3.4).

2.15.3.9 Specification

This defines the way in which the process is specified.

---

1 Although a process can be declared as variable instance, the present release does not fully support the use of variable instance processes. In particular, no "create" or "destroy" operation has been defined in this release of YSM.

For control processes (**PROCESS TYPE :** control), the allowed values are "bSTD", "state transition and action tables", "minispec" (unusual, but allowed — see §2.16.3.13.12), or "modification".

For data processes (**PROCESS TYPE :** data), the allowed value is "minispec" or "modification".

For process groups (**PROCESS TYPE :** group), the allowed values are "data flow diagram" or "modification".

For example:

> **SPECIFICATION :** minispec

There is exactly one specification used to define the 'function' of the process:

- Data flow diagrams are discussed in §2.14;

- Minispecs in §2.16;

- bSTDs in §2.24;

- state transition and action tables in §2.24.4.

2.15.3.10 Modified process

Sometimes a function (as described by a minispec) may be required several times, using different data flows. This can be handled by writing two minispecs. However, this would not be very efficient, because only the names of the data flows will have been changed. Building and testing the function (not to mention specifying) will not be very efficient.

Rather than force this redundancy, YSM allows one minispec to be declared as a modification of another one. In this case, the function is the same, but different names are used for input and output data flows. The internal function is the same in both cases.

Control processes can also be declared to be a modification of another control process, with only the incoming and outgoing event flow names changed. There would only be one bSTD required, but it would be used to specify two state machines (which appear as control processes on the DFDs). In fact, it is not often that a state machine can be re-used in this way — it usually requires other processes.

In more general situations, one process group can be declared to be a modification of another process group. There is only one lower-level DFD to specify the group. Any control processes seen on this diagram have a single corresponding bSTD. However, there are two (or more) processes with the same internal logic. They must have different inputs and outputs, otherwise, they would only be one, single process.

If a process is defined as a modification of another process, the name of the original process needs to be given, together with any changes. The same function is carried out (though using different data and event flows). (Changes in names of input or output data or event flows are discussed in §2.15.3.11.)

2.15.3.10.1 Example

This simplified example concerns a system to control access to a section of single-track railway. No startup protocol is defined here, nor does the example maintain a safe distance between trains travelling in the same direction. However, it correctly deals with the contention for this resource. The DFD below shows two functions that are almost 'mirror images' of each other. Each control traffic in one direction.

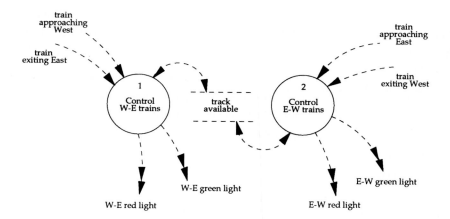

Only one of these two process groups is specified by lower-level diagrams. The diagrams below show how "Control W-E trains" is specified.[1]

The process "Control E–W trains" is defined to be a copy of "Control W–E trains", with only the input and output flows changed. This is given in a frame specification (some of the entries have been suppressed).

---

1 Note: the counter process group would be better modelled as an instance of enterprise object (held in the Enterprise Resource Library, but that is outside the scope of this release of YSM.

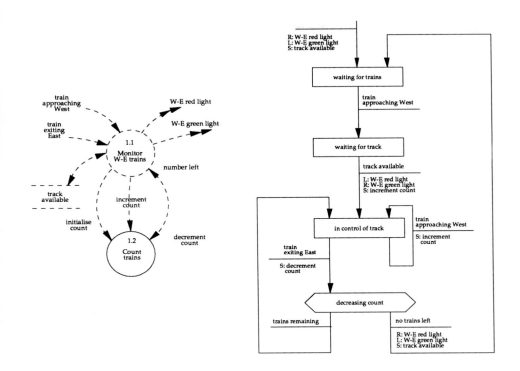

"Control E–W trains" is completely specified by this frame specification:

PROCESS :	Control E–W trains

MEANING :	Controls access to track for trains approaching from the East.

PROCESS TYPE :	group

PERSISTENCE :	continuous

SPECIFICATION :	modification

MODIFIED PROCESS :	Control W–E trains

CHANGES :	original	: modification
	train approaching West	: train approaching East
	train exiting East	: train exiting West
	W–E red light	: E–W red light
	W–E green light	: E-W green light

Both of these process groups have parallel logic for counting, state logic, etc., but it only needs to be specified once.

2.15.3.11 Changes

For each change of input or output, the original item name is given, followed by the new item name (the one used in the copy). The effect of these changes is *as if* the original specification (minispec, bSTD, or DFD) was edited with these changes to obtain a new specification. These items may only be components of incoming or outgoing data or event flows — data access flows may not be 're-routed'. For example:

CHANGES:	original	: modification
	train approaching West	: train approaching East
	train exiting East	: train exiting West
	W–E red light	: E–W red light
	W–E green light	: E–W green light

In this example, the input event flow "W train approaching" in the original process "Control W–E trains" corresponds to the event flow "E train approaching" in the process "Control E–W trains". This is a lowest-level event flow.

2.15.3.11.1 Allowed changes

Constants may be used instead of variable inputs (although this is rare). Access to entities may not be changed in a modification. The process that is defined as a modification will usually (but not always) have a different stimulus. When this is the case, the change does not have to be repeated under the "changes" entry. The new stimulus is assumed to be the replacement for the old one; any components of it replace the old components.

In a modification of a process that is specified by a minispec, renaming an input or output data item might appear to cause it to have the same name as a temporary item, local term or local function. If such minispec were constructed, it would contravene the 'unique name' rule (see §2.16.5.2). However, in generating a copy of a minispec, the modeller need not worry about this — these items are 'internal' to the original copy and the renaming does not compromise the behaviour of the process.

2.15.3.11.2 Exact copies of processes

Note: two processes on different DFDs are allowed to have the same name. In this case, the processes are exact copies or 'clones'. There is only one corresponding minispec (although each has a process specification). For example, "Display patient list" might be used many time in a hospital system. It would appear on several DFDs, each time with the same name and function. There is only *one* minispec that acts as a specification for these processes. The names of the data items that are input and output from the two processes will be identical.

2.15.3.12 Mean time between failures

The mean time between failures (MTBF) is a measure of the 'reliability' of the process. The average time between one failure and the next should not be less than this figure. This is a constraint on any implementation. Any acceptable design must be able to achieve this level of reliability. If the delivered MTBF is greater than this figure the system is acceptable.

Note: this parameter can be entered for any process, including process groups. The MTBF of a process group cannot be greater than the minimum of all the MTBFs for its included processes. This applies to the context process, as a special case.

See also §5.9, which discusses the System Essential Model performance aspect.

2.15.3.13 Mean time to repair

The mean time to recovery (MTTR) is an estimate of how quickly normal operation must be resumed after a failure. This parameter depends on the system type and business in which it used. For some systems (control systems, in particular), this may be critical, to avoid loss of life. In other systems, there is a penalty with the system not being available, but it is more of an inconvenience than a threat.

Sometimes it is possible to put a financial cost on being without the system. In these cases, this parameter can be used to do cost-benefit analysis of different implementations.

See also §5.9, which discusses the System Essential Model performance aspect.

2.15.3.14 Response time

For discrete processes only, this parameter describes how long the process can take to complete. Although perfect technology would allow it to complete in zero time, this is not, in fact, possible. This parameter is an upper limit of what is acceptable to the enterprise using the system. For a process that calculates the trajectory of a missile, this time may be very short and critical. For other processes (such as a 'payroll' function), the response time that is acceptable may be much greater.

See also §5.9, which discusses the System Essential Model performance aspect.

2.15.3.15 Minimum sample frequency

For continuous processes only, this can be used to estimate the minimum frequency at which the discrete equivalent of the process would have to be activated. It is only given if it is known (or believed) that the process is to be allocated to a digital processor — it would not make sense in an analog processor. Definition of how to calculate this sample frequency (which depends on the algorithm used, as well as the real-world dynamics) is outside the scope of YSM. (See, for example [Pow78].)

See also §5.9, which discusses the System Essential Model performance aspect.

2.15.3.16 Processing/activation

This is an estimate of how much processing is required for each activation of a discrete process. The equivalent for a continuous process, run in sample mode can also be given (provided it is anticipated that it will be allocated to a digital processor).

See also §5.9, which discusses the System Essential Model performance aspect.

2.15.4 Rules

2.15.4.1 Stimulus

1. If a discrete data process has a stimulus, it must be a discrete data flow.

2. If a discrete control process has a stimulus, it must be a discrete event flow.

2.15.4.2 Copies

1. A copy of a continuous process must also be a continuous process.
2. A copy of a discrete process must also be a discrete process.
3. A copy of a process that has a stimulus must also have a stimulus.
4. There must be no cyclic dependency in defining a process to be a copy of another process. That is, a process "A" is a copy of a process "B" if "A" is declared to be a copy of "B" in a process specification, or "A" is declared to be a copy of "C" and "C" is a copy of "A" (either directly, or indirectly).
5. No process group can be defined to be a copy of a process that is contained within it, or vice-versa.

2.15.4.3 Changes

1. The original and modification entry must be the same type of flow. They must both be data flows, or both be event flows.
2. The persistence of both the original and modification flow must be the same (either continuous or discrete).
3. The composition of compound modification flows must be the same as the composition of the original flows. In this context 'the same as' refers to the structure, not the name of the flows. The order of components in the composition (see §2.18.3.4) is significant in this respect.

2.15.5 Guidelines

2.15.5.1 Modifications

1. If the same function is used many times with modifications to its inputs and outputs, then consideration should be given to making it an enterprise operation. It should then be given a single general specification (see §2.17). It can then be used with different actual arguments (see §2.17.3.3.1).

2.15.5.2 Number of instances

1. If there is only a single real-world entity being controlled or dealt with, the number of instances is 1.
2. If more than one real-world entity is being controlled or dealt with, the number of instances depends on whether any of the following hold:

   - Each subsystem dealing with the entity is different, with different rules. In this case there must be modified logic to handle each. There will be specific processes for each, with different minispecs.

   - The system can be regarded as several copies of a single system. In effect, it splits into distinct, duplicate systems. This is dealt with by defining and checking one system model. It is then copied (or 'cloned') as many times as is necessary. Thus (for example), the same system installed at many user sites requires only modelling for one site.

- The subsystems are identical and yet they interact in some way. We need to distinguish between the instances (because of the interaction modelling requirement) and yet we wish to say the sub-systems are each dealt with by the same logic (because they are identical). For a lift system that handled four identical lifts, the central allocation of requests (dealt with by a process "Schedule lifts") needs to interact with and distinguish between the different instances of the lift. However, each of the four lifts will have the same individual control logic. We therefore define the system as containing four instances of the process group "Control lift".

2.16 Minispec

2.16.1 Purpose

Minispecs are provided to give a rigorous specification for each data process within the system. In certain circumstances, they are also used to specify control processes.

Minispecs must be:

- Precise: so that they can be converted to delivered system components that can be tested against the minispec. They are also used to test the delivered processes.

- Understandable: so that they can be understood by subject-matter specialists.

The most important criterion for a minispec is that the specification should state the rules that relate the outputs to the inputs. Minispecs allow both external and internal specifications. External specifications are used to define the effect of the process and for testing and formal proof of correctness. Internal specifications are used to describe how the process will be built. External specifications are descriptive; internal specifications are prescriptive.

Specification of the process may be carried out in a variety of ways. There are standard YSM ways of doing this, but no one range of styles and modelling tools will suit all situations. However, the modelling techniques in this section will be sufficient for all common situations.

The grammars described in this section may be used in a formal way when the application merits it. In other applications a relaxed format and syntax may be used if this is more appropriate for the project and reviewers.

2.16.2 Examples

2.16.2.1 Example 1

PROCESS : Update instructors' salaries with pay award

DATA FLOW INPUTS : %percentage pay increase%, %audit level%

DATA FLOW OUTPUTS : %high pay audit%

EVENT FLOW INPUTS : —

EVENT FLOW OUTPUTS : —

STORED DATA INPUTS :
      <Instructor> (**attributes:** age, salary)

STORED DATA OUTPUTS :
      <Instructor> (**attributes:** salary)

TEMPORARY ITEMS :

name	: values
%pay increase%	: currency

LOCAL TERMS :
      <Old instructor> ::= <Instructor> with age > 25.

LOCAL FUNCTIONS : —

FUNCTION (external) :

FUNCTION (internal) :
    1. For each <Old instructor> perform the following three operations:
        1.1. %pay increase% is <Old instructor>.salary multiplied by
            %percentage pay increase%.
        1.2. <Old instructor>.salary is increased by %pay increase%.
        1.3. If <Old instructor>.salary > %audit level%
            issue %high pay audit%.

2.16.2.2 Example 2

PROCESS : Schedule initial treatment

DATA FLOW INPUTS : %initial patient data%

DATA FLOW OUTPUTS : %insufficient treatment message%

EVENT FLOW INPUTS : —

EVENT FLOW OUTPUTS : —

STORED DATA INPUTS :
<Patient> (**attributes:** number, infection level, narrative history)
<Drug>　　　　(**attributes:** number)
<Recommended treatment>(<Drug> is recommended for <Patient type>)
　　　　　　　　(**attributes:** recommended period, recommended dose)
<Patient> is typed as <Patient type>

STORED DATA OUTPUTS :
<Patient>　　(**attributes:** infection level)
<Scheduled treatment> (<Patient> is scheduled to receive <Drug>)
　　　　(**attributes:** date of treatment, time of treatment, desired dose, treatment period)

TEMPORARY ITEMS : —

LOCAL TERMS : epsilon ::= angle, value = 0.0001radians

LOCAL FUNCTIONS : —

FUNCTION (external) :

Pre-condition 1
　　Matching <Drug> exists AND <Patient> with number = %treatment patient% exists

Post-condition 1
　　　　　　<Patient>.infection level = %new infection level%
　AND　"<Patient> is scheduled to receive <Drug>" exists
　AND　<Scheduled treatment>.date of treatment = todays date( )

Pre-condition 2
　　No matching <Drug> exists AND <Patient> with number = %treatment patient% exists

Post-condition 2
　　　　　%insufficient treatment message% exists
　AND　%insufficient treatment text% = <Patient>.narrative history

FUNCTION (internal) :

2.16.2.2.1 ERD for example 2

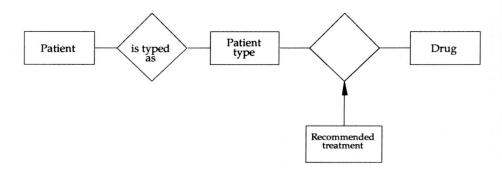

2.16.2.2.2 DFD for example 2

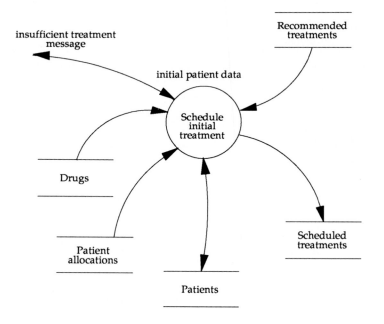

2.16.2.3 Example 3

PROCESS : Orient satellite for picture

DATA FLOW INPUTS : attitude

DATA FLOW OUTPUTS : rotation speed

EVENT FLOW INPUTS : —

EVENT FLOW OUTPUTS : satellite rotated for picture

STORED DATA INPUTS : <Required orientation>   (state variable : status) (attributes : required angle)

STORED DATA OUTPUTS : —

TEMPORARY ITEMS :   **name**   **:values**
                          desired angle   :orientation

LOCAL TERMS : epsilon ::= angle, value = 0.0001 radians

LOCAL FUNCTIONS : —

FUNCTION (external) :
initial actions:

FUNCTION (external) :
continuous actions:

FUNCTION (internal) :
initial actions: desired angle := <Required orientation>.required angle of <Required orientation> with status = due

FUNCTION (internal)
continuous actions: rotation speed.theta := f_4(attitude.theta, desired angle.theta) ;
                        rotation speed.rho := f_5(attitude.rho, desired angle.rho) ;
                        rotation speed.phi := f_6(attitude.phi, desired angle,phi) ;

    If   ABS(attitude.theta − desired angle.theta) < epsilon
        AND
                ABS(attitude.rho − desired angle.rho) < epsilon
        AND
                ABS(attitude.phi − desired angle.phi) < epsilon then
                signal(satellite oriented for picture)
    end ;

2.16.3 Components

2.16.3.1 Inputs and outputs of the minispec

The input and output data, event and access flows given in this frame format provide a link between the specification of the function and the other views such as data flow diagrams (see §2.14), data flow specifications (see §2.18) and access specifications (see §2.21).

Some of this is redundant with the above specifications, in particular, the components of data flows. It is completely redundant with the internal function specification and can be derived from it. However, it may be useful to draw up this list as an aid to writing the minispec. When the minispec is complete, these inputs and outputs can be automatically derived from the external (see §2.16.3.12) or internal (see §2.16.3.13) function specification.

In 'pencil and paper' environments, this should *not* be regarded as part of the specification. In automated environments, these inputs and outputs could be helpfully displayed.

2.16.3.1.1 Available and accessible items

These two concepts are useful in defining the allowed operations in minispecs:

- available item: an item whose 'value' can be used by the process. Available items include the process inputs, but also temporary items, local terms and local functions, where used.

- accessible item: an item that may be 'assigned a value' by the process. Accessible items include the outputs from the process, temporary items and local terms.

2.16.3.1.2 Communication with multi-instance processes

When data flows (and event flows) are sent to or from a process that has multiple instantiations (see §2.15.3.7), there is potential ambiguity about which instance of the process the flow relates to. As a general principle, each instance of the process only 'knows about' itself — it should not be aware that it has any siblings. However, supervisory processes may need to distinguish between the siblings.

For example, in a system that controls four lifts, there might be a scheduler process "Schedule lifts" and a four-instance process "Control lift". "Schedule lifts" is able to:

- generate data flows to the correct instance of "Control lift" with the operation "issue" (see §6.9.11);

- check which instance of "Control lift" generated a data flow with the operation "instance" (see §6.9.12).

As mentioned above, the individual instances of "control lift" are not be able to access this information.

2.16.3.2 Process

This gives the name of the process that is being specified. For example:

```
PROCESS: update instructors' salaries with pay award
```

2.16.3.3 Data flows inputs[1]

These are the data flows used by the process. For example:

```
DATA FLOW INPUTS : initial patient data
```

If an input is a data group, then all its components are available. In the example, "initial patient data" might have a data flow specification (see §2.18.3.3 that defined it as containing: "treatment patient", "drug number", and "new infection level". Within the minispec, the whole group may be used by referring to its name "initial patient data", or individual items may used by be referring to their names: "treatment patient", "drug number", or "new infection level". This means that data flow items not declared as being a data flow input can be used within the minispec. In this example, only "patient admission" might be listed as an input. Alternatively, all the component flows might be listed, depending on preference. With the correct support environment, this might not be an exclusive choice.

If there is more than one item in the list, they are separated by "," characters. For example:

```
DATA FLOW INPUTS : percentage pay increase, audit level
```

2.16.3.3.1 Data item delimiters

To stress the 'beginning' and 'end' of data items, delimiters may be used. (Data items include data flows, but they also include attributes and, possibly temporary items.). In some examples in this section, the items are delimited by enclosing them between "%" characters. Other delimiting characters could be used. Delimiters are not required if the grammar is well-defined, but if informal, 'relaxed' versions are used (see §2.16.6.5.5), the use of delimiters is recommended. For example:

```
DATA FLOW INPUTS : %percentage pay increase%, %audit level%
```

might be used for the above example.

2.16.3.3.2 Availability of input data flows

Discrete processes can only uses discrete data flows that are part (or all) of the stimulus to the process, together with continuous data and event flows. (Note: some discrete processes do not have a stimulus data flow — they may be triggered or run as the response to a temporal event.) Even then, it is not necessarily true that all the inputs will be available at a specific time. For example:

- An input data flow may contain optional items (see §2.18.3.3). In order to check whether these are present in a composite input, the standard YSM operation "present" may be used (see §6.9.14).

- Continuous data flows may only be available at certain times. In general, they will only be available when the source is active. This is discussed in §2.14.3.4.4. The standard YSM operation "available" (see §6.9.3) may be used to check whether they are available at a specific time.

---

1 Note: this entry is not present for control processes.

2.16.3.4 Data flow outputs[1]

This list is a summary of the contents of the 'output' data of the process. It gives the contents of the data flows that are seen as outputs on any data flow diagram that the process appears in. For example:

```
DATA FLOW OUTPUTS : high pay audit
```

The outputs are not necessarily always produced — these are the *possible* outputs. For example, an error message is a possible output and would appear in this list; hopefully it is not produced every time the process runs!

2.16.3.5 Event flow inputs

These are the event flows used by the process. For example:

```
EVENT FLOW INPUTS : auto available, manual request, manual_OK
```

Event flows corresponding to event groups (see §2.25.3) are treated in the same way as data groups. Within the process, continuous event flows are treated as boolean items. Discrete event flows can only be used in minispecs for control processes (see §2.16.3.5.2).

2.16.3.5.1 Use of continuous event flows in data processes

Data process can only use continuous event flows as inputs — they cannot use discrete event flows as inputs.

Even then, it is not necessarily true that all continuous event flow inputs will be available at a specific time. Sometimes a continuous event flow is only available at certain times. Continuous event flows are only available when the source is active. This is discussed in §2.14.3.5.3. The standard YSM operation "available" (see §6.9.3) can be used to check whether they are available at a specific time.

2.16.3.5.2 Use of event flows in discrete control processes

Discrete control processes can only use event flows as inputs. Continuous event flows can be used, as for data processes (see §2.16.3.5.1).

A discrete control process can use one or more discrete event flows that are its stimulus. A discrete event flow may contain optional items (discussed in §2.25.3. In order to check whether these are present in a composite input, the standard YSM operation "present" should be used (see §6.9.14).

If the stimulus is a single discrete event flow, then it will always be present when the process runs.

Note: discrete event flows are usually dealt with by a continuous control process that is specified by a bSTD. Continuous control processes cannot be specified by a minispec.

2.16.3.6 Event flow outputs

These are the event flows that are produced by the process. For example:

---

1 Note: this entry is not present for control processes.

```
EVENT FLOW OUTPUTS : auto_alarm
```

These are explicitly set by the standard operations "raise", "lower", or "signal".

The outputs are not necessarily always produced — these are the *possible* outputs.

2.16.3.7 Stored data inputs[1]

This shows stored information used by the process. (Not all such items are used every time the process runs.) For example:

```
STORED DATA INPUTS :
    <Patient>        (attributes: number, infection level, narrative history)
    <Drug>           (attributes: number
    <Recommended treatment> (<Drug> is recommended for <Patient type>)
                     (attributes: recommended period, recommended dose)
    <Patient> is typed as <Patient type>
```

shows that this process uses information corresponding to the following view of enterprise information:

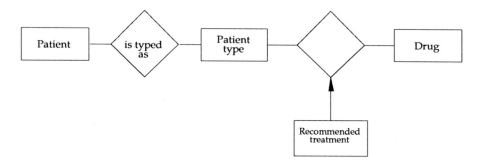

Different types of stored information may be included in this list, as described below, under "entities", "relationships" and "associative entities".

2.16.3.7.1 Entities

Entities may be accessed in various ways. However, all accesses involve at least one attribute or state variable, so references to an entity always require at least one attribute or state variable. This includes a 'match', which is usually a match on the identifier.

In the examples in this section, entity names are enclosed between "<" and ">" characters. Any other convenient way of distinguishing entities can be used. This is then followed by a list of actions carried out on the entity. Each type of action is enclosed in parentheses. Again, any convenient notation may be used. Allowed actions include creation, deletion and use of attributes and state variables.

---

1 Note: this entry is not present for control processes.

Attributes

An attribute may be used as an input to a process, by providing a value that is used to carry out calculations within the process. To identify an attribute, several alternative formats are available. The one shown above uses the term "attributes:", followed by a list of the attributes used. For example:

```
<Patient> (attributes: infection level)
```

refers to the attribute "infection level" of the entity "Patient". This should be read:

infection level of patient

Other possible notations include "<Patient>.infection level", "infection level(<Patient>)", "Patient.infection level", "infection level of patient", or "infection level of <Patient>". For use of several attributes, the convention adopted in this manual is convenient. For example:

```
<Patient> (attributes: number, infection level, narrative history)
```

Delimiters may also be used if desired. For example:

```
<Patient> (attributes: %infection level%)
```

```
<Patient>.%infection level%
```

or

```
%infection level% of <Patient>
```

for a single attribute, or, for a list of attributes:

```
<Patient> (attributes: %number%, %infection level%, %narrative history%)
```

State variables

Access to state variables is shown in a similar way to attributes. As an input, the process can check the state of a state variable. For example, a process that was used to set up a new occurrence of "Scheduled treatment" might need to check whether the drug being prescribed had been approved for general use. (Some drugs might still be in research, or currently restricted. They might even be obsolete and no longer allowed to be prescribed.) To test this, the process would need to check the state of the state variable "availability" of the entity "Drug". This would be shown:

```
<Drug> (state variable: availability)
```

2.16.3.7.2 Relationships

A relationship may be used as an input to determine which occurrences of entities participate in that relationship. For example:

```
<Patient> is typed as <Patient type>
```

states that the process uses a relationship "is typed as", which refers to the entities "Patient" and "Patient type". This should be read as:

> *Each patient is typed as a specific patient type*

This form of the relationship is referred to as the "relationship frame" (see §2.2.3.4.2). The relationship frame is given, rather than just the relationship name (in this case "is typed as"), because this is the form in which it will be used in the functional part of the minispec.

2.16.3.7.3 Associative entities

Attribute access is exactly as for entities (see §2.16.3.7.1

> <Recommended treatment> (**attribute**: recommended period)

An associative entity may also be referenced by a process to determine which occurrences of entities participate in the relationship (an associative entity is also a relationship). Whenever the relationship aspect of the associative entity is important to the process, it should be listed as "associative entity name", followed by the relationship frame that it defines (see §2.7.2.3.2). For example:

> <Recommended treatment> (<Drug> is recommended for <Patient type>)

indicates that the associative entity is "Recommended treatment" and it defines the relationship "<Drug> is recommended for <Patient type>".

This allows a formal check to be carried out against the way the relationship part of the associative entity is used in the minispec (see §2.16.3.10.2, §2.16.3.10.2 for examples).

2.16.3.7.4 Qualified access flows

If a process uses a qualified access flow, the convention used is that the qualifier appears in front of the entity name. For example, if the access flow was "(updated)infection level" and the entity it related to was "Patient", then in the minispec, this item is written "(updated)<Patient>.infection level" and read:

> *updated infection level of patient*

Qualified access flows are discussed in §2.14.3.1.7.

2.16.3.8 Stored data outputs[1]

This entry describes the stored data items that the process, creates, deletes or modifies. For example:

> **STORED DATA OUTPUTS :**
>   <Patient> (**attributes**: infection level)
>   <Scheduled treatment> (<Patient> is scheduled to receive <Drug>)
>   (**attributes**: date of treatment, time of treatment, desired dose, treatment period)

---

1 Note: this entry is not present for control processes.

Any entity, relationship or associative entity that is created or deleted is listed here. If an attribute is assigned a value, then it is also listed, using the format described for stored data inputs.

Additional to use of attributes (in this case in an update operation), the process might create or delete occurrences of the entity. Optionally, this might be denoted in some way. For example:

```
STORED DATA OUTPUTS :
    <Schedule treatment> (<Patient> is scheduled to receive <Drug>) (create)
```

indicates that the process can create an occurrence of the associative entity "Schedule treatment".

2.16.3.9 Temporary items

Temporary items are used within the process to hold intermediate values. They are data items, but they are not part of the input or output data flows. They are not attributed. They cannot be used, or referenced from 'outside' the process. Each temporary item required by the process must be defined by giving it a name and stating the values it can take. For example:

```
TEMPORARY ITEMS:
    name                : values
    %pay increase%      : currency
```

defines an item %pay increase%, which uses values from the ADT "currency". In a procedural specification, a value would be calculated and assigned to this value before it could be used. Temporary items are not used in external specifications (they are within the process and are *not visible*).

Very rarely, a set of values may be used, for example:

```
TEMPORARY ITEMS :
    name                : values
    %algorithm used%    : on-off, three-term, smith-predictor
```

This might be used to determine which of several algorithms were to be used and then as part of selection logic within the minispec. Although the use of such "value list" temporary items is allowed, they are generally better avoided.

2.16.3.9.1 Choice of name

The name of the temporary item should be chosen to reflect its meaning and must be distinct from that of any other input or output data item, local term (see below) or local function (see below). These temporary items are *not* accessible from outside the process.

Note: control processes can only use temporary items of type "boolean".

2.16.3.10 Local terms

A local term is a definition that may be used within the minispec. It consists of a name, followed by the definition of what that name means. It allows the inputs or outputs (data flows and stored data) used by a process to be restricted or 'filtered'. There are three types of item that may be given local names in this way:

- subsets of entities or relationships that satisfy a particular criterion;

- subsets of possible input data flows;

- constant values, taken from some abstract data type.

In each case, the name of the local term is given, followed by the definition. In this manual, the definition symbol "::=" is used, but any other convenient convention may be adopted. This should be read as:

   is defined as

The definition of the local term is equivalent to defining a subset of the original item, with a new name. Local terms are not functions or operation. They are just a 'filtering' of existing items. The filter selects those occurrences that satisfy some criterion. The item that is being restricted retains its original characteristics. Thus a local term defined as a restriction of a data flow input remains a data flow input; a local term defined as a restriction of a set of entity occurrences remains a set of entity occurrences; a local term defined as a constant remains a value of that abstract data type.

2.16.3.10.1 Choice of name
The name of the local term should be chosen to clarify the meaning of the minispec, but must be distinct from that of any input, output or temporary item (see §2.16.3.9) or local function (see §2.16.3.11) that is used by the process. These local terms are not accessible from outside the process.

2.16.3.10.2 Local terms for stored data items
It is often the case that a data process operates on certain restricted subsets of an entity or relationship. This subset may be defined by giving a descriptive name. This name can then be used throughout the minispec.

Example of local terms for stored data items
This example shows several types of local terms. The example relates to entities and relationships shown in the diagram:

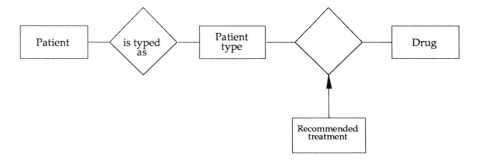

> **LOCAL TERMS:**
>   <Heart attack risk> ::= <Patient> with weight > 100 kg.
>
>   <This patient type> ::= <Patient type> for which
>   "<Patient> is typed as <Patient type>" exists.
>
>   <Appropriate treatment> ::=
>   <Recommended treatment> that references <Drug> and
>   <This patient type>

This defines three local terms: "Heart attack risk", "This patient type" and "Appropriate treatment", with the following interpretations:

1. "Heart attack risk" is defined as a selection of occurrences of the entity "Patient" for which the attribute "weight" has a value greater than 100 kg.

2. "This patient type" is defined as a selection of occurrences of the entity "Patient type". The ones that are chosen are those for which "<Patient> is typed as <Patient type>" is true. This selection is only allowed at points in the minispec where specific occurrences of <Patient> are bound (see §6.10.1.4).

3. "Appropriate treatment" is defined as a selection of occurrences of the associative entity "Recommended treatment". It acts as if it is an associative entity with the same meaning as "Recommended treatment" (except that only part of the set of occurrences of that associative entity are 'visible'). It will be bound whenever <Drug> and <This patient type> are bound. Expressions such as "<Appropriate treatment>.treatment period" are allowed (this accesses an attribute of the associative entity "Recommended treatment").

The above example shows that local terms act as a 'filter', selecting only those items that satisfy the definition criterion. A filter acting on an entity still gives an entity; a filter acting on a relationship still gives a relationship and a filter acting on an associative entity still gives an associative entity. Thus <Appropriate treatment> is defined as a selection of occurrences of <Recommended treatment>.

Example of local terms for data flow inputs

In a system that applied pay awards to individual employees, there might be a process that updated each employee with their new salary. If the data flow input to the process was "pay award", containing the components "employee name" and "percentage pay rise", then:

> **LOCAL TERMS :** %excessive pay award% ::= %pay award%
>   with %percentage pay rise% > 25

would 'filter out' from the incoming data flow "pay award" only those for which the component "percentage pay rise" was above a threshold. These selected data flows are now referred to by their name "excessive pay award".

Reducing minispec complexity by local terms — a trade-off
The advantage of using local terms is that complex selection logic need only be stated once in the definition and not repeated several times in the minispec. Local terms are still inputs and use of that local term is equivalent to the use of the corresponding input, even though it is not referred to by its own name within the body of the minispec.

For example, within a minispec that was responsible for allocating instructors to run courses, the definitions:

```
LOCAL TERMS:
        <Required course> ::= <Course> that <Scheduled course> references.

        %first day% ::= <Day>.date for <Day> that <Scheduled course> refers to

        %last day ::= %first day%.date + <Required course>.length

        <Required day> ::= <Day>
                with <Day>.date >= %first day and <= %last day%

        An <Available instructor> ::= <Instructor>
                who is not allocated on any <Required day>.

        A <Suitable instructor> ::= <Available instructor> for which
                "<Available instructor> can teach <Required course>" is TRUE.
```

might be used. Suppose the stimulus for the process was a data flow containing information about the scheduled course. The minispec would then reduce to:

```
If there is a <Suitable instructor> then
                create "<Suitable instructor> runs <Scheduled course>" for
                        one of the <Suitable instructors>
otherwise
                issue %no suitable instructor warning%.
```

This makes the function of the minispec very easy to understand — of course, most of the work has now gone into defining the local terms!

Care must be taken not to make the definition logic very complex, but this example is probably easier to understand than if no local terms were used at all.

2.16.3.10.3 *Local terms as restrictions of incoming data flows*
Local terms may also be defined in terms of incoming data flows. These may be used to define certain subsets of the possible inputs. For example, given an incoming data flow "pay rise" containing the components "number" and "percentage increase", the definitions:

```
LOCAL TERMS :
        %illegal pay rise% ::= %pay rise%
                with %percentage increase% > 50.

        %exceptional pay rise% ::= %pay rise%
                with 10 < %percentage increase% <= 50.

        %standard pay rise% ::= %pay rise% with %percentage increase% <= 10.
```

might be used in a minispec that audited pay rises.

2.16.3.10.4 Local terms for constants

A constant can be defined as a local term. It is then available item and can be used anywhere an available data item is allowed. It cannot be modified. It is not an accessible item. The exact format for defining constants is a matter of style, but, for example·

```
LOCAL TERMS : epsilon ::= angle, value = 0.0001 radians
```

was used to define a constant angle "epsilon", with a value of .00001 radians.

2.16.3.11 Local functions

Sometimes an operation may be required several times in a minispec. Each time it is used it takes values of available data items to produce results that may be used in further calculations or assigned to an output. The use of a local function may provide simplification. For example:

```
LOCAL FUNCTIONS :
        prescribed dose ::= <Patient>.weight * %dose rate%
```

As with local terms, the definition symbol "::=" is used in this manual, but any other convenient convention may be used. This should be read:

is defined as

This can be used at any point in the minispec where a specific occurrence of "Patient" is bound (see §6.10.1.4). It acts as an operation with two inputs — the weight of the "Patient" and the value of a data item "dose rate". To use this local function, we might have a temporary item "dose rate" and the following (rather unlikely) internal specification:

```
%dose rate% := 0.15 milligrams per kilograms ;
For each <Patient>
        create <Scheduled treatment> for
                        <Patient> and <Drug> with
                date of treatment := todays date( ),
                time of treatment := current time( ) + 2 hours,
                desired dose := prescribed dose ;
End ;
```

Within the loop, a single occurrence of the entity "Patient" is bound for each iteration. When the statement "desired dose := prescribed dose" is encountered, the attribute "weight" of that patient is multiplied by the value of "dose rate" and the result assigned to the attribute "desired dose" of the associative entity "Scheduled treatment".

The use of local functions *may* result in simplification of the minispec. However, an enterprise operation (see §2.17) can accomplish the same purpose. Furthermore, an operation can be used in many minispecs, whereas a local function can only be used in the minispec in which it is defined.

2.16.3.11.1 Choice of name
Local functions should be given names to reflect their meaning and clarify the minispec, but the names must be distinct from any input or output data item, temporary item (see §2.16.3.9), or local term (see §2.16.3.10). These local functions are *not* accessible from outside the process.

See also operation definition in §2.17.

2.16.3.12 Function (external)
The external function entry defines the function carried out by the process in terms of its effect. It does not prescribe *how* the function is carried out (see §2.16.3.13).

This external specification can be used for testing and formal proof techniques. It is also an alternative to the internal function for agreeing policy, where users prefer this type of specification. This has the benefit of suppressing the internal mechanism and stressing the 'effect' of the process.

The most common tool used in defining the function externally is pre- and post-condition specification. This is described in this section, but see also §2.16.4 for a description of other external modelling tools.

2.16.3.12.1 Pre-/post-condition specification principles
The pre- and post-condition modelling tool states what must be true after the process has carried out its function. It does this for each possible situation that may hold when the process is activated. For each, it asserts the conditions that must be true after the process has completed. In effect, this is a 'before and after' formulation of the behaviour of the process.

Each possible situation that may be true when the process is activated is referred to as a pre-condition. There may be any number of pre-conditions and each has a corresponding post-condition. The sequence in which these pre-/post-condition pairs is given is not significant.

2.16.3.12.2 Pre-conditions
The behaviour of the process is determined by its possible combinations of input values. Each separate combination is referred to as a pre-condition. If there are several, they are listed.

Each pre-condition is a boolean expression that can be evaluated when the process is activated. At any time the process is activated, the value of this expression is either TRUE or FALSE. The pre-conditions use values of any of the available inputs and/or the existence of occurrences of relationships or entities. The pre-condition often involves all or part of the stimulus (in the example below, the stimulus includes "treatment patient" and "drug number"). For example:

> **Pre-condition**
> Matching <Drug> exists
> AND
> <Patient> exists with number = treatment patient

In this there are two pre-condition clauses, connected by an "AND". Only if both are true is the whole pre-condition true.

"Matching <Drug> exists" is TRUE if there is an occurrence of "Drug" with the same value for the attribute "number" as the value of "drug number". This is an unspecified match entity operation (see §6.10.5.1).

"<Patient> exists with number = treatment patient" is true if there is one or more occurrences of "Patient" that have a value for the attribute "number" which is the same as the value of "treatment patient". This is a match entity operation (see §6.10.5).

2.16.3.12.3 Exclusivity of pre-conditions

The action of the process is often easier to understand if the pre-conditions are formulated so that no more than one is true, whatever the states and values of the process inputs. This is a simplification guideline. For formal proof of correctness of the process, this requirement can be relaxed, so that more than one pre-condition can be true at the same time.

2.16.3.12.4 Post-conditions

After the process has completed, the post-condition makes an assertion about the situation that holds.

Each post-condition must evaluate to a boolean value, so that after the process has completed, this condition can be evaluated. It value is either TRUE or FALSE. Information used in post-conditions may include values of all available inputs and also accessible data items or existence of occurrences of relationships or entities. In other words, post-conditions may include values of outputs. For example:

> **Post-condition**
>   <Patient>.infection level = %new infection level%
>   AND
>   "<Patient> is scheduled to receive <Drug>" exists
>   AND
>   <Scheduled treatment>.date of treatment = todays date( )

In this example, there are three post-condition clauses. For this particular post-condition to be TRUE, each of the three clauses must be TRUE.

1. "<Patient>.infection level = %new infection level%" is TRUE if the value of the attribute "infection level" for all bound occurrences of "Patient" has the value in %new infection level%. It is a convention that if an entity is bound in the pre-condition, then it is still bound in the post-condition. This is the case with <Patient>. It is a further convention that if a post-condition uses an entity that has already been bound, then there is an implicit 'for all bound occurrences'.

2. "<Patient> is scheduled to receive <Drug> exists" makes an assertion about an occurrence of a relationship. This uses the standard relationship frame format. (In this particular case, the relationship is an associative entity, as can be determined by looking at the "Stored data outputs" entry. However, for this clause, this is not significant.) This clause requires that a "create" must have been carried out within the process.

3. "<Scheduled treatment>.date of treatment = todays date( )" makes an assertion about an attribute of "Scheduled treatment". This relies on the convention that if one post-condition clause binds an entity, it is assumed bound for all following clauses in that post-condition. In this case, asserting that "<Patient> is scheduled to receive <Drug> exists" caused <Scheduled treatment> to be bound.

2.16.3.12.5 Pre-condition and post-condition pairs

Each pre-condition has a corresponding post-condition. A discrete process must satisfy the following:

> *If the pre-condition holds, then, if the process completely carries out its function, the corresponding post-condition must be true.*

This is usually referred to as a weak requirement. A stronger requirement is:

> *If the pre-condition holds, then, the process will complete in a finite time and the corresponding post-condition will be true.*

YSM uses this second interpretation of pre-/post-conditions. (The reason why the first form might be true, yet not the second, is that the process might not complete for certain input values. YSM requires that all processes complete in finite time. In fact for an essential model, the process completes in zero time.)

If a process is activated with one of the pre-conditions true and when the process completes, the corresponding post-condition is false, then the process has failed. Formal proofs of correctness may be applied to the internal function specification to show that this cannot happen. This is outside the scope of this release of YSM, but is covered in [Gor88], for example.

2.16.3.12.6 Minimal principle for post-conditions

Each post-condition makes an assertion about what the effect of the process is. Any significant effect is stated. Because there are an infinite number of effects that the process might have caused, a minimal principle is applied to post-conditions:

> *If a pre-condition is true, then any significant effects must be stated in the corresponding post-condition.*

In other words, *only those* effects stated in the post-condition should result from the action of the process. Nothing else should happen. There should be no 'side effects'.

2.16.3.12.7 Numbering pre-/post-condition pairs

Where there is more than one pre-/post-condition pair, the pairs may be numbered. This is conventional, but optional.

2.16.3.12.8 Use of null post-conditions

If none of the pre-conditions evaluates to TRUE, then no assertion is made about the effect of the process. For purposes of symmetry, pre-conditions are sometimes constructed so that exactly one pre-condition holds when the process is activated. If this is done, then a 'null' post-condition may be added. For example:

```
Pre-condition 1
        Matching <Drug> exists AND
        <Patient> exists with <Patient>.number = %treatment patient%

Post-condition 1
                <Patient>.infection level = %new infection level%
        AND     "<Patient> is scheduled to receive <Drug>" exists
        AND     <Scheduled treatment>.date of treatment =
                                                todays date( )

Pre-condition 2
        No matching <Drug> exists
        AND <Patient> with
                        <Patient>.number = %treatment patient% exists

Post-condition 2
                %Insufficient treatment message% exists
        AND     %Insufficient treatment text% =
                                <Patient>.narrative history

Pre-condition 3
        No <Patient> with
                        <Patient>.number = %treatment patient% exists

Post-condition 3
```

Post-condition 3 is null. Using the minimal principle, it means the process should 'do nothing' in this case.

If this convention is adopted, then *exactly* one of the pre-conditions holds. (Demanding that pre-conditions are mutually exclusive is optional — see §2.16.3.12.3)

2.16.3.12.9 Binding of stored data items in pre-/post-conditions

Where the pre-condition is satisfied by a single occurrence of an entity, any post-condition that refers to that entity, should be taken as referring to that specific occurrence of the entity. Where more than one occurrence of the entity satisfies the pre-condition, the post-condition must be read as "for each of the occurrences of the entity satisfying the pre-condition, the following is true ...". (This is a definition of binding —see §6.10.1.4.)

2.16.3.12.10 Use of qualified values in pre-/post-conditions
In some situations, it is necessary to refer to values that have changed in the post-specification. For example, to specify that all instructors' salaries had been increased by 10%, we might have:

> **Post-condition**
>   <Instructor>.(new)salary = 1.1*<Instructor>.(old)salary

This distinguishes between the value that held *before* the function had been carried out, and the current value. Strictly speaking, there is only one "<Instructor>.salary", so the 'old' value is no longer available when the post-condition check is carried out. However, in reading pre-/post-conditions, the reviewer should imagine that both are available at the time the post-condition is checked.

If qualified names are used for access flows in pre-/post-conditions, the qualifiers may also be used in naming the access flows on the DFD showing the process (see §2.14.3.1.7).

When access flows are named, their specification (in an access flow specification) might include a definition of what the qualifiers mean. However, if standard qualifiers such as "old" and "new" are used, this does not have to be spelt out in the access flow specification.

Rather than "old"/"new", "pre"/"post", or "before"/"after" may be preferred.

2.16.3.12.11 Pre- and post-conditions for continuous processes
For continuous processes, pre- and post-condition specification breaks down into two parts:

- **initial actions**: these can be specified using pre- and post-conditions. The initial actions are carried out once, when the process is enabled. They may be used to load initial values, set up store values for integrators, etc.

- **continuous actions**: these are carried out continuously, as long as the process is enabled.

In the example:

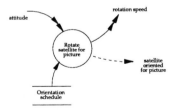

a continuous process generates an output ("rotation speed") that varies over time. This rotation speed causes the momentum wheels (terminator) to rotate the satellite. The attitude of the satellite is sensed by a measurement system (terminator) and used by the process as the continuous input "attitude". In order to determine the required rotation speed and when it has reached the desired attitude, the process must initially read the required attitude. This is held as an associative entity "Required orientation" in the store "Orientation schedule". This is shown below:

Note: Both rotation speed and attitude are defined in terms of a compound adt, with three components, "theta", "rho" and "phi".

PROCESS : Orient satellite for picture

DATA FLOW INPUTS : attitude

DATA FLOW OUTPUTS : rotation speed

EVENT FLOW INPUTS : —

EVENT FLOW OUTPUTS : satellite rotated for picture

STORED DATA INPUTS : <Required orientation>   (state variable : status) (attributes : required angle)

STORED DATA OUTPUTS : —

TEMPORARY ITEMS :   **name**   **:values**
desired angle   :orientation

LOCAL TERMS : epsilon ::= angle, value = 0.0001 radians

LOCAL FUNCTIONS : —

FUNCTION (external) :
initial actions: Pre-
         Post-
              desired angle := <Required orientation>.required angle of
                   <Required orientation> with status = due

FUNCTION (external) :
continuous actions: Pre-1
              ABS(attitude.theta − desired angle.theta) < epsilon
                   AND ABS(attitude.rho − desired angle.rho) < epsilon
                   AND ABS(attitude.phi − desired angle.phi) < epsilon

         Post-1
              rotation speed.theta = 0 AND rotation speed.rho = 0
              AND rotation speed.phi = 0
              AND satellite oriented for picture exists ;

         Pre-2
              NOT(pre-1)

         Post-2
              rotation speed.theta = f_1 (attitude.theta, desired angle.theta) ;
              rotation speed.rho = f_2 (attitude.rho, desired angle.rho) ;
              rotation speed.phi = f_3 (attitude.phi, desired angle,phi) ;

FUNCTION (internal) :
initial actions:

The functions f_1, f_2, f_3 are chosen to 'hide' the exact rotation algorithm used (which seemed too esoteric for this manual). They could either be expanded as explicit calculations 'in-line', defined as local functions, or (more naturally), defined as operations. Indeed, by allowing suitable parameters to be used, it is likely that a single operation could be defined with all three angular components.

The interpretation of the initial actions is as for a discrete process, with the interpretation that 'after' is now *immediately after the process is enabled*. The post-condition must hold at that time; it does not have to hold at subsequent times, even while the process is active.

For the continuous actions, the following must hold:

> *At any time that the pre-condition is true, the corresponding post-condition must also be true.*

(It does not make sense to refer to 'before' and 'after' in this case.) These post-condition must continue to hold at all time the process is active.

2.16.3.13 Function (internal)

This describes the 'internal' view of the function — in other words, how the function is constructed. The function is made by 'fitting together' standard building blocks.

The most common tool used in internal specification is "structured text", which is described briefly in this section. A fuller definition of the grammar used in structured text is given in §6.6.

Structured text gives a prescription of how the input items are transformed into the output items in a step by step, algorithmic manner. No other inputs or outputs may be used, but temporary items may be used to hold temporary values. The individual operations may use any available value and assign these values (or any value calculated using them) to any accessible item.

Other modelling tools are also allowed — see §2.16.4.

2.16.3.13.1 The 'building blocks' for functions

Processes are built up out of lower-level, primitive building blocks. These operations may be:

- **standard YSM operations:** including "signal", "initialise", "raise", "issue", etc. These are defined in §6.9. Issue is particularly useful (see §6.9.12).

- **data access operations**: operations such as "create", "delete", etc. These are defined in §6.10.

- **standard operations 'owned' by abstract data types**: as listed in §6.4.

- **operations**: additional to pre-defined operations of the above types. YSM provides a formal way of adding new operations by means of operation specifications. These are discussed in §2.17;

- **assignment statement**: a value is copied from one data item (data flow, attribute, local variable) to another.

Note: not all the above operations are allowed in particular types of process. For example "enable" is not allowed in a data process; data access operations are not allowed in control processes.

2.16.3.13.2 Building up the function from primitive operations

A convenient picture of how the function is built up from these operations is to think of them as 'off the shelf', 'black boxes'. Each has a name, with particular inputs and outputs. Each input or output has a specific abstract data type associated with it. Different functions is defined by selecting different boxes, or connecting them in different orders.

For example, a function that calculates the angle between two vectors is shown below, using a generic data flow diagram to show how these operations are connected together[1]:

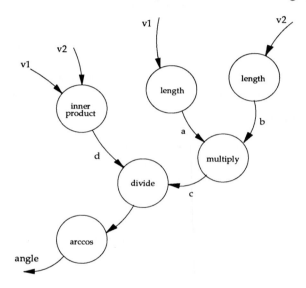

The building blocks used in this example are:

- "length" is an operation that takes a vector input and returns a real output. An example of how this might be defined as an enterprise operation is given in §2.17.3.4.2.

- "inner product" defined on two vectors. An example of how this might be defined as an enterprise operation is given in §2.17.3.4.2.

- "multiply" indicates that the standard operation "multiply" on two reals has been used. This returns a real value (see §6.4.10).

- "divide" indicates that the standard operation "divide" on two reals has been used. This returns a real value (see §6.4.10).

- "arccos" indicates that the standard operation "arccos" which takes a real as an input and returns a value of type angle (see §6.4.3).

---

1 This is not an 'ordinary DFD', but an implementation-type of DFD, showing how available functions are connected together to form a required function. It has therefore been shown with the names of the components being connected, rather than functional names for each process.

2.16.3.13.3 *Principles of structured text*

Structured text is a procedural or a 'one step at a time' way of specifying functions. It is a procedural statement of what needs to be carried out. It shows one particular way of carrying out the function.

For example, the above example could be written:

```
d := inner product(v1, v2)
a := length(v1) ;
b := length(v2) ;
c := a*b ;
angle := arccos(d/c) ;
```

This is a sequence of statements that are carried out. There are several other valid options for this function. For example:

```
a := length(v1) ;
b := length(v2) ;
c := a*b ;
d := inner product(v1, v2)
angle := arccos(d/c) ;
```

This illustrates one of the characteristics of structured text — it implies a particular procedural choice of implementation (although this can be avoided, to a large extent, by the "Parallel" construct — see §6.6.3). This is a definite disadvantage to structured text in the essential modelling stage — decisions about implementation might be made too early.

Note: the above example could be written:

```
angle = arccos( (inner product(v1, v2) /(length(v1)*length(v2) ) )
```

This avoids commitment to a specific sequence of calculation, beyond that implied by the precedence of the operators in the expression. If 'fed into' a compiler, some sequence would be chosen, but that is irrelevant from the point of view of policy. The above is selected as an example to illustrate structured text, rather than to make the statement that structured text is the best way of specifying this specific function.

2.16.3.13.4 *Dialects*

The assignment statements:

```
"a := b"
```

```
"the value of a is assigned to b".
```

```
"the current value of a is used as the new value for b".
```

all have the same effect, but are expressed using different vocabulary or syntax. Choosing one form, rather than another is more a matter for convenience of review, clarity, etc., rather than for correctness. Indeed, for the above equivalent statements, if one form was correct, then so must be the others. This concept may be used informally, in tailoring the minispec to the intended audience.

If the concept is to be used more formally, the term dialect is used to describe different variants of the same textual statement. One or more pre-defined dialects may be used within a minispec. These dialogues may be standard YSM dialects, or new dialects may be defined, as required for the enterprise. For many enterprises, the ability to define and use new dialects is not required, and general and standard dialects may then be used. (The standard YSM dialects are listed in §6.7.)

The primitive statements use available items and assign results to accessible items.

From these 'building blocks', the required function is built up using the following constructs:

- **sequence**: one statement, followed by another, followed by another, … ;

- **parallel**: several statements used at the same time;

- **selection**: one or more conditions are tested and, depending on whether they is true or not, different statements used;

- **iteration**: the same statements are used repeatedly. The number of repeats may be fixed (for example "do 6 times") or it may be variable (termination of the iteration depending on the satisfaction of a test condition). Note: iterations are not allowed in control processes.

A compound statement is a construct, treated as a unit. For example:

```
    If leap year then
            number of days := 29 ;
    else
            number of days := 28 ;
    end ;
```

is a simple selection (see §6.6.4.4). Regarded as a compound statement, it could be referred to as "S". The 'outer' construct is then defined as:

```
If month=September OR month=April OR month = June
        OR month = November then
    number of days := 30 ;
elsif month = February then
    S ;
else
    number of days := 31 ;
end ;
```

where the construct is a multi-way selection (see §6.6.4.6), with one of the chosen statements is compound (S) and the others simple.

2.16.3.13.5 Lexical conventions

In a procedural specification, it is important to identify what the individual statements are. Although the author may know what the intended components are, later users of the specification may misinterpret the specification if the 'beginning' and 'end' of individual statements is not clearly identified. For example:

> If the worker is hourly paid then
> If today = Sunday then
> rate of pay := 1.5
> calculate the pay for the day by multiplying the number of hours
> worked by the rate of pay

raises questions about whether a calculation is carried out for all days, or just Sunday. What are the statements carried out anyway? A simple rule like 'one statement to a line' clearly fails for complicated statements.

It is for this reason that some lexical conventions are adopted. These show the beginning and end of constructs (both individual statements and the sequence, selection and iteration constructs). The convention used is not significant and YSM allows any well-defined way of delimiting constructs to be adopted.

Note: it is possible to construct automatic parsers for different lexical conventions. One format can be converted to another.

Statement delimiters
The lexical convention mainly adopted in this section is that one statement is separated from the next by a ";" character. This is sometimes referred to as a "statement delimiter". Any other convenient symbol could be used. For simplification in this reference, each statement is written as 'ending' in a ";".[1]

Construct delimiter
The lexical convention adopted in this section is that each construct is introduced by a special format that is specific to that construct. For example:

> If ... then ... end

where the ... indicates that there is a variable component. The end of a construct is indicated by an "end". These are sometimes referred to as "construct delimiters". Any other convenient symbol could be used.

Numbered text
An alternative to the above is to use 'numbered text'. In this, each statement corresponds to a sentence terminated by a full stop. Each statement is numbered within a sequence. For example:

> 1. Use the values in %v1% and %v2% to calculate the inner product and store the result in %d%.
> 2. Calculate %a% as the length of %v1%.
> 3. Calculate %b% as the length of %v2%.
> 4. Multiply %a% and %b% together, giving %c%.
> 5. %angle% is computed as the inverse cosine of %d% divided by %c%.

---

[1] This is not quite the same as saying "any two statements are separated by a ";".

In this example, the textual dialect is used, as is conventional with numbered text. The '%' convention is also shown in this example. In practice, this is often used with the textual dialect; the procedural dialect, without '%' delimiters is often used if the text is not numbered. These are the two combinations used for illustration in this section.

2.16.3.13.6 Nesting constructs

One of these constructs may be used 'within' another (this is sometimes referred to as a nested structure). For example, a selection may be included within another selection, or an iteration. At the lowest level, the statements must be YSM, enterprise or data access operations, or assignment statements.

2.16.3.13.7 Conventions for showing nesting

When nesting occurs, it is recommended that indentation is used, with each 'level of nesting' indicated by an increase in the indentation. Ruling lines may also help to indicate structures as having a 'beginning' and an 'end'. For example, the old saying:

"Thirty days hath September, April, June and November,
All the rest have thirty one, save February alone,
which has 29 in leap years and 28 otherwise"

becomes:

```
If month=September or month=April or month = June
                         or month = November then
            number of days := 30 ;
elsif month = February
            If leap year = 0 then
                        number of days := 29 ;
            else
                        number of days := 28 ;
            end ;
else
            number of days := 31 ;
end ;
```

This will be referred to as the "bracketed format" in the description of constructs given below.

An alternative, equivalent format is to use boxed text. This requires more technical support (an ordinary word-processor will probably not support this, for example). For example:

```
If month=September or month=April or month = June
          or month = November then
          number of days := 30
elsif month = February
          ┌──────────────────────────────────────┐
          │ If leap year  then                    │
          │              number of days := 29     │
          │ else                                  │
          │              number of days := 28     │
          │ end ;                                 │
          └──────────────────────────────────────┘
else
```

2.16.3.13.8 Simple and compound statements

To define nesting more formally, the concepts of simple and compound statements is used. A simple statement corresponds to one of the primitive operations or assignment statements (see §2.16.3.13.1). For example:

```
number of days := 28 ;
```

2.16.3.13.9 Use of dialects

Because each dialect is really a different way of formatting the same information, it is possible to convert from one to another in a mechanical way. This is useful when one dialect is used for discussions with subject-matter experts and users, but another is used as a standard specification dialect. This conversion may be carried out informally by an analyst at the same time as they present or 'walk through' the minispec with a subject matter expert.

Use of YSM or user-defined operations

It is possible to have several dialect versions of each of these operations. The different versions of the standard YSM operations are described in §6.9. The way user-defined operations are declared to have dialogue variants is described in §2.17.3.5

2.16.3.13.10 Balance between rigour and understandability

Where the full rigour of a formal language is not required, then any informal equivalent may be used, although this may introduce ambiguity and lack of precision if not used carefully. This 'relaxation' should be carefully controlled by the analyst and used as required. Minispecs are of no use if they cannot be verified — a careful balance has to be struck between precision and readability. When an informal grammar is used, the use of delimiters to highlight data items is helpful.

Some guidelines on relaxation and presentation are given in §2.16.6.3.

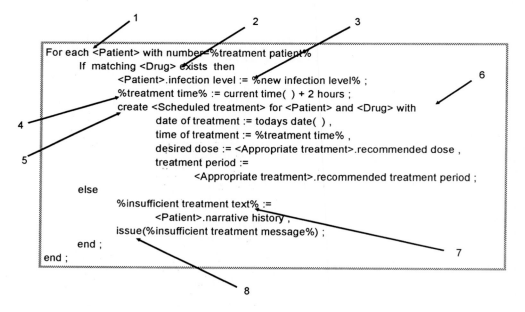

```
For each <Patient> with number=%treatment patient%
     If  matching <Drug> exists  then
          <Patient>.infection level := %new infection level% ;
          %treatment time% := current time( ) + 2 hours ;
          create <Scheduled treatment> for <Patient> and <Drug> with
               date of treatment := todays date( ) ,
               time of treatment := %treatment time% ,
               desired dose := <Appropriate treatment>.recommended dose ,
               treatment period :=
                    <Appropriate treatment>.recommended treatment period ;
     else
          %insufficient treatment text% :=
               <Patient>.narrative history ,
          issue(%insufficient treatment message%) ;
     end ;
end ;
```

2.16.3.13.11 Example of structured text

This example shows how data access operations, data items (inputs, outputs, temporary and local terms) and local functions are used. It uses a fairly formal version of a procedural dialect. It could be relaxed to make it seem less technical and more user-friendly. However, it is easier to see the underlying structure in this format.

Notes

1. This restricts the statements within the outer structure to those occurrences of <Patient> that match the treatment request. This is a "match entity" operation — see §6.10.5. (In fact, there is exactly one such patient.)

2. This checks whether there are any occurrences of drug that have the same number as that contained in the data element "drug identifier" that is contained in the input data flow "treatment request". In order for this shortened form of the instruction to be used, this data element must be specified as an attribute of "Drug" (see §2.18.3.7). This is an "unspecified match entity" operation — see §6.10.5.1. More generally:

```
If <Drug>.number = %drug identifier%
```

could be used. This expanded format is always acceptable, but some prefer the shortened form. In either case, the effect is to return "true" or "false" which is used as a selection condition. Further, <Drug> is now bound (see §6.10.1.4).

3. This assigns a new value to an attribute of the patient. Because this is inside a construct "for each <Patient>", the assignment is allowed (each iteration corresponds to a unique patient).

4. "treatment time" is a temporary item. (Definition of %treatment time% as a temporary item does not gain much in this example, but it is used to show how it would be done. In practice, this particular minispec would almost certainly be improved by *not* using a temporary item.) The value assigned to it is obtained by using an enterprise operation that returns a "time" (in fact, the time at this instant). Addition of another "time" (2 hours) then gives a time, which is assigned to the attribute. The attribute must have abstract data type "time".

5. "create" is a data access operation and in this case it is followed by an associative entity. For an associative entity, each entity that participates in the relationship must be uniquely identified. In this format, they are identified as following the "for" keyword. Both "Patient" and "Drug" are unambiguously identified by their position inside the two constructs. Because "create" often assigns other attributes, the "create" format used here allows a list of attribute assignments to be carried out. This list is given between "with" and the ";".

6. Each component of the list assigns a value to the attribute (on the left) from an expression (on the right). This assignment uses a standard operation "todays date( )", that returns a "day" value — this value is then allocated to the "date of treatment" attribute of "Scheduled treatment".

7. "insufficient treatment text" is a component of the outgoing data flow "insufficient treatment message".

8. "issue" is used to indicate a data flow is generated as an output. (This is optional, and setting values in any output items may be assumed to cause them to be "sent out" — however, some people feel using this aids clarity.)

2.16.3.13.12 Specifying continuous data processes
For continuous processes, structured text breaks down into two parts:

- **initial actions**: these can be specified using pre- and post-conditions. The initial actions are carried out once, when the process is enabled. They may be used to load initial values, set up store values for integrators, etc.

- **continuous actions**: these are carried out continuously, as long as the process is enabled.

In the example:

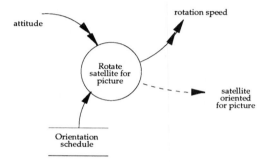

a continuous process generates an output ("rotation speed") that varies over time. (See §2.16.3.12.11 for a background to this example.) The frame specification for this is given in §2.16.2.3.

Note: Both rotation speed and attitude are defined in terms of a compound adt, with three components, "theta", "rho" and "phi".

The functions f_4, f_5, f_6 are chosen to 'hide' the exact rotation algorithm used (which seemed too esoteric for this manual). They could either be expanded as explicit calculations 'in-line', defined as local functions, or (more naturally), defined as enterprise operations. Indeed, by allowing suitable parameters to be used, it is likely that a single enterprise operation could be defined with all three angular components.

Another example
Where a process has no state memory, but generates event flow outputs from its event flow inputs in a way that does not vary over time, a data process is normally used, rather than a bSTD. To specify the data process a decision table or a set of boolean expressions may be used, defining the outputs in terms of the inputs. For example, the process "Determine correct auto-manual status" is a continuous data process:

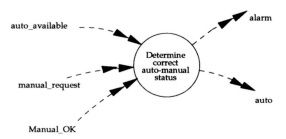

It is a data process, event though it deals with continuous event flows. These are treated as boolean items within the minispec.

PROCESS : Determine correct auto-manual status	
DATA FLOW INPUTS :	
DATA FLOW OUTPUTS : —	
EVENT FLOW INPUTS : auto_available, manual request, manual_OK	
EVENT FLOW OUTPUTS : auto, alarm	
STORED DATA INPUTS : —	
STORED DATA OUTPUTS : —	
TEMPORARY ITEMS : —	
LOCAL TERMS : —	
LOCAL FUNCTIONS : —	
FUNCTION (external) : **initial actions:**	
FUNCTION (external) : **continuous actions:**	(auto = auto_available $\wedge$ ~(manual request $\wedge$ manual_OK)) $\wedge$ (alarm = ~(auto_available))
FUNCTION (internal) : **initial actions:**	
FUNCTION (internal) **continuous actions:**	

Note: this process has no initial actions.

2.16.4 Alternative modelling tools for specification

In addition to pre-/post-conditions and structured text, other tools are sometimes used, as described in this section. Some of these are suitable for external specifications; some for internal specifications and some may be used for both.

2.16.4.1 Decision tables

An alternative, equivalent version of the pre-/post-condition example given previously is the use of decision tables, where the pre-conditions and the post-conditions are laid out in a table. For an example give previously, such a table would be:

Decision tables are particularly useful in specifying control processes, where the only items that may be used are boolean items.

The example shown here is laid out with pre- and post-conditions as columns. It has two situations where there are no post-conditions. Decision tables are good for checking all possible pre-conditions have been considered, even if there are no required post-conditions for certain pre-conditions.

\<Patient\> with number =%treatment patient% exists	Matching \<drug\> exists	Post-condition
Y	Y	1. \<Patient\>.infection level = %new infection level% 2. \<Scheduled treatment\> for \<Patient\> and \<Drug\> exists. 3. \<Scheduled treatment\>.date of treatment = todays date( ).
	N	1. %insufficient treatment message% exists with %insufficient treatment text% = \<Patient\>.narrative history
N	Y	
	N	

There are alternative layouts possible for decision tables. For example, the following format is able to deal with more pre- and post-conditions:

<u>Pre-conditions</u>				
1. \<Patient> with \<Patient>.number = %treatment patient% exists	TRUE	TRUE	FALSE	FALSE
2. Matching \<Drug> exists	TRUE	FALSE	TRUE	FALSE
<u>Post-conditions</u>				
1. \<Patient>.infection level = %new infection level%	TRUE	FALSE	FALSE	FALSE
2. \<Scheduled treatment> for \<Patient> and \<Drug>exists	TRUE	FALSE	FALSE	FALSE
3. \<Scheduled treatment>.date of treatment = todays date( )	TRUE	FALSE	FALSE	FALSE
4. %insufficient treatment message% exists with %insufficient treatment text% = \<Patient>.narrative history	FALSE	TRUE	FALSE	FALSE

The above examples used "TRUE" and "FALSE" as entries. This is conventional for external specifications. For internal specifications, "Y" and "N" are conventional (to indicate whether the action is carried out).

2.16.4.1.1 Using decision tables for discrete control specification

As discrete control processes generally compare exclusive discrete event flows to one or several combinations of continuous event flows, decision tables are often the best way of specifying them.

Consider "Correct cash problem", which is a discrete process group that is triggered when there is a 'cash' problem'. This group contains a control process ("Determine best source of funds") and two data processes. (As the process group is discrete, all three processes are discrete, of course.)

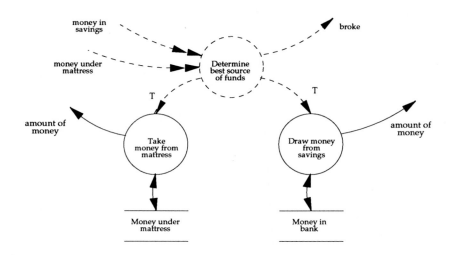

The inputs to "Determine best source of funds" are:

- "money in savings", which is a continuous event flow that is output by a process that monitored the status of a savings account.

- "money under mattress", which is a continuous event flow, output by a process that monitored this alternative savings account.

The following decision table could be used as an internal specification for "Determine best source of funds":

condition				
money in savings	TRUE	FALSE	TRUE	FALSE
money under mattress	TRUE	TRUE	FALSE	FALSE
action				
T: Draw money from savings	Y	N	Y	N
T: Take money from mattress	N	Y	N	N
S: broke	N	N	N	Y

2.16.4.2 Transfer function specifications

2.16.4.2.1 For continuous data processes

For continuous data processes, particularly in control systems, pre-/post-conditions is not the most appropriate modelling tool. This is because pre- and post-conditions do not highlight the behaviour of the process over a period of time (they are more suited to specifying what happens at a particular point in time).

In such situations, other methods are used to specify the relationships that must apply between the outputs and inputs of a process. For linear systems and controllers, transforms are often used to specify controller function. For example, a controller:

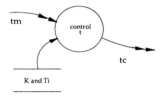

which keeps track of an input required value which changes according to a linear ramp function in a second order process might have the specification in terms of the transfer function:

$$G(s) = \frac{K}{T_i}\left(\frac{1 + s\,T_i}{s}\right)$$

where

K is a control parameter, T_i is a time constant,

and

$F(s) = \int_0^\infty f(t)e^{-st}\,dt$ is the usual Laplace transform definition (this would be assumed)

The output "tc" would then be defined in terms of the input "tm" and this response function by an equation in the s domain:

$$TC(s) = G(s)TM(s)$$

Note: K and T_i might be 'hard-coded' into the process. The version shown would allow them to be modified in real time.

___

2.16.4.2.2 *For discrete data processes*
The equivalent type of specification for discrete processes that sample continuous inputs would be by means of z-transforms. For example, the controller might be defined to have the z-transform:

$$G(z) = \frac{1 - K_1}{1 - K_2} \, \frac{1 - K_2 \, z^{-1}}{[1 - 2z^{-1} + K_1 z^{-2}]}$$

where

K_1, K_2 are controller constants

F(z) represents the z-transform of a continuous function f(t), which is defined as:

$$F(z) = \sum_{k=0}^{\infty} \frac{f(kT)}{z^k}$$

2.16.4.3 Network graphical
A very radical way of specifying minispecs would be to use a diagrammatic technique, such as the one shown in §2.16.3.13.2. This would show primitive operations connected together to obtain the required function and avoid the requirement for any procedural form of the specification.

2.16.4.3.1 *Technical support*
An advanced architecture would allow 'browsing' to select the operations that were to be connected.

2.16.4.4 Decision trees
This tool shows a sequence of decisions as a decision tree. Each time there is a decision, there is a two- or multi-way branch. (These decision trees are mainly used for selection logic.) In the example given below, the first decision is "are they a discount customer?" The next decision

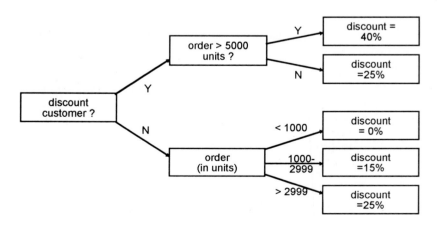

depends on the route taken from this first decision, but may be "Which of the three ranges does the number of units fall into — less than 1000, 1000 to 2999 or 3000 or more?" The right-hand 'leaf' in the tree corresponds to the actions to be carried out if all preceding branch conditions lead to that point.

2.16.4.5 Graphic presentations for procedural specifications

It is possible to define graphic icons for each of the standard building blocks and give the operations as text. Data access operations may be shown as text, or by special icons. Tools of this type include Nassi–Shneiderman charts and flowcharts. Although each of these tools have their devotees, as a general guideline, if the minispec is so complex to need such tools, it is generally better to partition it into simpler components, each of which can be specified clearly without such tools.

However, the analyst should be aware of the existence of these tools and should not refuse to use them in any circumstance. There are occasions when users demand the use of such tools, possibly because of company policy. YSM does not include these tools within the method, but does not prohibit their use — the responsibility for using and cross-checking the correctness of such tools is in the domain of the user and analyst, rather than the provider of the method.

2.16.5 Rules

2.16.5.1 Inputs and outputs

1. Data flows are inputs or outputs. They cannot be both.

2. A continuous data process cannot use any discrete flow inputs. This is because discrete inputs would not be available at all times and the process could not carry out its function.

3. A discrete process may not output continuous data flows, nor may it contain "raise" or "lower".

2.16.5.2 Local terms, temporary items and local functions

1. The names of all data items (inputs, outputs, local terms, temporary) and local functions must be distinct. This is a restriction on the choice of names for temporary items, local terms and local functions.

2. All local terms must be used at least once.

3. All input data items must be used at least once in the minispec body. Where a local term is a restriction of such an input, reading the local term counts as reading the input data item.

4. All output data items must be given a value at least once in the minispec.

5. All temporary items must be given a value and the value must be used.

6. In procedural minispecs, any temporary item must always be assigned a value before it is used. This must apply for all possible values of the input data items.

7. All local functions must be used at least once in the minispec.

8. For all possible values of the input data items, the inputs to any use of a local function must have a value each time it is used.

2.16.5.3 Enterprise and data access operations

1. For all possible values of the input data items, the inputs to any operation must be available each time it is used.

2. For all possible values of the input data items, the inputs to any data access operation must be well-defined each time it is used.

2.16.5.4 Structured text

1. For all possible values of the input data items, the value of any test condition in a selection must evaluate to "true" or "false". In other words, all data items used in that test condition must have been given prior values, or failing that, the condition's value is independent of their value.

2. For all possible values of the input data items, the iteration parameters (start and end values for counters etc.) must always be well-defined on entry to the iteration.

2.16.5.5 Pre-/post-conditions

1. For a pre-/post-condition formulation, for all possible values of the available inputs, each pre-condition must evaluate to TRUE or FALSE.

2.16.5.6 Concerning control processes

1. No data access operation may be used in a minispec for a discrete control process.

2. A discrete control process with stimulus (a discrete event flow) may use that flow value as a boolean value. It has the value "true". This is only of any use when the stimulus is an event flow that contains several event flows — this may then be used to determine which one occurred. Alternatively, the operation "present" (see §6.9.14) may be used.

3. The only inputs other than the stimulus that may be used in a discrete control process are continuous event flows, which should be considered to be of type boolean. They can be combined with the usual operations of "AND", "OR", "NOT". They may also be used with "rising" to generate discrete event flow outputs.

4. Temporary items used in control process specifications may only be specified as boolean.

2.16.6 Guidelines

2.16.6.1 Complexity

No minispec should be too 'complex'. While this is very difficult to quantify, the minispec should be simple enough to be reviewed with a degree of confidence that it embodies the required policy.

2.16.6.2 Inputs and outputs

Generally, if a data flow is declared as a group and the components are referred to in the minispec, but not the data flow as a whole, then only the name of the components is given. If the data flow is used as a unit, but its components are not referenced, then only the name of the whole flow is given, not the components.

If the flow is used as a whole and individual components are referenced, then either the whole data flow or its components may be shown in the inputs and outputs. Depending on the support environment, both might be visible. (It is not likely that this is cost-effective in pencil and paper support environments.)

2.16.6.3 Choice of specification tool and style

The function carried out by the process is defined here. YSM makes no restriction on the grammar used to do this, but there are some general principles that should always apply. These are:

1. The functional specification should be unambiguous. In other words, it should always give the same effect for given inputs. This can either be tested by using formal grammars or by giving the minispec to different analysts and subject-matter specialists, with some test data. If any of them interprets the specification differently from the others, then it is ambiguous. Note: the sole use of subject-matter specialists is counter-productive, as they will tend to subconsciously correct ambiguities as a result of their own knowledge.

2. The functional specification should be complete, in the sense that it deals with all possible values of the inputs and assigns values to all the required outputs.

Within these restrictions, any suitable specification tool may be used. Choices that have to be made are:

*   Whether to use an external or internal specification: external specifications are more suited to testing and quality assurance; internal specifications are more suitable for construction blueprints. (It is allowed to have both and even, check them against each other.)

*   Which modelling tool should be used.

*   The style and format in which the tool will be used.

2.16.6.3.1 *External or internal specification*

If the user understands and can verify an external specification, then build an external specification in the analysis phase. This will be used to verify the policy. If the algorithm used to carry out a function is being verified by the subject-matter specialist an internal, procedural specification should be used. If the user is not able to verify an external specification, then an internal specification should be used during analysis and an external specification constructed for testing.

If an internal specification has not been built during analysis and it is cost-effective to use this in program design (this will depend on the technical support environment and the complexity of the function), construct an internal specification during system design.

If the minispec is to be used as a specification of any delivered unit (in other words, it is part of an implementation model), construct an external specification before testing the unit. Unit testing will be against this external specification.

Note: if the same action has to be carried out on an indefinite number of iterations of a component of an incoming data flow, it is usually easier to use an internal, procedural specification.

2.16.6.4 External specification

2.16.6.4.1 Choice of modelling tool for external specification
Choose pre-/post-conditions, except in the following special cases:

1. There are a large number of pre-condition combinations. In this case, choose decision tables.

2. The process monitors or controls the environment. If it is continuous, use Laplace transforms (see §2.16.4.2.1); if it is discrete, use z-transforms (see §2.16.4.2.2).

2.16.6.4.2 Making pre-/post-conditions more 'user-friendly'
Pre- and post-conditions may be made to read more easily by some of the following techniques:

- Separating each pair from the next by a dividing line or several blank lines.

- Indenting or underlining to help break the specification into more obvious pre-/post-pairs.

- Giving each pair of pre- and post-conditions an identifying number or letter. The numbering sequence is not significant.

- Describing them as 'situations before' and 'situations after' or just 'before' and 'after', rather than 'pre-' and 'post-' or 'pre-condition' and 'post-condition'. This is particularly effective when walking through these specifications — 'before' sounds much less 'technical' than 'pre-condition' and consequently gets a better response.

2.16.6.5 Internal specifications

2.16.6.5.1 Local terms
Sensible use of local terms can often make a minispec more readable and (particularly where the same complex selection logic appears several times) reduce the checking that has to be done. However, trying to do everything in local term definition is usually counter-productive — a sensible balance is required.

A local term is often a good place to differentiate between an essential condition and the test for that condition. For example:

> For each <Patient> with %cause of removal% = null do the following …

is better expressed as:

> For each <Patient in study> do the following …

where the local term "<Patient in study> is introduced by:

> <Patient in study> ::= <Patient> with %cause of removal% is null.

The second states the requirement as the user will understand it. The local term definition clarifies exactly what is meant by " a patient being in the study".

2.16.6.5.2 Choice of modelling tool

With most technical environments procedurally oriented, structured text is the natural tool (see §2.16.3.13), but more sophisticated support environments would preferably support a more graphic approach (see §2.16.3.13.2).

2.16.6.5.3 Choice of style

Possible choices include very formal mathematical or text syntax, or a much less formal textual equivalent. This may be referred to as a choice of dialect. More formal styles are easier to check for correct syntax and may even support formal 'proving' techniques. For human verification, a less formal style is recommended.

In advanced support environments, automatic conversion between dialects may allow the specification to be captured in a formal, internal grammar, which can be automatically 're-laxed' for human consumption.

2.16.6.5.4 Structured text

1. Two subject-matter experts should be able to take the minispec with the same test case inputs and predict the same outputs. This test should be carried out without the assistance of the author. It is a test of clarity. If the formulation of the minispec is difficult for a subject-matter specialist to understand, a different presentation must be used.

2. A minispec should not contain any logic that could be removed (thus reducing in complexity), while maintaining the same behaviour. This is a test of simplicity. For example, the rules:

> increase every instructor's salary by 5%

and

> multiply every instructor's salary by twenty and add on the original salary; Now divide every instructor's salary by 20.

are equivalent. However, the first is preferable in terms of simplicity and clarity.

2.16.6.5.5 Making structured text more 'user-friendly'

Structured text may be made to read more easily by some of the following techniques:

- The addition of non-significant words such as "and then" to help make it more "user-friendly". (This is sometimes referred to as "relaxed structured text"). These may be 'stripped out' of the final agreed specification.

- Starting each statement on a new line. This should be taken for granted, except with users who are very 'computer-oriented'.

- Indenting by a fixed amount by each extra level of structure (this is highly recommended).

- Providing left-hand ruling lines to show the beginning and ends of structures (e.g. see §2.16.3.13.6). This is very helpful, but difficult to maintain neatly in most current technical support environments.

- Providing boxes around structures (e.g. see §2.16.3.13.6). This is also very helpful, but even more difficult to maintain in most current technical support environments.

- Numbering the components of structures. This may become tedious under maintenance, but some users find it very helpful.

- Providing a 'bullet' character before each statement.

2.17 Operation Specification

Note: this section need only be read if the standard YSM operations are insufficient and 'user-defined' operations are required.

2.17.1 Purpose

Many simple operations are already provided by YSM:

- there are pre-defined operations for each of the standard abstract data types (see §6.4);

- standard operations such as "signal" (see §6.9);

- a standard data access grammar (see §6.10).

These operations are likely to be sufficient for most enterprises.

YSM also allows new operations to be defined and added to those already available. Each operation that is added in this way requires an operation specification.

Each operation requires one or more inputs and returns one or more results. (Very rarely, an operation has no input arguments, for example, a random number generator. Other examples include 'environment functions' — for example "time of day( )".) The operation specification defines these inputs and outputs, the effects of the operation and the way it is used.

Some operations defined by the enterprise might be very general and applicable to many enterprises. Others may be more complex and very specific to the enterprise — for example, "inner product" defined on two "vector"s. This might be useful in an enterprise that did a lot of mathematical modelling, but it is unlikely to be used in a bank.

2.17.1.1 Appropriate rigour

YSM provides a rigorous framework for definition of these operations and checking that they are used correctly. This may be described as "strong typing" of a model. For many projects, this level of rigour may not be appropriate — in such cases, a more relaxed approach is equally acceptable.

Note: the Enterprise Resource Library is discussed in §4.1.5.

2.17.2 Examples

OPERATION : area product

MEANING : Calculates the area of a rectangular region bounded by two lengths at right angles.

ARGUMENTS :

argument	: type	: direction
x	: length	: in
y	: length	in
a	: area	: out

DEFINITION : a := real(x) * real(y)

USE :

algebraic:	"x*y" returns area.
procedural:	"product(x, y)" returns area.
textual:	"a is set to the product of x and y".

OPERATION : round

MEANING : Rounds to the nearest whole number

ARGUMENTS :

argument	: type	: direction
x	: real	: modify

DEFINITION : x := real(integer(x + 0.5))

USE :

procedural:	"round(x)" alters the value of x to the nearest real equivalent of an integer.
algebraic:	"~x" returns a real value that is equivalent to the nearest integer to x.

2.17.3 Components

2.17.3.1 Operation

Each operation has a name by which it is identified. This name is unique within the enterprise and is entered here to indicate which operation is being defined. For example:

```
OPERATION : area product
```

The following are also possible operation names: "integer product", "real product", "area product", "inner product". The first two are so commonly required that they are provided as support for the standard ADTs "integer" and "real" respectively, so users of YSM do not need to define them. However, the other two examples appear in this section as examples of defined operations. They are identified by these names.

Note: this name is not necessarily the name by which it will be referenced (see §2.17.3.5).

2.17.3.2 Meaning

This entry gives a description of the function carried out by the operation in general terms. For example:

```
MEANING : Calculates the area of a rectangular region bounded by two
          lengths at right angles.
```

This is more 'user friendly' than the definition of exactly what the effect of the operation is, which is given in the "definition" entry.

2.17.3.3 Arguments

The arguments are the inputs and outputs of the operation. Each argument is listed here, together with its associated typing and direction. The typing is defined by giving the name of an abstract data type that the operation acts on. For example:

ARGUMENTS:		
argument	: type	: direction
x	: length	: in
y	: length	: in
a	: area	: out

there are two input arguments, each of which must be given a value of type "length". The result is a value of type "area".

2.17.3.3.1 Formal and actual arguments

The names used for the arguments are only significant in the definition of the operation. These names are those used in the definition entry (see §2.17.3.4). Within a minispec the operation might be referenced using different names, for example:

```
floor area := area product(P, Q)
```

where "P" and "Q" are data items of type "length" and the data item "floor area" is of type "area".

The arguments "x", "y" and "a" are referred to as formal arguments. When the operation is used each of these must be replaced by values of the correct data type. The values used each time the operation is invoked are referred to as the actual arguments. (The process of associating the values of the actual arguments with the formal arguments is sometimes referred to as 'binding' the values.)

2.17.3.3.2 Arguments that are compound ADTs

The arguments in the operation "area product" are simple ADTs. Arguments may also correspond to compound ADTs. An example is shown in §2.17.3.4.2.

2.17.3.3.3 Generic operations

The type of the argument must be the name of an abstract data type, except that if the operation may act on more than one abstract data type the reserved term "generic" is used. For example:

OPERATION : swap

MEANING : interchanges two items of the same type.

ARGUMENTS :

argument	: type	: direction
x	: generic	: modify
y	: generic	: modify

DEFINITION :

Pre-

x, y have same abstract data type

Post-

x = y(pre) AND y = x(pre)

USE : procedural : "Swap(x, y)" interchanges x and y

2.17.3.3.4 Direction

The direction is indicated by one of:

- **in**: the argument value is used, but not changed by the operation.

- **out**: the argument is given a value, but the operation does not require the argument to have a value before it is invoked. In effect the value is 'created' by the operation.

- **modify**: the operation modifies a pre-existing value. "round" is a simple example of this type of operation (see preceding example).

Input values may be obtained from any available data item of the correct type, specifically:

- **an attribute**: (see §2.8.3.6) — this uses the current value for that attribute (the entity must be bound at the point the operation is invoked);

- **a non-attributed elemental data flow**: (see §2.18.3.10) most elemental data flows are attributed, but not all;

- **a temporary item**: (see §2.16.3.9) this item must have been given a value prior to the operation being referenced;

- **a constant**: one of the allowed values of the ADT (for example, "2.45" for a "real" constant). For some ADTs, units may be used, for example "1.6 metres" (see §2.10.4.3 for a discussion of these supported units);

- **an expression**: for example, "2.5 + a*b" is an allowed "real" expression, if "a" and "b" are both real.

The outputs of the operation may be assigned to any data item of the correct type that is accessible (see "minispecs" for a discussion of this). A "modify" argument has the same restrictions as an output argument.

2.17.3.4 Definition

The body of each definition is a specification of how the input arguments are used to determine the output value(s). Either structured text or a pre- and post-condition formulation may be used. For example, the definition of "area product" includes:

```
a := x*y
```

which multiplies the two input arguments together, giving the result. Input arguments are values that are used and output arguments have values assigned to them. In this case, implicit type conversion (see §2.17.3.4.3) is used.

Modified arguments are both used and assigned to. For example:

```
x := real(integer(x + 0.5))
```

rounds a real number to the nearest 'whole' real number. This is an example of explicit type conversion (see §2.17.3.4.3).

2.17.3.4.1 Recursive operations

An operation may refer to itself, when it is described as being "recursive". For example:

OPERATION : factorial

MEANING : Gives (number)*(number - 1) *...* 2 * 1

ARGUMENTS :	**argument**	**: type**	**: direction**
	x	: cardinal	: in
	y	: cardinal	: out

DEFINITION :	Pre-	x = 0
	Post-	y = 1
	Pre-	x > 0
	Post-	y = x*factorial(x-1)

USE :   procedural: "factorial(x)" returns cardinal.
        algebraic: "x!" returns cardinal.

2.17.3.4.2 Access to components of compound ADTs

Access to components of compound ADTs is carried out in the same way as access to components of compound data items. For example, for the operation "inner product", the individual components of the arguments are referenced:

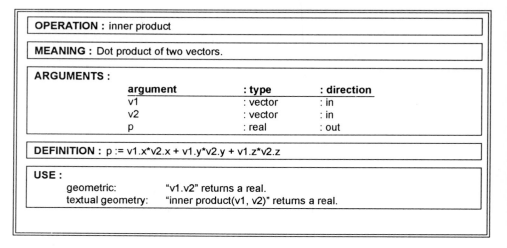

OPERATION : inner product

MEANING : Dot product of two vectors.

ARGUMENTS :	**argument**	**: type**	**: direction**
	v1	: vector	: in
	v2	: vector	: in
	p	: real	: out

DEFINITION : p := v1.x*v2.x + v1.y*v2.y + v1.z*v2.z

USE :
    geometric: "v1.v2" returns a real.
    textual geometry: "inner product(v1, v2)" returns a real.

(A possible definition of an ADT "vector" is as a compound ADT with three components, "x", "y" and "z" — see §2.10.5.)

2.17.3.4.3 Inheritance of operations

When the argument ADTs have been defined by restriction of another ADT, any operation supported on the parent ADT may be used. For example, "area product" is defined on two "length"s and returns an "area". Both of these ADTs are defined as restrictions of real. Any operations that are allowed on "real"s may be used. This may be carried out in one of two ways:

1. Explicit type conversion: the inheritance and restriction is explicitly invoked by using the name of the ADT. For example:

   ```
   a := area( real(x)*real(y) )
   ```

   The type conversion function "real" converts the "length"s back into the parent ADT — "real". The operation "*" is then used, giving a "real". This is then restricted to an "area".

2. Implicit type conversion: the inheritance and restriction is implied, without actually showing it. For the above example, we would have:

   ```
   a := x * y
   ```

Most people will prefer implicit type conversion as being much more 'user-friendly'. Explicit type conversion may be used if there is any ambiguity.

Type conversion for inherited ADTs is further discussed in §2.10.3.7.2.

2.17.3.4.4 Strong typing

The checking that actual argument ADTs match the formal argument ADTs is referred to as 'strong typing'.

Strong typing has benefits of more checkability, but there are costs in setting up such an environment. Its use is therefore not cost-effective in all situations.

Theoretical treatments of programming languages discuss strong typing in more detail. Most modern languages provide strong typing with explicit type conversion rules and techniques. Unchecked use of parameters is bad practice, at least in theory.

2.17.3.5 Use

For a given operation there may be many ways in which it may be "used". For example, if there was an operation "percentage increase" that increased a number by a percentage, we might allow any of the following formats to be used in referencing it:

```
increase the sum insured by inflation rate
```

```
percentage_increase(sum insured, inflation rate)
```

```
apply %inflation rate% percentage increase to %sum insured%
```

are three possible syntaxes for invoking this operation. Whichever form is adopted, the operation requires two inputs of type "number" and "percentage". (The type percentage might be defined as a restriction of "number" perhaps, but, in principle, they are two different types of data.)

The first form's name does not so explicitly imply it uses a percentage as the other two. However, as long as any use of it always provided it with a percentage, the three forms are equivalent.)

2.17.3.5.1 Dialects

Different access formats may be provided for each possible anticipated use of the operation. For example, an operation:

```
USE :
        algebraic:      "x*y" returns area.
        procedural:     "Product(x,y)" returns area.
        textual:        "a is set to the product of x and y".
```

defines three alternative ways of accessing the same operation that operates on two lengths, giving a real result.

These alternatives may just be listed, or they may be defined as being the required formats in any one of a number of dialects.

2.17.3.5.2 Named dialects

Each vocabulary, or set of legal expressions is termed a dialect and given a unique name within the enterprise. In the above example, the form of the operation for the "algebraic", "procedural" and "textual" dialects are given. There may be more than one way of using a given operation in the same dialect.

The environment might be one where great rigour is not justified and a more 'user-friendly' dialect is required, or it might be a very formal environment, with a very well-defined syntax for operations. There may even be dialects that can be automatically compiled.

Introduction to the dialect concept is given in §1.2.3.2.1 and the use of dialects in minispecs is described in §2.16.3.13.4. The standard YSM dialects are defined in §6.7.

2.17.4 Rules

1. An operation may have one or more arguments, but must have at least one.

2. There must be at least one output that is "out" or "modify".

3. The names of the formal arguments must be distinct.

4. Only information in the argument list or that relating to the environment may be used in the definition section. However, other operations may be used. For example, the operation "todays date( )" might be invoked in an operation that calculated elapsed days since a given day. (Note: cyclic dependency of operation *is* allowed.)

5. In the "use" entry, the pairs of opening and closing quotes must balance. Exactly one instance of each of the arguments must appear between each pair.

2.17.5 Guidelines

1. There should be at least one input argument. Random number generators and operations that return information about the environment (such as todays date) are exceptions.

2. Generic operations should contain explicit type checks as part of the pre-conditions.

3. The format in which an operation is used in a human readable dialect should be one that provides an understandable format in use. Thus the names should be meaningful and generally correspond to verb phrases.

4. The names of arguments should be chosen so that the use specifications are not ambiguous. Specifically, these names should not be the same as any other sequence of characters used in the description of use, as this can be very confusing.

2.18 Data flow specification

2.18.1 Purpose

This type of specification is used to specify data flows and their components. Any data flow seen on a DFD (of which the context diagram is a special type) has a corresponding data flow specification.

Where a discrete data flow contains several items that occur at the same time, the information that occurs each time the flow exists is termed a data group. Each of its components is a data flow, which may or may not be shown explicitly on data flow diagrams. These components also have data flow specifications.

A data flow that cannot be broken down into smaller components is referred to as an elemental data flow. These are specified as attributes, or (in rare cases) in terms of their meaning and values.

Data flows may also be grouped to provide clarification where a large number of flows would obscure the purpose of a high-level DFD. These are multiple flows — they are a notational convenience on DFDs. Both discrete and continuous data flows can be contained in multiple data flows.

2.18.2 Examples

DATA FLOW : patient admission note

MEANING : Information recorded at the time the patient took a room at the hospital.

STRUCTURE : group

COMPOSITION : patient name
+ [next of kin | contact person|responsible agency]
+ (referring physician)
+ 1{medical complaint + complaint priority}

DATA FLOW : patient name

STRUCTURE : element

PERSISTENCE : discrete

ENTITY : Patient

ATTRIBUTE : name

DATA FLOW : next of kin

STRUCTURE : element

MEANING : The legal next of kin. This person must be contacted in case of severe deterioration of the patient's condition (or death).

PERSISTENCE : discrete

ENTITY : —

TYPING : abstract

ADT : name and address

DATA FLOW : customer complaint

DATA FLOW : customer complaint response

MEANING : Dialogue relating to customer complaint about purchase.

STRUCTURE : dialogue pair

2.18.3 Components

2.18.3.1 Data flow

Each data flow has a unique name that is used to specify it. This entry identifies the data flow being specified. For example:

> **Data flow : patient admission note**

No component of a data flow may have the same name as any other data flow or component of a data flow.

If the data flow is a dialogue pair, then the first entry is the initiator and the second the response.

2.18.3.2 Meaning

For an data flow that is not an attributed data item (see §2.18.3.7), this entry gives the significance of the data item to the system.[1]

2.18.3.2.1 Qualified data flow

As described in §2.14.3.7.2, a data flow name may have a qualifier. If a qualifier is used, then the distinction between different qualifications of the same data item is given here. For example, in the specification of the data flow "actual treatment", we might have:

> **MEANING** : The actual treatment contains information about the treatment administered to the patient at a particular time.
>
> A (valid) actual treatment is one that has been checked for consistency. It has values for its components that lie within the expected ranges.

For this example, a semantic distinction between different occurrences of "actual treatment". By implication, "(valid)actual treatment" on a data flow diagram is more than just a treatment — its values have been checked to conform to some validity checks. This semantic distinction may be a useful one to make. However, it does not absolve the modeller from writing a minispec, showing how (valid)actual treatments are 'filtered out' of the sum total of all candidate actual treatments.

2.18.3.3 Structure

This entry has one of the following four values:

- **element**: a data flow that is a single, indivisible piece of data.[2] For a continuous data flow, the data flow represents a continuously changing quantity. For a discrete data flow, every time the data flow exists, this has a single value for an instant of time. At other times, it is not available.

---

1 Note: this is one of the rare examples where an entry in the frame format is not always present and its presence or absence can only be determined by fields not yet discussed. The order of fields in this frame specification is not critical, but it seems more appropriate to keep to the standard of a "meaning" entry being given immediately after the name of the item.

2 In this section, the term data, rather than information will be used. This will avoid some semantic awkwardness

- **group**: a discrete data flow consisting of several pieces of data. Every time a discrete data flow defined as a group exists, there is one occurrence of this 'packet' of data. (A continuous data flow cannot be defined to be a group, but must be defined to be multiple.) The structure of the packet will be defined (see §2.18.3.4);

- **dialogue pair**: the data flow can only be specified as a dialogue pair if two names have been given. In this case, the first name is the initiator; the second is the response flow. Both of these must be 'group' or 'element' data flows, each with their own data flow specification.

- **multiple**: as shown on the DFD, this is a single data flow. In reality, it is an overview representation of many flows. These do not have any temporal connection. (An analogy would be a 'multi-core' cable, containing many signal cables. Each individual data flow corresponds to one signal cable. The multiple items defined in this flow are specified as contained flows (see §2.18.3.5).

To define the type of data flow one of the above four keywords is entered. For example:

```
STRUCTURE : group
```

2.18.3.4 Composition

If the structure is 'group', the composition entry gives the contents of the flow and structure of one occurrence of the data flow. A multiple data flow does not have a 'composition' — its contents are just listed (see §2.18.3.5).

This composition of a data group is defined using the inclusion, selection, iteration and optionality constructs.

The composition is usually written out using a standard syntax for these constructs. This is described under each data composition structure. For example:

```
COMPOSITION :      patient name
            +      [next of kin | contact person | responsible agency]
            +      1{medical complaint + complaint priority}
            +      (referring physician)
```

2.18.3.4.1 Inclusion

Each of the components are listed, separated by "+" symbols, which should be read as "together with" or "and".

Two items that are "joined" by a "+" symbol occur at the same time. For example:

```
patient name + patient age
```

indicates that the two data items "patient name" and "patient age" occur together.

— for example, "a data flow contains a packet of information". A purist might demand that we refer to these as information flows, but it is probably too late for that.

2.18.3.4.2 Selection
This is used to indicate that exactly one of several possible components will be present at any one time. It should be read as "choose one, but only one, of this list". For example:

> [next of kin | contact person | responsible agency]

The first selection choice is shown by an opening bracket followed by a delimiter ("|" is used in this reference). The last choice is shown between the delimiter and the closing bracket. Intermediate choices, should they be present, are between two delimiters. There must be at least two choices, so there must be at least one delimiter!

See also §2.18.5.3 which discusses the difference between selection and contained data flows in a multiple data flow.

2.18.3.4.3 Iteration
One data item may consist of several iterations of another item. This is denoted by showing the item(s) to be repeated inside a pair of enclosing braces. This should be read as "iterations of". For example:

> {medical complaint + complaint priority}

indicates that there is an indefinite number of pairs of items. The first of each pair is the "medical complaint"; the second is the "complaint priority".

Bounds of iteration
Where there is a lower limit in the number of iterations, this is shown as a number before the opening brace. For example:

> 1{medical complaint + complaint priority}

If this lower limit is not shown, then by implication, it is possible that "zero iterations" may be present.

A number following the closing brace shows there is an upper limit, by essential policy, on the number of items that may occur. For example:

> 1{month total}12

states that at least one, but no more than twelve iterations of the item "month total" is present.

If the upper limit is omitted, it is assumed that there is no upper limit to the number of iterations.

Should the same number or data name appear on both sides of the braces, it is assumed that precisely that number must be present. For example:

> 7{daily total}7

states there must be exactly seven daily totals.

2.18.3.4.4 Optionality
Some components of a data group are not always present. These components are shown enclosed by parentheses. For example:

```
patient name + (patient age)
```

means "patient name" will always be present, but "patient age" may or may not be present. This should be read:

patient name, together with (optionally) patient age

2.18.3.4.5 Precedence

For complex structures, the groupings are unambiguous for the syntax recommended. They usually have an intuitive interpretation. For example:

```
[soup + salad | appetiser]
```

shows two items to choose from: the first is "soup and salad"; the second is "an appetiser". Were this written:

```
[soup | appetiser] + salad
```

one could choose between soup or an appetiser, but you would always get a salad. The position of the list delimiter or the closing paired bracket shows the precedence intended.

This precedence also applies to the other structure notation. For example:

```
{{name + (phone number)} + department}
```

shows that each "department" has some number of occurrences of "name" (optionally paired with "phone number"). There would be a number of occurrences of this 'department record'.

In case of problems in resolving the structure, the algorithm given in §2.18.5.1.1 will allow a 'parse tree' to be constructed.

2.18.3.4.6 Alternative modelling tools

YSM allows other modelling tools to be used to describe the structure of data items. Graphic modelling tools of various types have been used in the past and any automated support tool is likely to use some more interactive way of building up the structure and presenting it to the system modeller. This is allowed.

Note: a rather textual modelling tool — a parse tree is shown in §2.18.5.1.1.

2.18.3.5 Contained data flows

For multiple data flows, the component data flows are listed. To stress the fact that these are really independent flows, the list is just given by listing the component flows, separated by 'commas'. For example, a high-level DFD (or the context diagram) might show a single data flow "stuff" to represent many flows. This could be specified as follows:

```
┌──────────────────────────────────────────────┐
│ ┌──────────────────────────────────────────┐ │
│ │ DATA FLOW : stuff                        │ │
│ ├──────────────────────────────────────────┤ │
│ │ STRUCTURE : multiple                     │ │
│ ├──────────────────────────────────────────┤ │
│ │ CONTAINED DATA FLOWS : p, q, r, s, t     │ │
│ └──────────────────────────────────────────┘ │
└──────────────────────────────────────────────┘
```

For dialogue flows specified as multiple, the list is shown tabulated. For example, a high-level DFD might show a dialogue data flow, with names "user-input" and "user-output". Its specification might be:

```
┌──────────────────────────────────────────────┐
│ ┌──────────────────────────────────────────┐ │
│ │ DATA FLOW : user input                   │ │
│ ├──────────────────────────────────────────┤ │
│ │ DATA FLOW : user output                  │ │
│ ├──────────────────────────────────────────┤ │
│ │ STRUCTURE : multiple                     │ │
│ ├──────────────────────────────────────────┤ │
│ │ CONTAINED DATA FLOWS :                   │ │
│ │            report request   : : report   │ │
│ │            new course       : : imminent course warning │ │
│ └──────────────────────────────────────────┘ │
└──────────────────────────────────────────────┘
```

Note: as shown, flow of this type can contain dialogues and unidirectional flows. For example, the above multiple flow contains two unidirectional flows and one dialogue flow. Contained dialogue flows may themselves be declared as dialogue (in which case each component is specified), or multiple (in which case they are defined as above.

See §2.18.5.3 for a discussion of the difference between contained data flows in a multiple data flow and selection items in a group.

2.18.3.6 Persistence
For a data flow declared as elemental, this gives whether the flow is time–discrete or time–continuous. Allowed values are "discrete" and "continuous".

Note: a data flow declared as a group must be completely composed of discrete flows. Continuous flows can only be contained in multiple flows. If a multiple flow contains a continuous flow, then it is considered to be continuous.

2.18.3.7 Entity
For elemental data items, this gives the name of the entity of which the data element is an attribute. Most information is information 'about something' — this is the entity to which the data is attributed to. The allowed values are the name of the entity or "—" (which defines the data item as being 'non-attributed'. For example:

ENTITY : Patient

If an entity name is given here, then there will also be an entry for "attribute"; if the data is not attributed, then there will be an entry for "typing".

2.18.3.7.1 Distinction between attributed data and stored data

YSM requires that all stored data be attributed. This section says that '*most* data is attributed'. What does this mean? The distinction between stored data and attributed data is sometimes stated incorrectly.

An attribute of an entity is a property of an entity — values of that attribute describe the entity in some way. Not all these values need to be stored. However, they are still attributes. For example, in a system that controlled the temperature of a solution in a tank, the temperature of the solution is an attribute of the associative entity that relates the solution batch to time. This attribute is not usually stored. It might be stored if the requirement were to log the history of the solution.

As this example demonstrates, attributed data does not necessarily have to be stored. The attribute specification for each attribute defines whether it is stored or not (see §2.8.3.4).

Even if the data element is defined as an attribute that is stored, there is still a distinction between this element having a value and the store containing that attribute being updated. For example, the data element "patient name" is defined to be the same as the attribute "name" of the entity "Patient" (see example). This data element has a value at the point in time the data occurs, but the value of this attribute for a specific occurrence of "Patient" is not updated in stored data until a "create" or "update" data access occurs.

The values of attributed data elements that are part of data flows are not stored until a data process specifically transfers them to stored data using one of these operations.

2.18.3.8 Attribute

For attributed data elements, this gives the name of the attribute. Note: attributes of two different entities may have the same name, but for the same entity, the names of attributes are unique. For example:

```
ATTRIBUTE : name
```

All attributed data elements have a corresponding specification — see §2.8.

2.18.3.8.1 Entity reference

If the data element is a reference to a specific occurrence of the entity, then this is indicated by "<>" in the attribute field. For example:

```
DATA FLOW : referring physician
ENTITY : Physician
ATTRIBUTE : <>
```

This is provided to allow a decision on how an identification of a required occurrence is made, particularly in the user interface. For example, there might be several identifiers of the entity "referring physician" already determined. On the other hand, the only one defined might be "registration number" (perhaps allocated by the government). It is not likely that this would be the actual means of identification used in a 'user-friendly' system. A more likely choice might be to display a list of doctors in the area, allowing one to be selected. It would be wrong to insist that the data flow contains the attribute "registration number" in analysis. There is no way a potential patient, or indeed medical assistant would know this.

Entity references provide this type of deferment. They indicate that a unique referent to an entity occurrence is required, but do not commit to exactly how it will be achieved.

2.18.3.9 Typing

For non-attributed data elements, this states whether the values of the data element are defined in terms of an abstract data type or given as an explicit list of values. The allowed values are "abstract" or "value list". For example:

```
TYPING : abstract
```

Note: almost all data should be attributed. This makes it difficult to come up with examples of non-attributed data. The examples given below, purporting to be non-attributed data, could, in most cases, be defined to be an attribute. This might require that a new entity be added. Rather than search for obscure examples, we have 'gone with' the examples given. When the entity is identified, these examples would have to be re-specified as attributed data elements.

2.18.3.10 ADT

If the typing of a non-attributed data element is "abstract", then this entry gives the abstract data type used for its values. For example:

```
ADT : cardinal
```

```
DATA FLOW :  contact person

STRUCTURE :  element

MEANING :  A person to be contacted in case of severe deterioration of the patient's
           condition (or death)

PERSISTENCE : discrete

ENTITY : —

TYPING : abstract

ADT :  name and address
```

in the definition of "complaint priority" indicates that the values will be taken from "0, 1, 2, ...". The basic use of ADTs is to define a set of values that may be compared, assigned, etc.

An ADT may be compound. For example, both "contact person" and "next of kin" are typed as a "name and address (see §2.18.2):

These two specifications both use the ADT "name and address", which might be defined as:

ABSTRACT DATA TYPE : name and address

MEANING : Structure used to hold information about a particular person.
The information held is a "name", together with an "address".

STRUCTURE : compound

COMPONENTS : <u>component</u>   **: type**
name   : alphabetic
address   : address

Compound ADTs are discussed in §2.10.5.

Note: introducing the entity "Person", with subtype "Patient", together with the relationship "<Person> is contact for <Patient>" would avoid the requirement for this non-attribute data item.

2.18.3.11 Values

If the typing of a non-attributed data element is "value list", then this entry gives the list of values used. If the element has a specific value list, it may only be tested for equality with one of the values. For example:

DATA FLOW : initial infection assessment

STRUCTURE : element

MEANING : This is an assessment of how ill the patient is when admitted.

PERSISTENCE : discrete

ENTITY : —

TYPING : value list

VALUES : critical, very ill, stable, healthy

Note: the above is (fairly obviously), an attribute for "Patient" and should therefore be re-specified.

This type of specification often occurs for 'messages'. For example, "Solution strength update error" might have:

VALUES :	solution not found, threshold out of range, solution in use

When the message is produced, this is the possible values it could have.

2.18.3.12 Parameters

If an ADT has parameters, there are choices allowed for any specific application. These choices are specified here. In the example below, "length" has optional parameters for "minimum" and "maximum". They are specified to restrict the legal range for this specific element. If the chosen ADT's parameters are marked as required, they must be specified here. This field is only used for elements with "abstract" typing.

DATA FLOW : distance between workstations

STRUCTURE : element

PERSISTENCE : discrete

ENTITY : —

MEANING : The measure of length between two adjacent workstations on the assembly line. This is derived from the two workstation positions.

TYPING : abstract

ADT : length

UNIT : metre

PARAMETERS :
parameter	: value
minimum	: 1
maximum	: 500

RESTRICTIONS : —

Note: this non-attributed data item might be attributed to an associative entity referring to two workstation occurrences.

2.18.3.13 Restrictions

A restriction defines the attribute to be only a certain selected subset of the possible values of that ADT. The use of restrictions use is often an alternative to the use of parameters associated with the ADT, but restrictions are usually more specific constraints than were anticipated when the ADT was defined. For example:

DATA FLOW : complaint priority
STRUCTURE : element
MEANING : This is an estimate of how important the complaint is.
PERSISTENCE : discrete
ENTITY : —
TYPING : abstract
ADT : cardinal
RESTRICTIONS : complaint priority < 10

Note: if an associative entity "Complaint" is created, then this would need to be re-specified as an attributed data item.

Restrictions may be written down fairly informally, or they may use a more formal grammar. An example of an informal restriction is:

RESTRICTION : First character is alphabetic.

which could be used to demand that the first character of an alphanumeric was not numeric. Each restriction must evaluate to TRUE or FALSE and the allowed values for the attribute are those taken from the ADT that give TRUE for the restriction. A more formal definition of allowed restrictions is given in §6.5.

2.18.4 Rules

2.18.4.1 Composition

1. For optional components, there must be a closing parenthesis for every opening parenthesis.

2. For a selection, there must be at least two items in the list. The list must be preceded by a "[" and be followed by a "]". Each selection, apart from the last, is followed by a "|", to separate it from the next choice.

3. For an iterated item, there must be a closing brace following every opening brace. The bounds of iteration need not be specified; the default is zero to infinity (no upper limit to number of iterations).

2.18.4.2 Data elements

1. For a given value list, the values given must be distinct from one another (but may be the same as those of another or of an ADT).

2. For an attribute that is defined in terms of an ADT that has associated parameters, values of all non-optional parameters must be given.

3. For an attribute that is defined as a restriction of an ADT, the restriction may only use functions and operations that are defined on that ADT.

2.18.5 Guidelines

2.18.5.1 Composition

1. For iteration, the items within the brace represent one occurrence of an iteration. They should be named accordingly. In other words, if one occurrence is one total, the data item should be named "total", not "totals".

2. As deep nesting of structures can make a composition very difficult to understand, it is recommended that no more than one level be used within another. For complex structures, the use of named data items with separate composition definitions will make the structure easier to understand. For example, rather than:

```
dinner =
            [[[pate | pasta] | [chicken soup | onion soup]]
       +    [ice cream | cake] + entree | sandwich]
```

Try:

```
dinner =        [full meal | sandwich]
full meal =     [appetiser | soup] + dessert + entree
appetiser =     [pate | pasta]
soup =          [chicken soup | onion soup]
dessert =       [ice cream | cake ]
```

(The "=" represents definition of structure by means of a data flow specification.)

Note: If a selection is directly included within another (as here, it can always be replaced by a single level of selection. Thus the above becomes:

```
Dinner =
            [ [pate | pasta | chicken soup | onion soup]
       +    [ice cream | cake] + entree | sandwich]
```

2.18.5.1.1 Parsing data composition specifications

1. Parse the expression, starting at the left-hand end. (The algorithm can be made from the right, or even 'both ways at once'. However, a demonstration of left to right parsing should suffice.) On encountering any opening paired symbol (a "[", "{" or "(") find the matching right symbol and 'cross out the whole structure'. Note: the 'depth' of intervening pairs of the same symbol will have to be checked — the matching symbol is the first one at the same depth.

2. When the right hand end is reached, there are two possibilities. Either the whole line is crossed out (the first and last character are matched outer symbols of a construct), or there are several components linked by a "+" symbol.

3. If there are several components, draw the top of the tree, with a "+" symbol. Now link this to nodes for each component, linked by the "+" symbols in the original expression. Each node has the expression within the component.

4. If there is only one component, it must be a selection, iteration, or optional construct. For a selection, add a leaf for each of the components, as separated by "|" characters. Any nested constructs are 'crossed out' when scanning for these "|" characters, so there is no ambiguity. For an iteration or optional item, add a single leaf with the item enclosed between the delimiter pair.

This is now a 'top-level' decomposition. Each leaf node contains a simpler expression that can be parsed in the same way. The process stops when the leaves contain data elements.

The example below shows a parse tree for the composition:

{{ name + (phone number) } + department}.

It can be constructed in a top-down sequence using the algorithm.

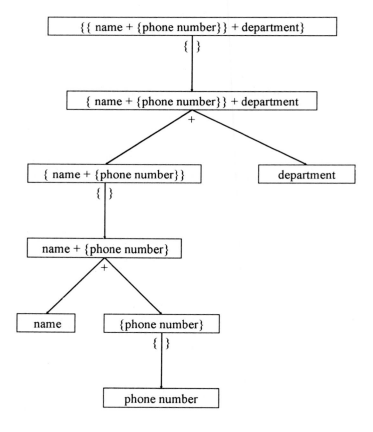

2.18.5.2 Data elements

1. An element should be named to describe its use in the system. A subject matter expert should be able to describe the meaning of an element from its name alone.

2. If a set of values is commonly used for different attributes or elements, then it is a good idea to create an ADT with these values, providing the values are used with the same meaning.

2.18.5.3 Difference between multiple flows and selection

This is a subtle distinction. Indeed, many consultants will not wish to use the multiple flow concept. This is an allowed position to take, although high-level dialogue data flows cannot be specified without using the multiple flow specification frame (or its equivalent).

The intended use is:

- the selection construct is used within a single 'packet' of data. It represents alternative components of the packet. For example, in a system to record payments from customers, sometimes there would be a credit card number, sometimes details about a cheque; there would always be a payment amount.

- a multiple flow represents a much 'looser' alternation. Contained flows occur at different times. The multiple flow is only used to give an overview of many flows, which have no link together in time.

This is a rather relative distinction. For example, if a more detailed examination of "Record payment" was carried out, it might be decomposed into different data flows — "cash payment", "credit card payment", "cheque payment". This would imply that a higher-level DFD showing "payment" would require that "payment" be declared as multiple.

2.19 Function–Entity Table

2.19.1 Purpose

The function–entity table is used as a high-level modelling tool to visualise the relationships between functions carried out and the information required to support those functions.

This tool is useful in:

- **Strategic planning**: function–entity tables may be used to study dependency between enterprise economic units or several possible systems. These studies are strategic planning activities.

- **Project modelling**: the table may be used to provide an overview of information usage by functions within one system.

2.19.1.1 Enterprise and system use of function–entity tables

A function–entity table used as part of the Enterprise Essential Model is referred to as an enterprise function–entity table.

A function–entity table used as part of the System Essential Model is referred to as a system function–entity table. Note: this is an optional view for this model. It may be used as a temporary, working view or as a retained view, depending on the support environment.

The prefix enterprise or system may be dropped if no ambiguity results.

2.19.2 Example

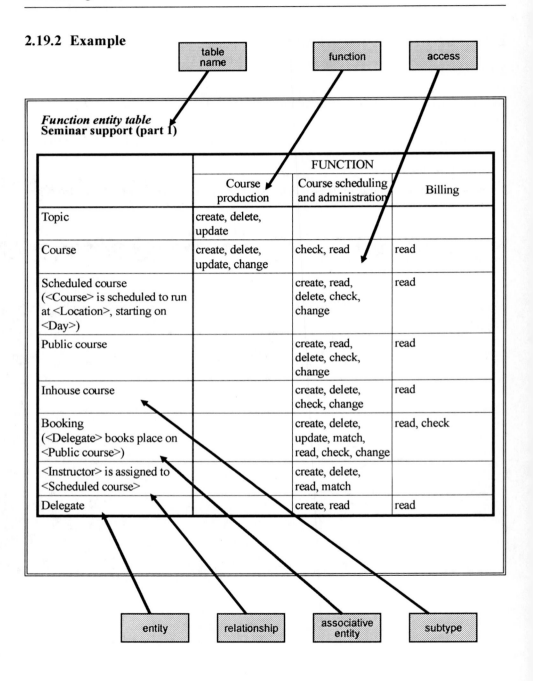

table name → **function** → **access**

Function entity table
Seminar support (part 1)

	FUNCTION		
	Course production	Course scheduling and administration	Billing
Topic	create, delete, update		
Course	create, delete, update, change	check, read	read
Scheduled course (\<Course\> is scheduled to run at \<Location\>, starting on \<Day\>)		create, read, delete, check, change	read
Public course		create, read, delete, check, change	read
Inhouse course		create, delete, check, change	read
Booking (\<Delegate\> books place on \<Public course\>)		create, delete, update, match, read, check, change	read, check
\<Instructor\> is assigned to \<Scheduled course\>		create, delete, read, match	
Delegate		create, read	read

entity **relationship** **associative entity** **subtype**

2.19.3 Components

2.19.3.1 Access

This gives a view of the use of the entity or relationship by the function. Note: this action is a possible use by the function — it will not always occur. Any more detailed model of that function must explicitly show when that access occurs.

The allowed accesses are:

- **create**: the function creates a new occurrence of the entity or relationship.

- **match**: the function needs to know if a specific occurrence of the entity or relationship exists.

- **read** (for entities only): the function needs to know the existing values of attributes that have been previously assigned in a create or update.

- **update** (entities only): the function may change or assign values of attributes to an entity.

- **delete**: the function may destroy one or more occurrences of the entity or relationship;

- **check:** examine a state variable to determine a correct response;

- **change**: change the state of a state variable.

To save space in tables, the following abbreviations are sometimes used:
"C" (create), "?" (match), "R" (read), "U" (update), "D" (delete), "S_c" (state check), "S_a" (state change). (Any other agreed abbreviations could be used.)

Note: if the column shows a function allocated to a system, then an entity–event table showing all events that are the responsibility of that system provides a more detailed view of the information usage of that system. Each row of the function–entity table becomes a row of the entity–event table. For example, the row "Course" in the function–entity table shows which functions make which accesses to this entity; the row "Course" in the entity–event table shows which events cause accesses to that entity. If the functions shown in a function–entity table are event–response processes, then the function–entity and entity–event tables are identical.

2.19.3.2 Associative entity

If an associative entity used by the group of functions being considered, then its name is given here and the row gives the use that each function makes of it. It is suggested that the relationship frame of the associative entity is given in parentheses below the associative entity name. This helps in understanding the dependency of the associative entity on the entities that it refers to. For example:

```
Scheduled Course
(<Course> is scheduled to run at <Location>) starting on <Day>
```

indicates that the associative entity "Scheduled course" acts as a relationship and refers to the "Course", "Location" and <Day> entities.

The entries in this row give the accesses of this associative entity by the different functions.

2.19.3.3 Entity

If an entity is used by any of the functions, then its name is given here and the row gives the use of this entity by the functions. For example:

Course

This row gives the use of the entity "Course" by the given functions.

2.19.3.4 Function

Each column shows the information usage corresponding to one function. The name of this function is shown at the top of the column. For example:

Seminar support

When the table is used in strategic planning, the function will either be a planned system, or an economic planning unit. The example at the beginning of this section is of this type.

When the table is being used to show how one system's functions are organised around the information they use, the function will be a process group or single data process. For example:

Function entity table
Course booking system (part only)

	FUNCTION

	Record firm booking
Patient	read
Drug	match
Recommended treatment (<Patient type> should have amount of <Drug>)	read
<Patient> is typed as <Patient type>	match
Scheduled treatment (Patient> will receive <Drug>)	create, update

gives the accesses made by the function described in §2.16.2.2. This is a single data process. This modelling tool might also be used to show groups of functions corresponding to those seen in a high-level data flow diagram.

2.19.3.5 Relationship

If a relationship is used by any of these functions, then its name is given here and the row gives the use made of it by the functions. For example:

<Instructor> is assigned to <Scheduled course>

This is a relationship "is assigned to" that is defined to refer to the entity "Instructor" and the associative entity "Scheduled course". The full form of the relationship name is given, including the entity references, because several relationships may have the same name, provided they involve different entities. (For further details on this see §2.6.)

2.19.3.6 Subtype

An entity that is a subtype of another entity is not distinguished in any specific way in this table. However, there is an important consistency criterion — any access to the subtype must also be an access to the supertype if that supertype appears in the table. For example:

Inhouse course

is a subtype of "Scheduled course" and down any column (corresponding to a single function) the accesses of "Inhouse course" are always found in the accesses to "Scheduled course". There may, however, be some accesses to the supertype that are not relevant to this subtype. For example, "Course scheduling and administration" does not read "Inhouse course", although they do read "Public course" (which is another subtype of "Scheduled course").

2.19.3.7 Table name

Each table restricts attention to a group of functions and the information model components that they use.

No special identification needs to be given to it unless it is retained as permanent view. A table used as part of the Enterprise Essential Model must have name that is different from that of any other enterprise function–entity table.

If the table is used to study the relationship between functions and their information usage within the scope of one system, the name should be the name of the system.

Because it may not be convenient to show all of the entities and relationships in one table (it might become very large), the table may be split into sections. In this case, an identifying sequence number might be added, as in the example:

Name : Seminar support (part 1)

2.19.4 Rules

1. Each entity, associative entity or relationship shown in the table must have an access. This means each row must have an access.
2. Entities may have create, delete, update, match, read, check or change accesses.
3. Relationships may have create, delete and match accesses.
4. If a relationship appears in a table, then all entities to which it relates must also be shown in that table. If the table is divided into sections, this rule applies to each section.
5. Each section may only show an entity, associative entity or relationship once. In other words, rows may not be duplicated in a single section of the table.
6. If a row is repeated in different sections, then all occurrences of the row must have identical accesses.
7. If a subtype of an entity is shown in the table, then for each function, the accesses to the supertype must include the accesses made to the subtype. There may be additional accesses, corresponding to the access to occurrences of the supertype that are not that subtype.

2.19.5 Guidelines

1. Each function shown in the table generally accesses one of the information model components. This means that each column has an access. For tables that are divided into several sections, this guideline does not apply to individual sections — the function may not need to access any of the information components shown in that section.
2. Repeated rows may be tagged with a "*".

2.19.5.1 Organising the table

Listing the entities and relationships in a random order does not lead to an especially useful table. For small tables this will not be so important, but for large tables, the interdependencies of the information model components (entities and relationships from either the Enterprise Essential Model or the System Essential Model) will become less easy to identify, because the reader will have to keep switching attention from one portion of the table to another.

Associative entities and relationships create dependency in the information model components by linking to other entities. One of the uses of this tool is to identify a group of such relationships (or associative entities), together with the entities to which they refer and check that the entities correctly support the relationships. To do this, the relationship and the entities to which it refers should be kept close together in the table.

Additionally, for a specific entity, the match and read accesses should appear in columns to the right of the create, delete, update entries. The functional dependency that this implies is best seen by reading the table from left to right (in other words the function in one column depends on the correct data access being carried out by the function in the preceding column).

Because some entities may participate in many relationships, this criterion may be difficult to achieve for all these relationships. When this occurs, the table should be split into several sections, each section of which provides a coherent grouping of entities and relationships that meet this criterion. In order to ensure that all the entities to which a relationship refers are seen in one section, it will often be necessary to repeat entities (including associative entities) on several sections. Where this is the case, an identifying "*" tag may be used to identify such entities. This is more important for pencil and paper environments. For automated support environments, selection of components in a section of the table is more of a 'filtering' from a single function–entity table. This could be carried out automatically.

Some entities do not participate in any relationships which are important to the functions being considered. The positioning of this type of entity in the table is of no great consequence, but listing them separately at the beginning or end of the table (or in a separate section) will keep the table simple.

2.20 Data Store Specification

2.20.1 Purpose

The data store specification defines which entities and relationships are included in the store. It is used to allow cross-checking between a DFD and an ERD. The DFD shows stored information using a named store. This store contains information relating to one or more entities and/or relationships. The data store specification defines exactly which entities and relationships this store icon represents.

Note: the store specification can define the store as a collection of other stores. However, as each of these is defined as containing entities and/or relationships, the effect is that any store is defined as containing entities and relationships. Store do not exist, except as an indication that a process on a DFD needs to access attributes of entities and relationship occurrences.

(The use of information by functions may also be modelled using the function–entity table — see §2.19)

2.20.2 Examples

DATA STORE : Solutions

ENTITIES : Solution

RELATIONSHIPS : —

STORES INCLUDED : —

DATA STORE : Teaching assignments

ENTITIES : —

RELATIONSHIPS : <Instructor> is assigned to teach <Scheduled course>

STORES INCLUDED : —

DATA STORE : Current solutions

ENTITIES : —

RELATIONSHIPS : —

STORES INCLUDED : Scheduled solutions
Solution in tank

2.20.3 Components

2.20.3.1 Name
This gives the name of the store. Within a system, each store has a unique name, which serves to identify it. For example:

> **DATA STORE :** Solutions

Note: If the store contains information about a single entity, this name is usually chosen to be the plural of the entity name.

2.20.3.2 Entities;
This entry identifies any entities that are 'in' this store. For example:

> **ENTITIES :** solution

All stored attributes of this entity may imagined to be 'in' this store. It does not hold attributes that are declared as derived or temporary.

State variables are also imagined to be 'in' this store. Changing the state of a given state variable is seen as an access to the store containing that entity.

2.20.3.3 Relationships
This entry identifies any relationships that are 'in' this store. The full relationship frame for relationship must be given. (There could be several "is assigned to" relationships, for example. For example: <Patient> is assigned to <Patient type>; <Patient> is assigned to <Doctor> ...) For example:

> **RELATIONSHIPS :** <Instructor> is assigned to teach <Scheduled course>

this must correspond to a binary relationship. Each occurrence of the relationship refers to one occurrence of the entity "Instructor" and one occurrence of the "Scheduled course".

2.20.3.4 Stores included
Rather than list all the entities and relationships in a store, it is sometimes helpful to group them together into named groups. These groups appear explicitly as stores on lower-level DFDs. Each has a specification in terms of the entities and relationships they contain. On higher-level DFDs a store is shown, with a specification showing it to contain the lower-level stores. For example:

> **STORES INCLUDED :** Scheduled solutions
> Solution in tank

Note: this form of the specification is optional. It could always be replaced by a specification that explicitly listed all of the entities and relationships in the store.

2.20.4 Rules

1. At least one of the "entity", "relationship", or "stores" entries must be present.

2.20.5 Guidelines

2.20.5.1 Choosing stores

1. If a data store contains more than just one entity or relationship, it should be for one of the following reasons:

   - It includes a relationship and the entities it associates. For example, "Patient", "is assigned to" "Doctor" might be combined in a store called "Assignments".

   - It includes all subtypes for one supertype

   - It includes relationships that are between the same entities.

   For higher-level DFDs, this guideline is relaxed and a store may contain groups of entities and relationships that are used by one high-level function.

2.20.5.2 Names of stores

1. A data store name is usually a plural noun, representing the group of items inside the stores.

2. A data store that includes only one entity should use the plural of the entity name.

3. A data store that is a relationship should use a plural noun form based on the relationship name. Generally speaking, this will be the plural of the name that would be used if the relationship was an associative entity.

4. A data store that contains several stores should be given a general name that is derived from the reasons the stores were grouped together. This name should give a good 'clue' to what is in the store. If this guideline cannot be satisfied, show the stores separately.

2.21 Access Flow Specification

2.21.1 Purpose

This provides specific detail on what is being accessed from the data store. All access flows have specifications; even those that are not given specific names on the DFD need a specification of this type. (Naming access flows is optional.)

Depending on technical support, the use of this modelling tool is optional — information contained in this specification is entirely redundant with that contained in the specification of access to stored data given in the minispec (see §2.16.3.1).

2.21.2 Example

```
STORE : Bookings

PROCESS : Record firm booking

ACCESS FLOW : —

MEANING :

COMPOSITION : <Booking>      (<Customer> makes reservation for <Delegate> to
                             attend <Public course>)
               create(attributes: date firm booking),
               update(attributes : date firm booking),
               check(state variable: status),
               change(state variable: status)
```

2.21.3 Components

2.21.3.1 Store
This is the name of the store that the flow is to. The flow being specified is identified by the flow name, process name and access flow name.

2.21.3.2 Process name
This is the name of the process whose access to the store is being specified.

2.21.3.3 Access flow
This is a name for the flow, as it appears on a DFD. Naming access flows is optional.

2.21.3.4 Meaning
This gives a general description of the use of the store by the process. purpose of the flow. Usually, the meaning is obvious from a listing of the composition.

2.21.3.5 Composition
A description of the components of the access flow. This is given as a simple list. Where attributes are involved, these may be specified. For example:

```
COMPOSITION :
        <Booking> (<Customer> makes reservation for <Delegate> to
                        attend <Public course>)
                match,
                create (attributes: date firm booking),
                update(attributes : date firm booking),
                check(state variable: status),
                change(state variable: status)
```

This indicates that the access flow includes match, create, update, check and change accesses. A match operation is used to check if there is an existing booking that is being updated. Sometimes a create access occurs, assigning values to the attribute "date firm booking"; at different times, an existing occurrence of "Booking" is updated with a value for "date firm booking". The state variable "status" is both checked and (sometimes) changed by the process.

2.21.4 Rules

1. An access flow's composition may only include references to entities and relationships held in the store.

2. All attributes included in an access flow must be attributed to the entity shown.

2.22 System Event Specification

2.22.1 Purpose
An event is something that:

- occurs outside a system boundary,
- the system must respond to.

These events are 'external' to the system. They may occur within other parts of the enterprise, or completely outside the enterprise. In either case, they are external events for the system in question. The system may be thought of existing to respond to the collection of these events.

Within the enterprise, there may be many events, each of which has an enterprise event specification (see §2.23).

The system event specification describes how a specific system deals with an event. It describes event-detectors, inputs and outputs associated with the event and required response time. System event specifications are part of the definition of the scope of the system — see §5.6.4.

2.22.1.1 Event lists
The enterprise event list shows a consolidated list of all such events. System event lists are a subset of this list, restricted to those dealt with by that specific system.

The only information given in the system event list is the event number, event meaning and possibly a grouping of events (as implied by the event numbers).

2.22.1.2 Event tables
An event table shows all the events to which the system must respond in one view. It may also be used to relate system components to those events. This may be filled in as different components of the system are identified and organised. It serves as a consistency link between those components. It is used to ensure that the model is consistent, complete, and organised around the events.

2.22.1.3 Format of specification and technical support environment
The events that the system deals with must be identified and each specified in detail. This can be done in several ways — the most appropriate way depends on the technical support environment.

The event specification used in this section uses a frame specification (see §2.1.6). Alternative tabular presentations may also be useful in 'pencil and paper' environments.These tabular presentations might also be an optional view using an automated tool. In automated environments, the minimal views are the event list and a specification for each event; any required table format may be generated, suppressing information that is not required.

2.22.1.4 Event specification support for event table
If a table layout is used in a pencil and paper environment, then each event will require a specification that specifies all other items not given in the table. The components of this are as described for the full event specification.

2.22.1.5 Example explanation

See §2.14.1.3 for a description of the background to this example. The same example is also used for describing data flow diagrams in §2.14 and behavioural state transition diagrams in §2.24.

2.22.2 Example system event specification

SYSTEM : automatic car painting system		

EVENT NUMBER : 2

MEANING : Solution drops below minimum strength

SYSTEM SIGNIFICANCE : The system must ensure solution is correctly replaced.

DATA FLOW :	inputs	:	outputs
	solution strength	:	solution change

EVENT FLOW :	inputs	:	output
	—	:	: flush tank

STORED DATA :	inputs	:	outputs
	<Solution>.use	:	—
	<Solution>.concentration	:	

DETECTION MECHANISM : stimulus

STIMULUS : solution strength

EVENT–DETECTION PROCESS : Monitor solution strength

EVENT–RESPONSE PROCESS : Inform operator of solution change

USE IN CONTROL PROCESSES :	condition	: control process
	falling(solution correct)	: Control car oxidation

RESPONSE TIME :	median	: 3 minutes
	maximum	: 15 minutes

2.22.3 Example system event list

Event
1. Car Moves In Oxidation Area
1.1 Car enters mix bank
1.2 Car enters bath
1.3 Car has bathed long enough
1.4 Car has bathed too long
2. Solution drops below minimum strength

2.22.4 Example system event tables

Event		Detection Mechanism	System inputs	System outputs	Event-detection process	Condition/ control process	Event-response process
1 Car moves into oxidisation area	1.1 Car enters mix bank	oxidiser use request		move into bath		oxidising request/ Control car oxidation	Post oxidising request
	1.2 Car enters bath	car in bath				car in bath/ Control car oxidation	
	1.3 Car has bathed long enough	time		move out of bath		MINS(state)= 2/ Control car oxidation	
	1.4 Car has bathed too long	time		free car		MINS(in soup) = 5/ Control car oxidation,	
2 Solution drops below minimum strength		solution strength	tank level	solution change, flush tank	Monitor solution strength	falling(solution correct)/ Control car oxidation	Inform operator about solution change

The diagram below shows another alternative layout for the table. It uses fewer columns compared with the preceding version:

Event or event group		Inputs : Outputs	Detection process → event flow	Response process
1 Car moves into oxidisation area	1.1 Car enters mix bank	oxidiser use request(stimulus) : move into bath		Control car oxidisation(C), Post oxidising request(D)
	1.2 Car enters bath	car in bath(stimulus) :		Control car oxidisation(C)
	1.3 Car has bathed long enough	: move out of bath		Control car oxidisation(C)
	1.4 Car has bathed too long	: free car		Control car oxidisation(C)
2 Solution drops below minimum strength		solution strength(stimulus), tank level : solution change, flush tank	Monitor solution strength→ solution correct	Control car oxidisation(C), Control solution replacement(C) Inform operator about solution change(D)

In this table:

- the inputs are separated from the outputs by a colon;

- (stimulus) after an input denotes it is a stimulus;

- (D) after a process indicates that it is a data process that responds to the event;

- (C) indicates that it is a control process that responds to it (by using it as a condition);

- → should be read as "generates". Thus the data process "Monitor solution strength" generates the event flow "solution correct". (Data processes of this type are event detectors.)

In this format table, the absence of a named stimulus flow indicates that the system must regard the event as a temporal event.

2.22.4.1 Preliminary versions of the event table

A partial event table may be used as a tool to focus on events without using the full specification or table. This is useful in the early stages of building the environmental aspect of the System Essential Model. The full table or event specifications is then generated from this preliminary version.

For example, a table showing:

- event,

- stimulus,

- preliminary, unstructured description of the systems response,

- A behaviour change column, with entries that are "Y" or "N"

is a convenient format. This preliminary table is then converted to the final specification format as the Essential Model is completed. For example:

Event or event group	Detection Mechanism	Description of response to event	Change of behaviour
1. Car moves into oxidisation area (group)			
1.1 Car enters mix bank	oxidiser use request	move car into oxidisation bath	Y
1.2 Car enters bath	car in bath	Wait for car to be oxidised	Y
1.3 Car has bathed long enough	time	Move car out of bath	Y
1.4 Car has bathed too long	time	Try to recover car that is 'stuck'	Y
2. Solution drops below minimum strength	solution strength, tank level	Make sure solution is changed	

2.22.5 Components

2.22.5.1 System

This gives the name of the system that deals with the event. Note that several systems may need to respond to the same event — there is a specification for each of these systems.

2.22.5.2 Event number

Each event is given a number, which serves as a shorthand identifier for the event within the system.

If the events are not organised into groups, they are simply numbered "1", "2", "3", ... If the events are organised into groups, the events are numbered in sequence within that group. Within group 3 (for example), the events would be numbered 3.1, 3.2, 3.3, 3.4, ...

Within a group, the numbers themselves have no significance; they represent neither sequence nor priority. Note: any grouping implied by this number is 'system specific' — the grouping used at the enterprise level (see §2.23.3.1).

2.22.5.2.1 Event group

An event group is a collection of events that share some commonality. This commonality may be:

- the origin of the event, i.e. the same terminator;

- the object of the event (e.g. all the events related to, or affecting "customer deposits")

- a common response to the events (e.g., such as modifying the attributes of a common entity. For example, the event group "Car moves into oxidation area" (see §2.22.3) was formed because all events in that group affect the same external entity, the car body.

Grouping events is a way of organising large numbers of events. It is purely an organisational convenience and does not imply any specific system organisation or behaviour. However, where responses to events have been organised into groups (see §5.7.1), renumbering the events to show the parallel grouping is helpful.

2.22.5.3 System significance

This gives an unstructured description of the significance of the event to this specific system. Several different systems may deal with the same event — presumably the significance of the event will be different for each of them.

This entry acts as an overview description of what the system has to do when the event occurs.

2.22.5.4 Meaning

This gives a description of the event in real-world terms.

An event is an incident occurring at a specific point in time. It occurs in the system's environment and it is relevant, in the sense that the system must respond to it.

2.22.5.4.1 Relevance of event

For an event to be considered a system event, there must a requirement that the system should respond to the event to achieve the purpose of the system. The required system response may be to transform data, change behaviour (achieved by changing state in a state machine) or both. If the system can ignore the event and not compromise the integrity or purpose of the system, then it is either not an event, or this system does not need to deal with it. In either case, it should not be specified for this system.

The event "Airplane takes off" is certainly an event to some system, but probably not to a word processing system.

2.22.5.4.2 External nature of event

For an event to be of interest, it must occur 'outside' the system. If a system produces an output data flow, then the generation of the output does not constitute an event — this activity is within the system boundary and, furthermore, it is solely the responsibility of the system to decide whether this event occurs or not.

2.22.5.4.3 Events must require responses

The receipt of an output data flow by a terminator is not an event either. For example, suppose an insurance claims system generates a data flow "suggested claim settlement", which is dealt with by an external "Claims adjustment auditor":

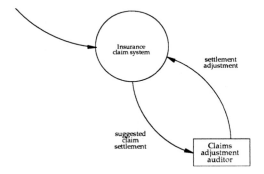

When the system generates this output, there is an event in the environment that affects the auditor — he/she receives the suggested claim settlement. However, this should not be included in the event list, because there is no requirement that the system respond to this event. (In fact, the system would not even know that this event had occurred, unless the auditor always acknowledges receipt of these suggested claim settlements when they are received.

2.22.5.4.4 Validations are not usually events

The fact that the system has validated a flow is an incident; however, that incident occurs within the system and should not be included in the event list. On the other hand, validations carried out by terminators *are* events, provided the system needs to respond to the validation.

In the auditor example, if the auditor agrees the settlement, or not, depending on the information contained in "suggested claim adjustment" (and possibly other available information, not specifically generated by the system), then it *is* correct to model "Auditor confirms settlement" and "Auditor denies settlement" as events. They occur outside the system and require a response by the system. Although the system indirectly caused these events to occur by generating "suggested claim settlement", the system has no control over whether the claims adjustment auditor confirms, denies (or ignores) the claim settlement.

The flow "settlement adjustment" is compound, containing two flows that are the stimuli for each of these two events.

If it was significant that the auditor had not responded in a given time, there would be a third event "Auditor fails to respond to suggested claim settlement". This would be a temporal event. (This type of event is sometimes referred to as a timeout.)

2.22.5.4.5 Distinction between event and stimulus
The event is the thing that happens in the system's environment. The detection of a flow by the system is *not* an event. The system may know that the event has occurred because of the presence of a flow. If so, this flow is the stimulus for the event, but it is not the event.

When an event is detected using a stimulus, it is important to clearly identify which event caused that stimulus to occur. This event should not be described as "x_request generated" (supposing the name of the stimulus to be "x_request") — the reason why that flow was generated and submitted to the system should be described.

In the example, "oxidiser use request" is such a flow. It is wrong to describe an event as "Oxidiser use request is detected", or even "Oxidiser use request generated". However, an event "Car enters mix bank", with a stimulus of "oxidiser use request" *is* correct.

2.22.5.4.6 Boundary conditions
Many systems must continuously perform some behaviour over the period of time when the system is considered to be running. For example, in the satellite system (see §5.7.3.3.6, the process "Integrate acceleration" must be continuously running while the satellite is in orbit. If the System Essential Model is limited to the normal operations of the system, then it may seem that one of the following is true:

1. There are some system behaviours which are not a response to an event.

2. Events are not only time discrete, but may have a persistent quality.

Neither of these is true. Analysis of the boundary conditions of system behaviour will result in identification of time–discrete events that cause a continuous system response. This response will be permanent, that is, it will neither be enabled nor disabled.

In the satellite example, there is a complex sequence of steps during the launch phase of the satellite that culminates with the satellite being inserted into orbit and powered up. It is the last event in this sequence which is the first event in the life of the system being modelled. The event "Satellite inserted in orbit" has a required response "Integrate acceleration".

2.22.5.4.7 Identifying entities in event descriptions

Events usually refer to one or more entities. To highlight this, the notation used to refer to 'one occurrence of an entity' may be used. For example, a system that dealt with bookings from outside customers to attend a scheduled course might include:

5.1 <Customer> makes provisional booking on <Scheduled course> for
<Delegate>.

5.2 <Customer> makes firm booking on <Scheduled course> for <Delegate>.

5.3 <Customer> converts status of existing <Booking> from provisional to firm.

2.22.5.5 Data flows

2.22.5.5.1 Inputs

Any data flow that is used in detection of the event, or in producing the response to the event. This data flow must be an input from the environment of the system. It will appear as (or be part of) an input data flow on the context diagram. For example:

DATA FLOW : **inputs**
solution strength

indicates that this data flow is required. (In fact, it is used to detect the event.)

2.22.5.5.2 Outputs

Any data flow that is produced as part of the response to the event. This data flow must be an output to the environment of the system. It will appear as (or be part of) an output data flow on the context diagram. For example:

DATA FLOW : **outputs**
solution change

indicates that this output is produced as part of the response to the event.

2.22.5.6 Event flows

2.22.5.6.1 Inputs

Any event flow that is used in detection of the response, or in producing the response to the event. This event flow must be an input from the environment of the system. It will appear as (or be part of) an input event flow on the context diagram. For example:

EVENT FLOW : **inputs**
—

indicate that this event does not require any event flow inputs, either to detect the event, or to determine the correct response.

2.22.5.6.2 Outputs

Any event flow that is produced as part of the response to the event. This event flow must be an output to the environment of the system. It will appear as (or be part of) an output event flow on the context diagram. For example:

```
EVENT FLOW :      outputs
                flush tank
```

indicates that an event flow "flush tank" is part of the response to this event.

2.22.5.7 Stored data

2.22.5.7.1 Inputs
Any stored data that is used in detection of the event, or in producing the response to the event. The notation used is described in §2.16.3.7. For example:

```
STORED DATA :      inputs
                <Solution>.use
                <Solution>.concentration
```

indicates that the state variable "use" and attribute "concentration" of the entity "Solution" is used. The "use" state variable defines which solution is currently in use (only one occurrence of the "Solution" may have the value "in tank" at any one time).[1]

2.22.5.7.2 Outputs
Any stored data that is created or updated in response to the event. The notation used is described in §2.16.3.7.

2.22.5.8 Detection mechanism

The detection mechanism is what the system uses in order to know that the event has occurred. All events must be recognised. The system can detect that the event has occurred using one of the following techniques:

- **a fixed point in time**: an example of this would be "Time to produce daily report". This time is part of the policy of the system;

- **a scheduled time**: an example of this would be "Time to take picture", when this time had been previously input as part of a system input. This type of event can be detected by comparing stored data against the current time;

- **a delay**: after some other event (for example, "Two seconds after a certain event … ");

- **the arrival of a discrete flow:** (either data or event);

- **the change in value of a continuous data flow:** to detect this type of event, an event–detector data process is required;

- **the change in value of a continuous event flow**: from true to false (or false to true).

Events of the first three types are referred to as temporal events. For these, the detection mechanism entry contains the word "time". For the last three types of event (which can be described as non-temporal events), the detection mechanism entry contains the word "stimulus". For example:

---

1 Grouping state variables together with stored information is optional. They could be shown in a separate entry.

> **DETECTION MECHANISM :** stimulus

2.22.5.9 Stimulus

If the detection mechanism for the event is "stimulus", then the name of the flow used is entered. For example:

> **STIMULUS :** solution strength

A flow used in this way is referred to as the "stimulus for the event" — note that it does not cause the event, but allows the system to recognise that the event has occurred.

If there is a stimulus flow entered, then it acts as an input.

2.22.5.9.1 Stimulus flows and event detection

Sometimes, for non-temporal events, the arrival of an input may only suggest the event *might* have occurred. An event–detection process may have to carry out further checks, using this input and other available items to decide whether the event has actually occurred. The input flow that causes the system to carry out the check is still regarded as the stimulus in these cases and it is listed under the "stimulus" entry. See §2.22.5.9 for a further discussion of 'event detection'.

2.22.5.9.2 Flows that are stimuli for several events

It is possible for a flow to be the stimulus for more than one event. For example, a continuous data flow "temperature" might be associated with the events "temperature too high", "temperature above critical level", ...

2.22.5.10 Event–detection process

An event–detection process is a process which determines when an event occurs. It may do this in two ways:

- by comparing the value of a system input flow input with given pre-determined values;

- by comparing the current time against a fixed (hard-coded) or previously scheduled times.

If there is a data process that detects *and* responds to the event, the name of the process is entered under data response and also under event detection. If the event is detected by a control process, then there is no entry for event–detection.

2.22.5.10.1 Processes that detect more than one event

Sometimes a process may detect several events. For example, a temperature may be monitored to detect the event "temperature too high", "temperature above critical level", ... In these cases, the name of the process is entered as an event–detection process for each event.

This is optional and creating separate event–detection processes for each event may simplify the model.

2.22.5.10.2 Temporal event detection

To detect temporal events, the system must use one of the following techniques:

1. Compare (in data process) the current time to a constant . For example:

   ```
   time of day( ) = midnight
   ```

2. Compare (in data process) the current time to a previously scheduled time, held as stored data. For example:

   ```
   <Picture sequence>.start time = current time ( )
   ```

3. Wait (in a control process) a specified time after a given event. For example, the event "time to let pedestrians cross" (see §2.24.3.4.6) is detected by:

   ```
   secs(state) = 3
   ```

2.22.5.10.3 Non-temporal event detection

Non-temporal events can be detected by several means. Often, the presence of a discrete flow arriving from the environment will ensure detection, as the time that flow comes in is exactly the time at which the event occurs. This is the stimulus (see above). Sometimes additional checks must be made in order to be certain that the event has occurred.

2.22.5.10.4 Example discussion

In the example given in §2.22.4, "oxidiser use request" is a discrete data flow, which is generated when the event "Car enters mix bank" occurs. Its presence is detected by the event–detection process "Post oxidation request", which signals to the event store "oxidising request". A control process, monitoring this event store, can now be 'aware' that the event has occurred. Because "Post oxidation request" is also part of the response to "Car enters mix bank", it does not appear in the "event–detection" entry.

Sometimes, when "oxidiser use request" occurs, the solution required (which is contained as part of "oxidiser use request") does not match that currently in use. This means that "Check solution is correct" must be part of the response to this event. "Check solution is correct" generates one of two discrete event flows — "solution is correct" or "solution is incorrect". If solution is incorrect, then the solution must be changed, so "Change solution" is also part of the response to this event.

The event "Solution drops below minimum strength" is detected by the event–detection process "Monitor solution strength". This process continuously compares the value of the continuous data flow "solution strength" with a stored attribute and raises or lowers the continuous event flow "solution correct strength".

2.22.5.11 Event–response process

An event–response process performs a required function of the system directly because the event has occurred. There may be more than one process involved in this response. Initially, there is a single (or no) event–response process. However, during system modelling, it may be decomposed. The single event–response process name is entered. The components of this process are shown on a child DFD.

In the example (see §2.22.4), event "2: Solution drops below minimum strength", requires "Inform operator of solution change" as part of its response.

2.22.5.12 Use in control processes

This entry defines any control processes that 'handle' the event. For each such control process, the condition is given (as it would appear in the behavioural STD).

The condition is specified, together with the control process in which it occurs. In the example at the beginning of this section, a "/" was used to separate the condition from the name of the control process in which it occurs. Any other suitable notation could be adopted (or there could be a separate entry, depending on the support environment).

This entry is only filled in if the event causes a change in the behaviour of the system. One or more control processes must contribute to this state behaviour. The name of each control process that detects the event or changes state (or both) is entered.

2.22.5.12.1 Types of condition

Discrete event flow inputs

If the event is 'signalled' by the occurrence of a discrete event flow, the name of that event flow should be entered for "condition". For example, event 1.2 (see §2.22.4) the discrete event flow "car in bath" is a system input. This will be directly detected by the control process.

Continuous event flow inputs

If the event is detected by examining a continuous event flow, the condition should be a "rising" or "falling" function using the flow. An event occurs at a moment in time; if a continuous event flow is used, the point in time at which the value changes is the point in time of event occurrence.

In the example (event 2) "falling(solution correct strength)" corresponds to the moment the solution falls below the strength threshold.

Timed delays

If the event is defined as a set time since a preceding event, and there will be no other events intervening for that control process, the condition will be the state clock. For example, in the example (event 1.3) the state clock is used to determine when the car has been in the bath long enough.

If the event is defined as a set time since the occurrence of a previous event dealt with by that control process, but there may be intervening events causing other states of behaviour, the condition will be a named clock.

For example, the named clock "in soup" is used to detect when the car has been in the bath too long (event 1.4) as an elapsed time from when the car entered the bath. (The time it entered the bath is event 1.2.) This might occur if the machinery to move the car out failed. It is a 'time out' type of event. Because there is an intervening event (1.3: Car has bathed long enough), a named clock must be used to keep time across states. The state clock only keeps time within one state and is reset on change of state. Named clocks must be used to time across states (see §2.24.3.4 for further discussion of clocks.)

Detected by data process

If the event is detected by a data process using a data flow, an event flow or event store will be used to 'tell the control process that the event has occurred. The "condition entry" is the event flow or event store name, usually a shortened name for the event.

For example, event 1.1 (see §2.22.4) is detected by "Post oxidising request", using "oxidiser use request". "Post oxidising request" signals to the event store "oxidising request", which is used by "Control car oxidation". (See §2.14.2 for the corresponding DFD.

2.22.5.13 Response time

This gives parameters describing how 'quickly' the system must respond to the event. In this release of YSM, parameters are used to describe this timing in a 'global' sense. All responses are completed in this time. For example, a median response time is given in which all responses will be completed for 50% of the occurrences of the event.

These responses include all changes of state of state machines and completion of any discrete data process that is a response to that event (including triggered processes).

Later releases of YSM will allow separate parameterisation of each of the responses. Some data flows (errors, alarms, etc.) may be more critical than others (logging, reports, etc.), even though they are all produced by a response to the event. This level of granularity in performance specification is not covered by this release of YSM.

2.22.5.13.1 Choice of parameters

YSM does not demand that particular parameters have to be used to characterise the required times for responses of the system. The two shown in this manual (median and maximum time) are ones that are easy to understand. More sophisticated analyses of response characteristics may require other types of parameter. YSM allows these to be added to the specification.

2.22.5.13.2 Median response time

The median time is the time that is an estimate of the 'most probable' response time. More precisely, it is defined as the time which there is a 50% chance of all processing being complete. There is an equal probability that it will take more or less time.

Many people will refer to this time as the 'average time', but strictly speaking, they are not the same and relationship between the two depends on the probability distribution. (See [Wea68] for a discussion of this.)

2.22.5.13.3 Maximum response time

The "maximum" entry gives the practical maximum time that should occur between the occurrence of the event and the completion of the system's response. Over a large number of trials, the system should respond in less than this time at least 95% of the time.

2.22.6 Rules

1. Each event must occur outside the system's boundary.
2. The system must need to respond to an event in order to achieve its purpose.
3. Each event must have a unique meaning — if the meaning entry is the same for two events, then they are the same event.
4. The detection mechanism for each event must be stated.
5. If the detection mechanism is "stimulus", then there must be an entry for the stimulus flow.
6. Any flow used as a stimulus must be an input.

7. Every event must have at least one entry in either event–response process or a condition/control process. (otherwise the event would have no effect on the system.)

2.22.7 Guidelines

2.22.7.1 Suggested wording for event description

To help ensure that events are correctly identified as occurring in the environment and not within the system, it is helpful to try to find a standard format description for the event. The following are suggestions which often (but not always) are helpful :

1. Temporal events: the name of the event should be "Time to ...", where "..." is replaced by a description of what the system does at that time. A list of function should not be given, but rather one overall summary name. For example:

   > Time to produce daily reports

   Event names of this type are a little 'self-fulfilling' in the sense that it is obvious what the system does when the event occurs. However, they are often very easy for the end-user to relate to. If the system does not produce outputs, the above format does not usually work very well (because there is no visible output).

2. Non–temporal events: the event should be described using a sentence in which the terminator (or some other external entity associated with the event) is the subject, followed by an active verb. For example, in the event list given in §2.22.3, all events have either "Car" or "Solution" as the subject.

2.22.7.2 Forming event groups

1. An event group should be at a more general level of wording or higher level of abstraction than the events it groups.
   For example, the event group "Car moves in oxidation area" uses the more general verb "move" and object, "oxidation area", which are then broken into more detail by the events "Car enters mix bank, car enters bath, etc.

2. Event groups may be defined as containing other event groups — in other words, the levelling of the events may be repeated (at least in principle) indefinitely.

3. A good grouping for events deals with several events that affect one entity (and/or controlled device). If the group of events thus formed is large, then it may be broken down into smaller groups. The following are not good reasons for grouping events and should be avoided:

   - grouping all events together that occur at the same time or in sequence, but deal with different entities (temporal cohesion);

   - grouping events together because they cause similar types of processing to be carried out (logical cohesion).

4. A subject-matter expert should be able to 'guess' most of the events in an event group from the name of the event group and the purpose of the system.

2.23 Enterprise event specification

2.23.1 Purpose

Within the enterprise, there may be many events. Some correspond to events outside the enterprise that are treated as events by one or more systems. Other events are part of the processing carried out by one system. They are not events as far as that system is concerned. However, they may be events for other systems. Each of these is described using an enterprise event specification.

The enterprise event list shows a consolidated list of all such events.

The enterprise event specification gives parameters relating to the event, rather than how it is used by a system. It describes how often the event occurs and what it means. It also lists the systems that deal with this event.

System event specifications are used to define the way a system deals with the event — see §2.22.

2.23.2 Example

```
EVENT NUMBER :  6.3.2

MEANING :  Solution drops below minimum strength

SYSTEMS USING :  automatic car painting system,
                 plant safety monitoring system

EVENT FREQUENCY :   average      : 1/hour
                    maximum      : 5/hour
```

2.23.3 Components

2.23.3.1 Event number

This is a unique, identifying number for the event within the enterprise. This is not the same as the number used within a specific system (see §2.22.5.2).

Events may be grouped, as for system event groupings. The enterprise grouping of events is for the convenience of understanding the events. It does not relate to any grouping used by a system.

2.23.3.2 Meaning

This gives a description of what the event means. In a sense, it is the 'event name'.

2.23.3.3 Systems using

This gives a list of all systems that use this event. There are two shown in the example. Note that each of these systems has an Essential Model including a system event specification for this event.

2.23.3.4 Event frequency

This gives the average rate at which the event is expected to occur, together with the (worst case) maximum. These are used to determine the required system processing load.

2.23.3.4.1 Average event rate

If the given time interval is chosen at random many times (say 100 times) and the number of times the event occurs is counted, then the average event rate is determined by dividing the number of events by the number of intervals tested.

2.23.3.4.2 Maximum event rate

If a large number of time intervals are chosen at random, then for 95% of these intervals, there would be fewer events occurring than this maximum value. This value is, in effect, a practical maximum — in general, it is impossible to guarantee that the event frequency will never be greater than a given value (however large) for a given time period.

2.23.3.4.3 Time period for average and maximum event frequency

The time units should be chosen to be the time over which 'the events can be averaged out'. This will be of the order of, but slightly greater than the minimum of all the 'maximum response time' for all systems dealing with this event.

For example, if only one system deals with the event and the maximum response time is 5 seconds, then the time period for the event frequency might be 1 minute. Choosing a time period of 1 second would make no sense (as it is unlikely that there would be more than 1 event in this time). If another system had response times for the same event that were of the order of an hour, then the event frequency time unit chosen for the first system is still satisfactory. The average (and maximum) number of events per hour can be obtained by multiplying the number of events per minute by 60.

If the time unit had been chosen as one hour, however, it is not true to say that the rates for one minute can be obtained by dividing by 60. The calculation is valid for the average, but not the maximum. Periods of a few minutes in which the event frequency might be significantly higher than the long term average would be masked.

2.24 Behavioural State Transition Diagram

2.24.1 Purpose
The behavioural state transition diagram (bSTD) highlights:

- the modes of behaviour of a system or portion of a system,
- what causes the system to change the modes of behaviour,
- the actions that must be carried out to cause this change.

The bSTD is used to define the effects of events and conditions on the behaviour of the system.

2.24.1.1 State machines
The bSTD is used as a view of a state machine. State machines of various types have been used in many areas of software engineering. Within YSM a specific type of state machine 'driven' by conditions is used.

2.24.1.1.1 Inputs
There are two types of discrete input that can be used directly as an input to the state machine:

1. discrete event flows: which act directly as conditions (these event flows appear on DFDs showing the state machine as a control process);
2. clock events: specific delays may be detected using clocks. These clocks are not shown — they are 'hidden' within (or owned by) the state machine.

The behaviour of the state machine may be modified by continuous event flows or clock periods. Both of these correspond to a time-continuous status which is TRUE or FALSE.

2.24.1.1.2 Outputs
The outputs from the state machine are discrete (discrete event flows, triggers and event store initialisations) or continuous (continuous event flows or enable/disables). These outputs are generated (in the case of discrete outputs) or changed (in the case of continuous outputs) as actions carried out at certain times. At other times, discrete outputs do not occur and continuous outputs do not change.

2.24.1.1.3 Machine type.
YSM uses a modified form of the Mealy state machine[Mea55]. Mealy machines associate actions with transitions. Mealy machines of various types have been applied to many types of problem. (Shields gives a good, but theoretical treatment of Mealy machines [Shi87].)

2.24.1.2 Alternative presentations
As an alternative to the bSTD, a tabular presentation may be used (see 2.24.4).

2.24.1.3 Example explanation
This example is also used for describing data flow diagrams (see §2.14 for background).

2.24.2 Example

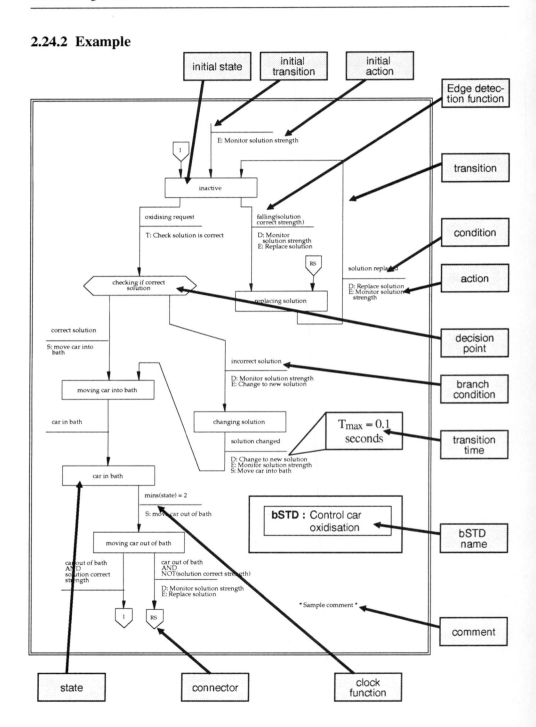

2.24.3 Components

2.24.3.1 Action

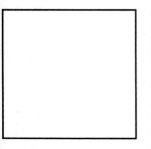

Actions generate or change the outputs of a state machine. These outputs may be continuous (continuous event flows, enables, disables) or discrete (discrete event flows, triggers and event store initialisations). Additionally, actions may be used to reset clocks, which are 'internal' to the state machine (these act as a discrete output).

Actions are associated with transitions, and take place the moment a transition occurs. Continuous outputs are changed at this point in time; discrete outputs are generated at this time. At other times continuous outputs are not changed and discrete outputs are not generated.

Actions are of eight types. Each is expressed as a standard operation, followed by the model component that is the object on which the operation is applied. The permitted operations are:

- **Begin:** start a named clock at zero time,
- **Disable:** stop a continuous process or prevent a discrete process from running when its stimulus occurs,
- **Enable:** start a continuous process or allow a discrete process to run when its stimulus occurs
- **Initialise:** create and initialise (or re-initialise) an event store. An event store is considered empty upon initialisation,
- **Lower:** set a continuous event flow to FALSE,
- **Raise:** set a continuous event flow to TRUE,
- **Signal:** generate a discrete event flow,
- **Trigger:** start a discrete process which runs to completion.

The exact form of the action depends on the style (or dialect adopted). Within this manual, the convention adopted is that the operation is abbreviated to a single letter, followed by a ":", followed by the model component that it acts on. For example:

> D: Change to new solution

The abbreviations are thus "B:", "D:", "E:", "I:", "L:", "R:", "S:", "T:".

Any other format is allowed, although a house style is advised.

2.24.3.1.1 Status of outputs
The outputs of the state machine obey the following rules.

Triggered processes

These are triggered at the point in time when any transition that has an associated action including a 'trigger' of that process occurs.

Enabled/disabled processes

If the state machine enables and disables a process, then that process is not active when the state machine is enabled. If the initial action is to enable the process, then the process immediately becomes active. Otherwise, it becomes active at the point in time when any transition that has an associated action including an 'enable' for that process is carried out.

Once the process is enabled, the process may be alternately disabled and enabled.

When the state machine is disabled, all processes enabled/disabled by it are automatically disabled.

Note that a state machine may be enabled at the same time as other processes that it does not enable and disable (see §2.14.3.17.3).

Discrete event flows

These are generated at the point in time when a transition generating them using a signal action is made. This discrete event flow does not exist at any other time.

Continuous event flows

These are not available if the state machine is not enabled. Any process using them should either:

- explicitly check for the existence of the event flow using the standard operation "available",

- be checked to prove that it cannot be active when the state machine is not enabled.

Failure to check this will lead to systems that are unreliable and behave in an unpredictable way.

On enabling the state machine, all continuous event flows output by the state machine have the value FALSE. The continuous event flow retains this value until a "raise" operation referencing this flow occurs. The value then becomes TRUE and remains TRUE until explicitly lowered.

When the state machine is disabled, all continuous event flow outputs cease to exist. They do not have a value — they are neither TRUE nor FALSE.

2.24.3.1.2 Multiple actions

More than one action may be associated with a transition. Actions are placed in a list, one per line. There are three ways in which these can be performed: "in order", "in any order" and "in parallel". In fact, for an essential state machine (part of an Essential Model), the actions take zero time, so "in parallel" and "in any order" are equivalent. (In implementation state machines all three will be allowed. This will be discussed in later releases of YSM.)

Line delimiters

Special delimiters may be used to show the end of one action and the beginnings of the next. The bSTD dialect uses ";" for this. If each action starts with a standard sequence of characters, delimiters may be omitted if this does not lead to ambiguity. (In the convention used in this manual, where each action commences with one of the following strings: "B:", "D:", "E:", "I:", "L:", "R:", "S:", "T:", providing no model component name includes a ":", any sequence of actions can be parsed unambiguously.)

The assumed default sequence that the actions are performed in sequence, unless the contrary is stated. For example, in TIC-TAC-TOE (see §5.7.6.2), to indicate that a machine move must be made before the game is checked for completion, the transition actions would read:

```
In order:
        T: Determine machine move
        T: Evaluate outcome
```

but, using the default convention, the following is equivalent:

```
T: Determine machine move
T: Evaluate outcome
```

If the actions can proceed in parallel, then there are two cases.

1. The 'parallelism' continues to the end of the action list. For example:

```
T: Accept deposit ;
In parallel:
        S: thank you
        T: Update account
        T: Print receipt
```

The action delimiter is required before "In parallel", but is omitted for the other actions. "Accept deposit" occurs before the other three actions, which could occur in any order (or at the same time in a perfect machine).

2. The parallelism does not extend to the end of the action list. In this case, there must be a way of denoting the end of the list. The recommended way of dealing with this is to indicate the end of the structure with a standard delimiter. The bSTD dialect (see §6.7) uses "end" (preceded by the action delimiter):

```
T: Accept deposit ;
In parallel:
        S: thank you
        T: Update account
        T: Print receipt ;
end
        S: transaction complete
```

2.24.3.2 Branch condition

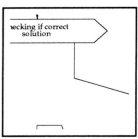

Each decision point must have two or more transitions from it to new states. Branch conditions are used to determine which of these transitions should be made. Each branch decision is a condition involving discrete event flows generated by a triggered data process or signals from other processes. Any signals from other processes must be a direct response (dialogue event flow) to signals sent out on entry to the decision point. Event flows from triggered processes and response event flows may be combined with continuous event flows and clock intervals.

For example, the branch condition:

> correct solution

is one of two possible discrete event flows generated by the data process "Check solution is correct". "Check solution is correct" is triggered whenever there is an oxidising request to determine whether the solution needs to be changed.

For initial decision points (see §2.24.3.11.1), the situation is slightly different. The branch condition does not have to include any discrete flows (although it can) — branch conditions that only involve continuous event flows are allowed.

2.24.3.3 bSTD name

The name of a bSTD is the same as the control process it specifies. This name identifies the bSTD.

> **bSTD:** Control car oxidisation

2.24.3.4 Clock function

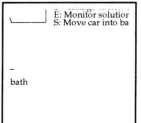

Clock functions can be used to determine the time on a clock. The functions provided are "secs", "mins", "msecs", "microsecs", "hours" and "days". Each of these returns a cardinal, giving the number of that time unit since the clock was started. Although the clock functions return a cardinal value, this may only be used to form a boolean expression.

There are two types of clock that may be used in a state machine. One special clock, the state clock, is always available; one or more named clocks may be declared and used as required.

The clock does not appear in any view, other than this bSTD. The clock is 'owned' by the state machine. It is not visible from outside the state machine and cannot be accessed, except by the state machine.

2.24.3.4.1 State clock

Each state machine has a state clock. It gives the time since the current state was entered. It starts from zero when the state machine is enabled and is restarted from zero every time there is a transition to a new state.

The state clock is distinguished by its name "state". Although it is used very much like named clocks (see below), it cannot be reset by an action. It is automatically reset on change of state.

2.24.3.4.2 Named clock

A named clock is more general than the state clock. Each state machine can have several named clocks (or none). Named clocks allow times to be recorded across states. A named clock acts as a stop watch; from the moment it is started by the action "BEGIN", it will keep track of time until it is restarted.

On enabling the state machine, named clocks are not set until an explicit "BEGIN" occurs. The clock has no value until it has been set (see §2.24.3.4.5). If the clock "T1" has not been set, neither:

```
secs(T1) >= 5
```

nor

```
secs(T1) < 5
```

(for example) would be TRUE and therefore no transition would be made.

2.24.3.4.3 Clock events

At the instant in time when a clock shows a certain time, this may be used to generate a condition. For example:

```
secs(state) = 3
```

gives a condition which is TRUE at exactly the point in time when the state clock has the value 3 seconds. This occurs exactly three seconds after the state is entered.

2.24.3.4.4 Clock intervals

A clock may be used to generate a condition that is TRUE over a period of time (a status). For example:

```
secs(state) > 5
```

is TRUE for all times after the first initial five seconds in a state. This clock interval may be treated like any other status and used to build up complex conditions. For example:

```
(secs(state) > 4 AND request) OR falling(reaction_OK)
```

would cause a transition if the continuous event flow became FALSE or a request occurred after four seconds in the state. If a request occurred any earlier, it would be ignored.

2.24.3.4.5 Initialising clocks

Named clocks may be reset to zero as an action. For example:

> B: pedestrian wait

The state clock is automatically set to zero on entering a state. A transition that does not cause a change of state does not affect it — only transitions to different states reset it.

2.24.3.4.6 Pedestrian crossing example
As an example of using a named clock, consider the example shown below:

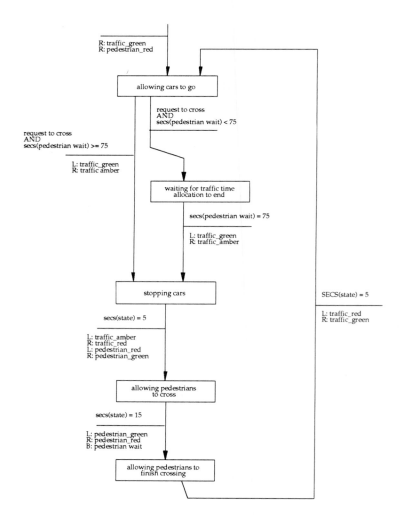

This allows cars to proceed until there is a request from a pedestrian to use the crossing. If there are no pedestrians, the cars are allowed to proceed for an indefinite time. In order to prevent heavy demand from pedestrians to cause the system to keep changing the lights, a pedestrian may have to wait if the lights have only just changed. However, no pedestrian should have to wait more than 75 seconds.

If the pedestrian requests the crossing at time that is greater than 75 seconds after the last time pedestrians had priority, the request is *immediately* responded to (after allowing cars to stop). If the request occurs any earlier, the state machine waits until the time interval has elapsed and *then* stops the cars.

(Note that the initial transition logic shown is not a 'safe' design for traffic. A better version would have a state in which neither pedestrians nor traffic was allowed to proceed as the initial state.)

2.24.3.5 Comment

A comment may be used to highlight any helpful data, such as aliases, ties to preceding or anticipated implementations, etc. The comment is not specified with additional textual specification.

2.24.3.6 Condition

A condition is the recognition by a state machine that a particular event or set of events have occurred. All transitions (other than the initial transition) have an associated condition. When the state machine is in the state that the transition is from and the condition becomes TRUE, the state machine changes state.

A condition must be only TRUE for a moment of time, in other words, it is time–discrete. It is discrete because of the presence of a discrete event flow, the change in value of a continuous event flow, or the detection of a specific point of time.

State machines change state when conditions become TRUE. They are, in a sense, 'driven' by these conditions.

2.24.3.6.1 *Statuses as modifiers of behaviour*

State machine behaviour may be modified by time–continuous situations, referred to as statuses. A status is a situation which is TRUE for a length of time. It is a range of time, or the presence of a continuous event flow. Statuses cannot be used directly as a conditions, but may be used as part of a condition. Allowed statuses are defined in §2.24.3.6.3.

2.24.3.6.2 Permitted conditions

Conditions act as an 'event recogniser' of a more restricted nature than event–detection processes (see §2.22.5.9). They may be built up of certain limited low-level component, using boolean expressions (boolean expressions are treated in §6.4.4). When the conditions become more complex, parentheses are used to clarify and avoid any possible ambiguity. There are six types of conditions:

1. Discrete event flow

   When the discrete flow occurs, the condition becomes TRUE and the transition is made. For example:

   ```
   car in bath
   ```

   would wait for the occurrence of the discrete event flow "car in bath" that is an input to the state machine.

   Note that a discrete event flow input cannot be 'negated', because the absence of a discrete event flow is not an event. Therefore, if "car in bath" is a discrete event flow, the expression:

   ```
   NOT(car in bath)
   ```

   is illegal.

2. Change in value of a status

   At the point in time a continuous event flow changes value, a condition can be generated by means of the edge detection functions "rising" or "falling". For example:

   ```
   falling(solution correct strength)
   ```

   generates a condition at the point in time when the continuous event flow became FALSE. More generally, it may be applied to any status. For example:

   ```
   falling(auto_available AND auto_requested)
   ```

   generates a condition at the point in time when the continuous event flows "auto_available" and "auto_requested" changed from both being TRUE to one or either of them being FALSE.

   The use of "falling" (or "rising") applied to a continuous event flow only generates a condition at times when the event flow is available. If that event flow is generated by a process (or terminator) that is not active, it has no value. It could not therefore change from TRUE to FALSE. The presence (or otherwise) of an event flow can be checked using the standard operation "available". For example:

   ```
   falling(available(motor_on))
   ```

   generates a condition at the moment of time that the continuous event flow "motor_on" ceased to become available. (The operation "available" returns a continuous boolean value — in effect, an anonymous continuous event flow. It cannot be used directly as a condition, but must be combined with an edge detection function, as here.)

3. Event store wait

This waits for the moment at which there is an event 'available' in the event store. As soon as it is available, the condition becomes TRUE and the transition takes place. The event is 'consumed'. For example:

> oxidisation requested

would cause a transition as soon as an "oxidisation request" is signalled to the event store. If there is an event in the store when the state is entered, no wait ensues — the transition is made immediately. (This is sometimes referred to as 'drop through'.)

Note that no explicit "wait" action is needed on entry to a state with such a transition out of it. 'Waiting' on the event store is implicit in that the name of the event store is shown as a condition of a transition out of the state.

4. Clock events

These are generated by comparing a clock value with a constant value (attributes cannot be used). Either the state clock or a named clock may be used. For example:

> mins(in bath) = 3

would generate a condition three minutes after the named clock "in bath" is started.

5. Alternative conditions

Two conditions may be given as alternatives. For example:

> (mins(state) = 2) OR mins(in soup) = 3)

generates a condition as soon as either the state machine has been in the current state for two minutes, or the clock "in soup" has the value three minutes.

6. Condition AND status

As well as being used with edge detection functions, continuous event flows may be used to modify the behaviour of the state machine. Transitions may be defined to be allowed only if a continuous event flow is TRUE (or FALSE). For example:

> (car out of bath) AND (solution correct strength)

could be used with the discrete event flow "car out of bath" and the continuous event flow "solution correct strength". In a sense, the transition is only permitted if the solution is currently the correct strength. The event "car out of bath" 'drives' the transitions; the continuous event flow 'permits' it (or not).

More generally, a status may be used instead of the continuous event flow.

2.24.3.6.3 Statuses

A status is a situation that persists over a period of time. There are six types of status:

1. Continuous event flow

The name of a continuous event flow may be used as a status. The status is TRUE at the times when the event flow is TRUE; it is FALSE when the event flow is FALSE or undefined (see 2.24.3.1.1). For example:

> solution correct strength

is a continuous event flow used by the state machine. It is used as an input to the edge function "falling" and also part of a compound condition.

2. Clock interval

   A clock interval may be used as a test. This gives the value TRUE for one period of time and FALSE at other times. For example:

   > mins(state) > 2

   is FALSE for the first two minutes in a state and TRUE thereafter. The value used must be a constant and the allowed operators are "<", "<=", ">=", and ">".

3. Status OR status

   At any time when either (or both) of the component statuses is TRUE, the compound condition is TRUE. For example:

   > (mins(state) > 2) OR solution correct strength

4. Status AND status

   At any time when both of the component statuses is TRUE, the compound condition is TRUE. For example:

   > (mins(state) > 2) AND solution correct strength

5. NOT(status)

   This 'inverts' the value of a status. For example:

   > NOT(solution correct strength)

   is TRUE at any time that "solution correct strength" has the value FALSE.

6. Available(continuous event flow)

   This can be used to check whether a given continuous event flow is available (see §2.24.3.1.1 for a discussion of why it might not be). For example:

   > available(plant running)

   could be used to check whether a continuous event flow was being generated or not. For safe control systems, checks of this type are obligatory. It cannot be assumed that continuous event flows are always present, with one of two values. The absence of a flow is not the same as saying the flow has the value FALSE. In a sense, as far as a receiver of a continuous event flow is concerned, all continuous event flows have three possible values: TRUE, FALSE and DON'T KNOW.

tion replaced
_____

eplace solution
lonitor solution
ength

2.24.3.7 Connector

Connectors are used to reduce graphic complexity of the state transition diagram. They allow a transition to be broken (graphically) in between the states that it connects. (The states must both be in the same state machine, i.e. the same bSTD.)

To describe a connector, the terms 'current state' and 'new state' will be used to describe the state before and after a transition takes place.

The current state is shown with a transition to the connector. The condition that causes the transition and associated actions are shown on this part of the transition. This is because there may be different conditions and actions associated with two transitions, even though they be to the same state.

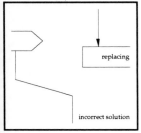

Another copy of the connector (with the same name) is shown connected to the new state. There is no associated transition or action.

The result is equivalent to showing a transition drawn between two states.

2.24.3.8 Decision point

A decision point is a point in time at which the state machine does not know what the correct behaviour should be. To determine the required state, the state machine triggers a discrete data process. The event flows from that data process are conditions determining the next state. Decision points are not states; they take no time.

There are two types of transition associated with the decision point:

- transitions to the decision point. There are made when a condition in the current state becomes true. They have a condition that satisfies the rules from transitions between two states. The actions may only be to trigger data processes or initiate actions that provide a result that is used in the decision point. These actions may be to trigger discrete processes or to send signals to other processes. In certain circumstances, more than one process may be triggered 'in parallel' if complex checks need to be carried out). The actions cannot include enables/disables or raise/lower type actions.

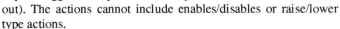
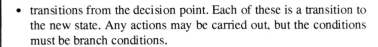

- transitions from the decision point. Each of these is a transition to the new state. Any actions may be carried out, but the conditions must be branch conditions.

These rules are slightly modified for initial decision points (see §2.24.3.11.1).

2.24.3.9 Edge detection functions

These functions allow the perception of the moment a continuous event flow changes. The "rising" function detects the moment at which a continuous event flow changes from FALSE to TRUE. The "falling" function detects the moment at which a continuous event flow changes from TRUE to FALSE. The detection of this moment is a condition.

For example:

falling(solution correct strength)

detects the precise moment when the continuous event flow "solution correct strength" changes from TRUE to FALSE, or, in other words, the exact time when the solution fails to be sufficiently strong.

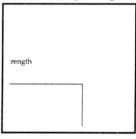

rength

2.24.3.10 Initial action

At the time the initial transition occurs, it is assumed that any process controlled by the state transition diagram is inactive, and that any continuous event flow raised or lowered by the state transition diagram is FALSE. Explicit actions on the initial transition need to occur if these defaults are not the case. In other words, to ensure that "Monitor solution strength" is active even when no cars are present to go into the bath, it must be activated on the initial transition.

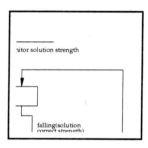

itor solution strength

falling(solution correct strength)

2.24.3.11 Initial state

Usually a control process starts up in a specific mode of behaviour and then there is an initial state. This state corresponds to behaviour defined by the initial transitions.

2.24.3.11.1 Initial decision point

In some situations, the initial action carried out by the state machine is to check its environment to determine which state it should be in . In these cases, an initial decision point is used. The data processes are triggered as initial actions.

It is also allowed to have an initial decision point, with no initial actions. The first action of the bSTD is to determine the required start-up state, using continuous event flow inputs.

For example, the bSTD below shows part of the logic to control an oil pump and fan. These are auxiliary plant that must be switched on before main plant is started. The responsibility of the bSTD is to ensure that they are started-up correctly. When the control process is enabled, it cannot assume that the oil pump and fan are off — they might be on. The first decision point determines the state that matches the plant status (hence the term 'match state' that is sometimes applied to this technique).

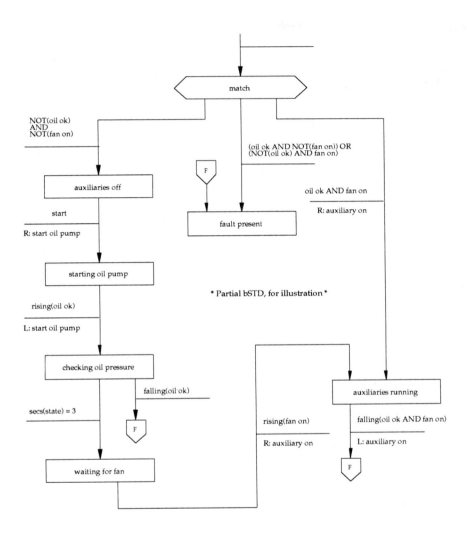

* Partial bSTD, for illustration *

2.24.3.12 Initial transition

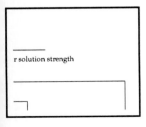

r solution strength

An initial transition identifies the 'start up' activities required to place a state machine in its initial state. These are shown as actions. It has no condition; the condition is assumed to be 'when activated'. Initial transitions are always shown, but they may have no actions.

There is only one initial transition. It indicates the state of the state machine when enabled.

2.24.3.13 State

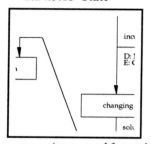

A state is a unique set of behaviours that persist over time. These behaviours result from the combination of the behaviour of the continuous processes and the status of continuous event flows which are under the control of the state machine. Note that for any state, it must be possible to determine unambiguously the value of any output continuous event flow and the status of any enabled/disabled process (see §2.24.5.7.1).

For two possible states to really be the same state, it is not enough that two states have the same behaviour. They must also have the same 'new state' for each of the possible conditions that may occur.

2.24.3.13.1 Final state

A final state is a state from which no transition can be made. When a state machine reaches a final state, its behaviour remains the same until it is disabled.

A state transition diagram may have more than one final state. A state transition diagram may have no final states. The state machine usually generates a discrete event flow to indicate that it has reached the final state. Often, there is another part of the system that needs to know that the final state has been reached.

2.24.3.14 Transition

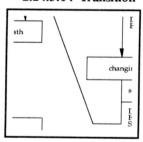

A transition is an instantaneous switch of behaviour. It takes place when the condition associated with the transition becomes TRUE.

A transition may have associated actions. If so, all these actions are carried out at the time the transition occurs (in perfect technology this takes zero time).

After the actions have been carried out, the state machine then takes on its new state of behaviour. The state after the transition (new state) may be the same or different from the state before the transition (current state).

2.24.3.15 Transition time

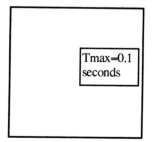

This gives the maximum allowed time that a transition is allowed to take. It allows the required performance of the state machine to be captured. There is a transition time for each possible transition of the bSTD.

This may be shown as a 'pop-up' item on a diagram, a diagram annotation, or a separate specification, depending on technical support.

2.24.4 Behavioural state transition and action tables

YSM allows an alternative presentation of the state machine, using tables. The successful use of these may depend on automated support, but the method allows them. The state machine may be described using a pair of tables, as shown below. Alternatively, the tables may be combined into one table, showing both transitions and actions.

2.24.4.1 Behavioural state transition table

state/ decision point	CONDITION								
	oxidising request	falling (solution correct strength)	mins (state) = 2	car out of bath AND solution correct strength	car out of bath AND NOT (solution correct strength)	car in bath	solution changed	correct solution	incorrect solution
inactive	checking if correct solution	replacing solution							
checking if correct solution (decision point)								moving car into bath	changing solution
moving car into bath						car in bath			
car in bath			moving car out of bath						
replacing solution							inactive		
changing solution							moving car into bath		
moving car out of bath				inactive	replacing solution				

2.24.4.2 Components
As for bSTD.

2.24.4.3 Behavioural state action table

state/ decision point	CONDITION								
	oxidising request	falling (solution correct strength)	mins (state) = 2	car out of bath AND solution correct strength	car out of bath AND NOT (solution correct strength)	car in bath	solution changed	correct solution	incorrect solution
inactive	T: Check solution is correct	D: Monitor solution strength E: Replace solution							
checking if correct solution (decision point)								S: move car into bath	D: Monitor solution strength E: Change to new solution
moving car into bath									
car in bath			S: move car out of bath						
replacing solution							D: Replace solution E: Monitor solution strength		
changing solution							D: change to new solution E: Monitor solution strength S: move car into bath		
moving car out of bath					D: Monitor solution strength replacing solution E: Replace solution				

2.24.5 Rules

2.24.5.1 Actions

1. There must be at least one action on the state transition diagram. (A state machine with no actions is a 'black hole' and would serve no useful function.)

2. An inactive processes cannot may be disabled. (An inactive process is one that has not previously been enabled.)

3. An active process cannot be enabled 'again' until it has been disabled. (An active process is one that has previously been enabled.)

4. It is illegal to "lower" a continuous event flow which has the value FALSE.

5. It is illegal to "raise" a continuous event flow which has the value TRUE.

2.24.5.2 Clocks

1. The state clock cannot be used in an action. It is automatically restarted on every change to a new state. It may not be named — there is only one state clock per state machine. This is equivalent to saying no named clock with the name "state" may be 'begun'.

2. A named clock can only be read in a condition if it has been previously started. This must be true of all possible 'routes' to this condition from the initial state. (A named clock may be started multiple times within a state transition diagram.)

3. If a clock is started (as an action) then it must be read (as part of a condition).

4. "B: state" is illegal as an action (this would reset the state clock). This rule is equivalent to the rule 'named clocks cannot be given the name "state"'.

2.24.5.3 Connectors

Definition: each connector links one or more current states to a given new state. Each pair of connectors (one defining current sate and one defining new state) is equivalent to a transition.

1. All connectors must be named. The name may be an abbreviation.

2. Each named connector must be linked to at least one current state.

3. Each named connector must be linked to exactly one state as a new state.

4. Any named connector must appear at least twice. Exactly one of these must be to a new state.

5. Connectors cannot be used on initial transitions.

2.24.5.4 Decision points

1. A decision point must be preceded by an action on the transition which triggers a discrete data process (there may be more than one process triggered) or generates a signal. Each triggered data process must produce at least one event flow that is used in one or more of the branch conditions associated with the decision point.

2. If an action prior to a decision point is to generate a signal, then it should be part of a dialogue event flow with another process.

3. At least two transitions must lead from a decision point.

4. When the processes triggered at the decision point are completed and the response to an event dialogue is received, exactly one branch condition of those leading from a decision point must be TRUE.

5. A transition may not be made both from and to the same decision point.

2.24.5.5 Edge detection functions

1. An edge detection function may only be applied to a continuous event flow.

2.24.5.6 Initial transitions

1. Processes controlled by the state machine are assumed to be inactive in the initial state unless they have been enabled on the initial transition.

2. There must be exactly one initial transition per state machine.

2.24.5.7 States

1. A state must be named.

2. The name of the state is unique to the state machine. It may appear only once on a state transition diagram.

3. A behavioural STD must have at least one state.

4. A state must have a transition leading to it.

5. Each state must be accessible by a sequence of transitions from the initial state. (This is a stronger requirement than the previous rule and demands that the STD is 'connected'.)

2.24.5.7.1 Behavioural consistency

A state must always represent the same behaviours. This is known as behavioural consistency. To ensure behavioural consistency:

1. Any continuous event flows under the control of the state machine must have the same value (TRUE or FALSE) no matter how the state is reached.

2. Any continuous process under the control of the state machine must be in the same activation status (enabled or disabled) no matter how the state was reached.

2.24.5.8 Transitions

1. A transition, except the initial transition, must always have a condition.

2. A transition may be made to and from the same state. If it does, however, actions must be associated with the transition.

2.24.5.8.1 Transition conditions

1. If a state transition diagram sets a continuous event flow as an action, it may not use that same continuous event flow in a condition.

Exclusivity of conditions

Only one transition condition associated with a state may be TRUE at any given point in time. For example, to define one condition as:

```
car out of bath
```

and a second condition (either on another transition or as an 'OR' combination with this one) as:

> car out of bath AND (solution correct strength)

is illegal. If "solution correct strength" is TRUE when "car out of bath" occurs, two separate transition conditions would be TRUE at the same time. The two conditions:

> car out of bath AND (solution correct strength)

> car out of bath AND NOT(solution correct strength)

is legal as there will be no point in time when both are TRUE.

2.24.5.8.2 *Correct use of continuous event flows in conditions*
The use of persistent event flows as conditions for making a transition is not allowed. A first-cut bSTD that has this defect should be replaced by one in which the transitions are spelt out more precisely. In the example below, x is a discrete event flow and y is a continuous event flow:

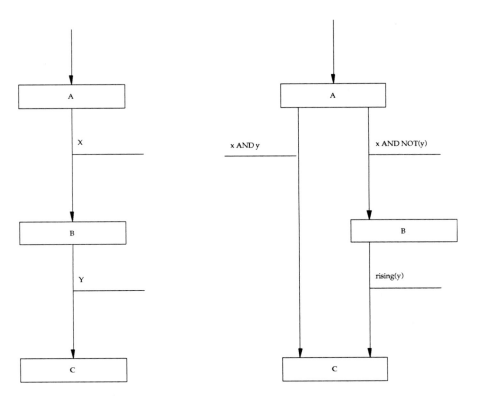

The left-hand diagram should be replaced by the one on the right. Transitions from states are always at a discrete point in time. [1]

2.24.5.9 Inputs and outputs

1. If an event flow is output by the state machine (as part of an action), it is illegal to use it as part of a condition.

2. If a process is disabled by the state machine, then it must also be enabled in at least one action.

3. If a continuous event flow is lowered by the state machine, then it must also be raised in at least one action.

---

[1] A convention could be adopted to allow conditions to be continuous, but YSM has decided to opt for the simpler version of the state machine in which conditions are always spelt out explicitly.

2.24.6 Guidelines

2.24.6.1 Comments

1. Comments should be used sparingly as they may clutter the diagram. Instead of placing lengthy comments on the diagram it should be added to the process specification of that diagram's parent process if it relates to the behaviour of the state machine. Comments may also be added in the 'meaning' entry of event flow specifications or processes that the control process controls.

2.24.6.2 Conditions

1. Too many "AND's" and "OR's" will lead to high complexity when considering conditions. There should be no more than three logical operators in any one condition.
   To avoid this complexity, either create event detectors which will inform the state machine of the occurrence of a collection of conditions via a discrete flow, or partition this state machine in smaller state machines controlling smaller areas of the system that talk to each other via discrete flows.

2. Parentheses should be used show the groupings of compound conditions even if they are not strictly required to ensure the correct evaluation sequence.

3. Event flow names beginning with "not" (e.g. "not true") should be avoided to prevent incomprehensible conditions such as "(NOT(not true))".

2.24.6.3 Named clocks

1. Named clocks should only be used when it is necessary to keep track of the time since a condition was detected across multiple states.

2. The name of a named clock should describe what is being timed. For example, a clock which times the length of time a door has been open might be called "door open". Reference to this clock would now be (for example) by: "mins(door open)".

2.24.6.4 States

1. The name of a state should describe the collective set of activities going on in the system while the system is in that state.

2.24.6.5 Partitioning

If a bSTD becomes very complex, it *can* be partitioned. This can be done to produce two state machines which work in parallel, communicating by means of event flows.

A bSTD which is fairly sequential (or a simple cycle) can always be broken into two state machines that activate each other in sequence. However, the benefit of doing this is debatable, because the initial bSTD would have been simple, even though there were a large number of states.

Since the complexity of the bSTD may be reflecting the real-world complexity of the controlled entity, decomposition of the state machine goes hand in hand with decomposition of that machine into smaller 'sub-machines'. If the controlled entity is separable into relatively independent components, it usually helps understanding and implementation if a parallel decomposition of the state machine is applied.

2.24.6.5.1 Partitioning for safe implementation

One important partitioning guideline is outside the scope of this release of YSM, but is very important for anyone designing control systems. That guideline is that it must be possible to *safely* implement and test the control system. This leads to a requirement that each state machine must be a unit of control that can be enabled and disabled in testing without compromising the safety of plant and/or personnel. It truly is an analysis requirement that this can be achieved and not just an design decision. It is a user requirement that the plant can be safely sequenced up to full running, even in the testing situation. The 'big bang' approach to testing control logic is not allowed (it may not even be legal in certain circumstances.)

2.24.6.6 Simplification

There are an infinite number of ways a state machine can be designed, even for quite simple problems. The best design is the one that is simplest and easiest to understand. The full theory of equivalence of state machines is outside this release of YSM, but the following discussion gives some idea of the issues involved. (For a more detailed treatment, see [Shi87].)

Two state machines are defined to be behaviourally equivalent if, when they are given the same inputs, they exhibit the same behaviour. This behaviour corresponds to the outputs, which are event flows and prompts.

A relatively trivial equivalence is to take a state machine and rename the states. This does not change the behaviour of the state machine. Indeed, the names of the states are entirely 'local' to the state machine. They cannot be 'seen' outside the state machine (although the state machine is more understandable if these names correspond to externally visible behaviour). Renaming the states does not therefore affect behaviour. The names of the states should be chosen to make the behaviour of the state machine easier to comprehend.

There are other situations in which two state machines have the same behaviour, but not at first sight, the same structure. The example below shows two versions of a state machine that controls a (very simple) vending machine. They have the same behaviour. In fact the one on the right may be reduced to the one on the left. The one on the left cannot be reduced to any simpler form.

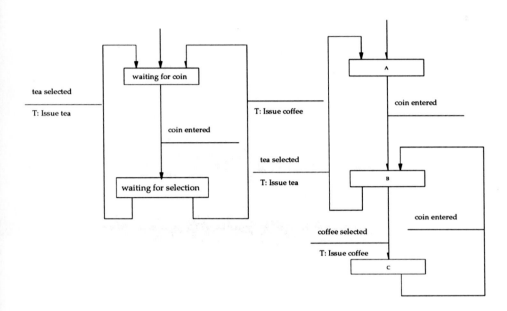

(To demonstrate the state machines are equivalent, pick any random sequence of the events that drive the state machine. 'Feed them into' the two machines and check that the same outputs occur. This is not a rigorous technique, but has an intuitive nature. In this example, there are only three events.)

Although no rigorous approach is given here, YSM recommends that the simplest state machine of the class of behaviourally equivalent state machines is chosen.

2.25 Event Flow Specification

2.25.1 Purpose of event flow specification

An event flow specification states the meaning of the flow and declares any other event flows that may be included in the flow.

2.25.2 Examples

EVENT FLOW : non-zero

MEANING : No trains left in section.

STRUCTURE : elemental

PERSISTENCE : discrete

EVENT FLOW : decrement count

EVENT FLOW : number left

MEANING : Dialogue used to keep track of trains in a section of track.

STRUCTURE : dialogue pair

EVENT FLOW : decrement count

MEANING : Decrease count of trains in section.

STRUCTURE : elemental

PERSISTENCE : discrete

EVENT FLOW : number left

MEANING : Result of decrementing count of trains.

STRUCTURE : multiple

CONTAINED FLOWS : non-zero, zero

2.25.3 Components of event flow specification

2.25.3.1 Name
A unique label for the event flow. For dialogue event flows and some multiple event flows, there are two entries.

2.25.3.2 Meaning
A description of the purpose of the event flow.

Note: for discrete event flows, this is quite often achieved by giving a reference to the specific event that this event flow indicates as having occurred.

2.25.3.3 Structure
The allowed values are[1]:

- **elemental**: a single, indivisible event flow;

- **dialogue pair**: a pair of event flows that play the roles of 'initiator' and 'responder'. Each has an event specification (note that a dialogue pair can consist of a discrete event flow and a continuous event flow, as well as the more obvious combinations);

- **multiple**: a grouping of event flows, chosen to simplify high-level diagrams.

2.25.3.4 Persistence
For elemental event flows, this entry defines whether the event flow is:

- **continuous**: a single continuous event flow, which has the values TRUE or FALSE at all times that it is available;

- **discrete**: a single discrete event flow, which is asserted from time to time;

2.25.3.5 Event flows included
For event flows declared as multiple, this gives a list of contained event flows. If these are the flows contained in a single event flow, then these are just given as a list. For a paired event flow, the list is tabulated.

Note: the occurrence of one of these is independent of another for discrete flows. Discrete event flows are always asynchronous.

As an example, consider the interface between a control system and a section of a coal fired power station that grinds coal up before burning it (a PF mill). Part of the controlled equipment relates to primary air fans and oil pumps (loosely called the 'auxiliary plant'). The interface to this can be shown as a multiple event flow, with the names "auxiliary control signals" and "auxiliary status". Each of the contained event flows has a specification, which are also shown. Note that this interface is a 'mix' of continuous and discrete event flows.

These specifications are shown below:

---

1 Note: event flows cannot be declared to be a group. This is because of their independent nature.

EVENT FLOW : auxiliary control signals

EVENT FLOW : auxiliary status

MEANING : Interface between the control system and the auxiliary plant for PF coal mill.

STRUCTURE : multiple

CONTAINED EVENT FLOWS : oil pump controls : oil pump status
primary air fan controls : primary air fan status

EVENT FLOW : oil pump controls

EVENT FLOW : oil pump status

MEANING : Interface between the control system and lubricating oild pump.

STRUCTURE : dialogue pair

EVENT FLOW : oil pump status

MEANING : When TRUE, the oil pump is running. When FALSE the oil pump is not running.

STRUCTURE : elemental

PERSISTENCE : continuous

EVENT FLOW : oil pump controls

MEANING : Used to control the lubricating oil pump.

STRUCTURE : multiple

CONTAINED FLOWS : turn oil pump on, turn oil pump off

EVENT FLOW : turn oil pump on

MEANING : Output to pump motor to start it.

STRUCTURE : elemental

PERSISTENCE : discrete

2.25.4 Rules

1. A discrete multiple event flow can only include other discrete event flows. Only one of these occurs on each presence of the flow.

2. A continuous multiple event flow can only include other continuous event flows. It is assumed all are present while this flow persists.

2.26 Event Store Specification

2.26.1 Purpose of event store specification

The event store specification describes the store and declares its maximum size.

2.26.1.1 Examples

> **EVENT STORE :** track available
>
> **MEANING :** The section of track is 'free' and may be claimed for use by trains in either direction.
>
> **MAXIMUM NUMBER OF EVENTS :** 1

> **EVENT STORE :** oxidising request
>
> **DESCRIPTION :** A signal from the mix bank that there are cars waiting to move into the bath.
>
> **MAXIMUM NUMBER OF EVENTS :** 4

2.26.2 Components of event store specification

2.26.2.1 Name

A unique label for the event store.

2.26.2.2 Meaning

This gives the significance of one event held in the store.

2.26.2.3 Maximum number of events

The limits of the resources or of the number of requests on the resources. This is a specified constant number. In one of the examples, the number of car bodies that can be in this one occurrence of mix bank at a time is specified as four.

2.26.3 Rules

2.26.4 Guidelines

1. An event store name should be singular.
2. An event store should be named for the resource in contention, followed by either "request" or "available".

3: The Enterprise Essential Model

Table of Contents

3.1 Introduction

3.1.1 Purpose

The Enterprise Essential Model (EEM) is a model of the information, events and functions used by an enterprise. It serves to integrate systems, so that the enterprise can truly be thought of as 'consisting of several systems'. The Enterprise Essential Model is the enterprise equivalent of the System Essential Model. Each System Essential Model is a subset of the Enterprise Essential Model.

The Enterprise Essential Model has no assumed implementation and is a conceptual, or essential model.

3.1.2 Scope of this release of YSM

The current release of the YSM is mainly concerned with 'shared information resources'. This is referred to as the 'the enterprise information aspect'. Future releases of YSM will cover other aspects of the Enterprise Essential Model, but this chapter (apart from this introduction) are specifically concerned with this information aspect.

This concept was first developed for information. It now includes other shared conceptual units. The full enterprise model is the Enterprise Essential Model; the enterprise information is one aspect of the Enterprise Essential Model.

3.1.3 Uses of the Enterprise Essential Model

The EEM allows system models to be built, with consistent access to enterprise information, functions and events. Specific uses of the EEM include:

- strategic planning: planning and initiation of system projects requires the EEM. It allows the potential impact on other systems to be visualised and the most cost-effective solutions identified.[1]

- use of the same information by more than one system in a consistent way;

- identification of events that effect several systems: this is important mainly in determining correct system boundaries.[2]

- modelling the interfaces between different systems within the enterprise boundary.

Regarding each computer-based application as an independent system is a strategy that will lead to long-term inefficiency and operating problems. Information and functions are likely to be duplicated with possible inconsistencies. Development resource will also be much greater than if an integrated design strategy is used. There should be a long-term plan for how all conceptual resources are used by the company.

---

1 This will be covered in later releases of YSM.

2 The strategic use of events will be covered in later releases of YSM.

3.1.4 Links to other models

3.1.4.1 Links to System Essential Models
Each system has an Essential Model. The structure of these is described in §5. The Enterprise Essential Model is the sum total of all of these; each System Essential Model is a subset of the Enterprise Essential Model.

This relationship between the Enterprise and System Essential Models is discussed in §4.

3.1.4.2 Links to the Enterprise Resource Library
The Enterprise Resource Library contains resources that *may* be used by the enterprise, as contrasted with the Enterprise Essential Model, which contains all informations and functions that *are* used by the enterprise.

Most of the components of the Enterprise Resource Library (as currently envisaged by YSM) are potential implementation unit types. Two specification types held in the Enterprise Resource Library are covered in this release of YSM. These are:

1. ADT specification: each abstract data type that is used (or may be used) by the enterprise has a corresponding abstract data type specification.

2. Enterprise operation specification: each anticipated operation that will be carried out on an abstract data type is specified by means of an enterprise operation. These are the only operations that can be correctly used on data items that are declared to be of a specific abstract data type.

3.1.4.2.1 Distinction between the Enterprise Essential Model and the Enterprise Resource Library
The Enterprise Resource Library contains 'useful' components, that may be used to construct systems. The Enterprise Essential Model is the definition of all information and function used by the enterprise.

The Enterprise Resource Library is described in §4.1.5.

3.1.5 Model assumptions
The Enterprise Essential Model concentrates on the meaning of the information, functions and events, rather than on any implementation techniques used to support these requirements. It may be assumed to be supported by perfect technology. This ideal storage and processing environment has the following characteristics:

- Unlimited storage capacity.

- Zero instruction time.

- Information is organised using entities, relationships attributes and state variables, with conceptual mechanisms that allows access to them (see §6.10). There are no periods of time over which stored information is not available.

- Information is never lost, corrupted or deleted, without a deliberate access from an activity to request its deletion.

- Derivations, relationship rules, participation constraints and other integrity rules are defined in the EEM. They are either automatically invoked when any access is carried out by a system model, or their corresponding logic is duplicated in each system model. This is an option in the way YSM is used. The first is preferable, but rather radical.

- Operations are carried out with infinite precision, in no time.

3.1.6 Strategic nature of Enterprise Essential Model

Building an Enterprise Essential Model requires a major commitment of resources running over a longer time scale than any single system. However, it is important that some effort is made in this direction to avoid the problems resulting from uncoordinated development of several computer applications. These problems include duplication of effort, inconsistent information being kept on several subsystems, etc.

In some organisations, the strategic nature of information is clearly recognised and it is not difficult to obtain resources to manage this as a corporate resource. It is less common for functions to be defined once and then re-used, but some companies already use this approach. The re-use of procedures and subroutines is taken for granted — program libraries have been used for many years. These implementation units are part of the Enterprise Resource Library (covered in later releases of YSM).

In other organisations, the philosophy may still be a 'just in time' approach to systems development. In these organisations, obtaining and allocating resource to build the Enterprise Essential Model may be more difficult. At the very least, it is recommended that system teams try to coordinate their information modelling and share resources to integrate their information usage.

Enterprise Essential Models are of great importance in strategic planning. Only by examining the 'larger picture' can valid decisions be taken as to *which* projects should be initiated and *when*. These decisions should be made to support the strategic goals of the company. In many case, this will be chosen to achieve the greatest cost-benefit for the company. (Strategic planning will be covered in detail by later releases of YSM.)

3.1.7 The Enterprise Essential Model as a shared resource

The Enterprise Essential Model describes the sum total of the information, functions and events used by an organisation. Each user may be aware of some part of this model. There are therefore multiple views, each corresponding to the subset of the enterprise activities that a particular user is involved in. (In this context, the term user means any agent that carries out part of the work of the enterprise. It includes both people and application systems.)

3.2 Structure of the Enterprise Essential Model

3.2.1 Distinction between system and enterprise views

Many of the modelling tools used in the Enterprise Essential Model are also used as views in the System Essential Model. To distinguish these two uses of the same modelling tool, an identifying prefix is added. Thus an enterprise entity relationship diagram (enterprise ERD) is an ERD used as part of the Enterprise Essential Model; a system ERD is an ERD used as part of a System Essential Model. (In this chapter if no ambiguity results, this prefix will be dropped.)

3.2.2 Aspects of the Enterprise Essential Model

The model contains two aspects:[1]

- the enterprise information aspect: describes the information used by the enterprise (see §3.2.3);

- the enterprise performance aspect: this describe event frequencies and number of occurrences of information aspect components.

3.2.3 Enterprise information aspect

This aspect describes the information used by the enterprise. This information is *all* information (not just stored information). The ERD is used to highlight the 'static' characteristics of the information. It provides the major semantic insight into what the information *means*.

Entity state transition diagrams are used to highlight the 'dynamics' of what happens to affect this information and when.

Text specifications are used to define the lower-level detail. In particular, this detail includes the attributes of entities and participation rules of relationships. The way these views fit together is shown below, with links between the components shown by lines between them.

Note: abstract data type and enterprise operation specifications are part of the Enterprise Resource Library.

---

1 Future releases of YSM will identify other aspects and views. This release of YSM does not give much detail on the performance aspect, but future releases, covering design will elaborate how this is used in more detail.

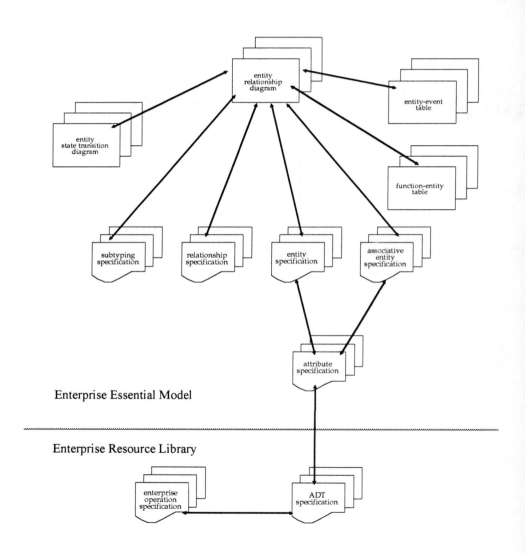

Enterprise Essential Model

Enterprise Resource Library

3.2.3.1 Structure of the information aspect
The modelling tools used for these views are described in more detail in §2.

- **Entity relationship diagram**: this is the main modelling tool used to declare the components of the enterprise information aspect. It shows entities, relationships, associative entities and subtypes. The EIA may contain many ERDs.

- **Entity–event table**: this shows the interaction of enterprise events with entities and relationships at a high-level.

- **Function–entity table**: this shows all functions within the enterprise and their use of information. It may show great detail, listing individual functions within systems. More important, however, are those functions corresponding to complete systems. A table showing these functions shows the shared information interface between systems.

- **Entity state transition diagram (eSTD)**: each entity may have one or more state variables. For each of these, an entity STD shows the sequence in which events occur and changes to this state variable.

- **Entity specification**: each entity that is not an associative entity has a corresponding entity specification, which includes a definition of the attributes of that entity. Each of these attributes has a corresponding attribute specification.

- **Relationship specification**: each relationship has a corresponding relationship specification.

- **Associative entity specification**: each associative entity has a corresponding associative entity specification. This specification includes a definition of the attributes of that entity. Each of these attributes has a corresponding attribute specification.

- **Subtyping specification**: Each subtyping of an entity into subtypes has a corresponding subtyping specification.

- **Attribute specification**: each attribute of an entity has a corresponding attribute specification. This entity must appear on at least one ERD.

3.2.3.2 Visibility on enterprise ERDs
The main information components appear on ERDs. Some may appear on more than one, as described below.

3.2.3.2.1 Entities on ERDs
Entities may appear on more than one ERD. They must appear on at least one. There is no special notation used to show an entity appears on several ERDs.

3.2.3.2.2 Relationships on ERDs
A relationship must appear on one ERD. They do not usually appear on more than one ERD, although this depends on the technical support environment. In automated environments, this poses no problem. In pencil and paper environments, the redundant specification of the relationship on several ERDs suggests that this should be avoided.

3.2.3.2.3 *Associative entities on ERDs*

An associative entity may appear on one or more views as an entity only (a 'pure' entity). In their role as a relationship, they follow the same rules as those given in §3.2.5.2.

3.2.3.2.4 *Subtypings on ERDs*

Subtypings are usually only shown on one ERD, although this raises problems when there are very large numbers of subtypes. In those cases, the subtyping may appear on several ERDs, each showing some of the subtypes. For automated environments, there are no problems in showing the subtyping in multiple views.

Each of the subtypes must appear on at least one ERD.

3.2.3.3 Completeness of enterprise information aspect

The enterprise information aspect can never really be described as 'complete' — it is dynamic and changes over time. However, at any time all of the correspondences described in §3.2.3 should be true. In other words, each entity should have an entity specification, each state variable declared in an entity specification must have an eSTD, ... These rules are given in §3.7.2.

It is very important is that the completeness checks apply to the portion of the EEM that is used to support a specific system. Only if all components required by that system are properly specified can the system be regarded as completely modelled.

3.2.4 Performance aspect

This contains views that refer to events and views that relate to stored information.

3.2.4.1 Enterprise events

These events are external to the systems that they are events for, but they are not necessarily external to the enterprise:

Some events are truly external to the enterprise — they 'happen' in other organisations, individuals or devices that the enterprise deals with.

Other events occur within the enterprise. They correspond to the completion of a process within one system; to another system, they act as events.

The enterprise event list shows events of both types. Information such as 'event frequency' is also given in event specifications.

3.2.4.2 Required storage

The enterprise performance aspect also includes the required storage for information. Views used are:

- **entity specification**: describing how many occurrences of each entity are expected and the required storage for each.

- **relationship specification**: describing how many occurrences of each relationship are expected and the required storage for each.

- **associative entity specification**: describing how many occurrences of the associative entity are expected and the required storage for each.

3.3 Information modelling principles

As described in the introduction, YSM captures system and enterprise requirements in an implementation-free format before considering how any specific hardware and software architectures may be utilised. For information requirements, this activity corresponds to identifying the required information, without paying any attention to its physical representation, etc. For stored information, the model deliberately suppresses information about how or where it is stored, database, direct access devices, etc.

Information requirements of this type are specified by means of an information model. YSM refers to this as the 'enterprise information aspect'. The process of identifying these requirements, organising them into a coherent model and then cross-checking that model is referred to as information modelling.

3.3.1 Semantic approach to information modelling

When verifying and understanding the models, a high-level view of what the information is and what it means is much more important than how it is represented. YSM thus uses techniques which concentrate on the meaning of the information. This is referred to as semantic information modelling. This approach originated in the work of Chen and others in the 1970s [Che76]. Flavin integrated semantic information modelling with other Yourdon techniques [Fla81] and YSM has significantly refined these concepts and techniques.

The main semantic tool used in YSM for information modelling is the ERD. This highlights entities and relationships, which represent specific semantic units within the model. In addition, this modelling tool provides a partitioning of the model into components which are then specified with other tools. For example, each entity has an entity or associative entity specification; each relationship has a relationship specification; each entity has an eSTD (or equivalent tabular representation); etc. The notation, semantics and rules for the enterprise ERD are given in §2.2.

Because of the significance of entities, relationships and attributes in the semantic approach to information modelling, this approach is sometimes referred to as entity–relationship–attribute modelling (ERA modelling).

3.3.1.1 Localisation of model features

A good model should aid understanding by having well-defined semantic components. Any fact should be localised — in other words it should be recorded in one well-defined area of the model. This aids understanding and verification by allowing us to 'window' or select one small portion of the model for further study. In YSM this is partly achieved by the modelling tools provided and their rules. For example, an entity is specified in only one place — the entity specification; this entity may be referred to in many other places. This is not something with which information modellers need to directly concern themselves with — it is a feature of the method.

One area where the modeller must be aware of this principle is in the definition of information views, as provided by multiple ERDs. A poor choice of such views will require the reviewer to keep 'switching attention' from one diagram to another, with a severe penalty in understandability and review quality. This is further discussed in §3.5.7.2.

Even within one view, a well-thought out diagram will aid, but a poorly laid out diagram will hinder understanding. Guidelines for layout are given in the corresponding sections of §2.

3.3.1.2 Semantic precision
Each of the model building blocks (entities, relationships and attributes) should be used as precisely as possible, which requires understanding of the intent of each of these 'building blocks'. As is always the case with semantics, however, this is rather intangible and it is difficult to come up with what is 'right' and what is 'wrong'. However, the information modeller should strive to conform to the general intent of the model building blocks as far as possible — this is very important in standardising communication.

An information model cannot be defended by stating 'it holds all the required information'. Even if it can be guaranteed that the information may be retrieved in a form required by the enterprise, this is not enough. The information model must be organised in a way which conforms with semantic interpretation of entities, relationships, attributes, abstract data types and so on. This allows it to be understood and thus used and maintained.

Because achieving semantic precision is difficult, it is very important to obtain a consensus on the most appropriate model component to use. Such consensus is best achieved in walk-throughs and presentation sessions involving users and information modellers. In this way, the individual style and bias of analysts may be minimised.

See §2, particularly §2.2, for further discussion of modelling tool semantics.

3.3.2 Fact-based information modelling
The main aim of information modelling is to capture and abstract the most important aspects of real-world facts. Each occurrence of a real-world fact is likely to be a very simple statement that can be verified by the analyst in discussions with the user and subject-matter expert. For example, such facts as:

Each employee has a personnel number

Each manager is responsible for a budget line

etc., are about the business enterprise, not about any system or database plan. These facts are not created by the analyst, but the analyst is responsible for formulating an abstraction and organisation for these facts.

By basing the information model on real-world facts, prior commitment to a particular use or organisation of data within the enterprise is avoided. Equally important, it is a *neutral* modelling strategy, which helps to avoid colouring models with the preconceptions of one particular person.

These facts may be organised in different ways by choosing to allocate them to different ERDs, naming attributes differently, and so on, but the underlying model is not affected by these organisational changes.

3.3.3 Relativist approach to information modelling

A particular real-world fact may be built into the model in one of several ways. There is no single 'correct' way of doing this for a particular fact — rather, the way chosen should be the one that most honestly reflects the *use* made of this fact by the enterprise. This is termed the relativist approach to information modelling.[1]

Adopting a relativist stance is *not* the same as saying 'any of the model constructs can be used, irrespective of meaning', which would contradict the semantic precision philosophy. The relative approach says the use may vary, depending on circumstances. For a given use of the real-world fact, there is one way which models that use most truly, using the semantic constructs provided.

3.3.4 Time entities

YSM requires that all information be attributed. That means that any data item used by a system must describe an entity or a relationship between entities. There can be no 'tables of values' or the like. In analysis, all information is about something. Each value is the value of an attribute for an occurrence of that entity. Sometimes this leads to abstract 'space' or 'time entities' being required.

For example, a record might be kept of the temperature at noon each day. This can be regarded as an attribute of the entity "Day". Each occurrence of "Day" has a value for this attribute. The continuous variation of temperature over time has been replaced by a set of values, taken at discrete times. The temperature has been sampled.

Sometime the time period may participate in relationships or associative entities. If the power generation company were committed to keeping records over a set time period (every minute, or every half-hour), then the diagram might be:

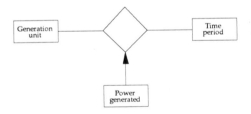

3.3.4.1 Scheduling

Some systems carry out a lot of 'scheduling'. It may be convenient to use a 'time period' entity in these systems. This makes it easier to visualise the relationship between the different entities and a given time period. In this manual, a training centre example has been used for some information modelling examples. The entity "Day" was chosen as a time period entity that would be used in the organisation. A slightly larger ERD of this part of the enterprise is given below, highlighting the access to the "Day" entity:

---

1 The absolutist approach would be to say that 'there is only one correct way of modelling a real-world fact. This representation of the fact would be used for any enterprise, irrespective of context'.

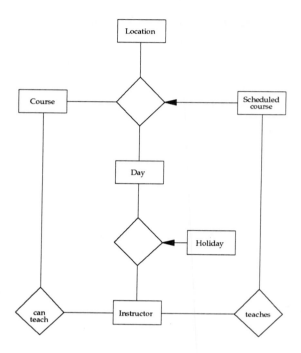

As an alternative, the entity "Scheduled course" could be converted to a binary associative entity, with an attribute "start date". This is an equally valid approach. YSM regards this as a 'relative' issue — it could have been done either way.

3.3.4.2 Spatial entities
In a system that calculated airflow over a wing, there might be a requirement to calculate certain characteristics at each point of the wing for different air speeds. One of these might be "air pressure". This would be modelled:

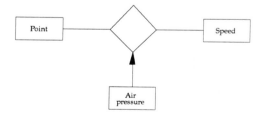

Note: in this case both entities are continuous, but are replaced by discrete equivalents. For example, there would be set of "Speeds" that the model would use.ing continuous quantities as entities[1]. The continuously varying "Point" would be replaced by a finite element mesh (or its equivalent).

1 The sampling described above is an artifact of imperfect technology. For example, conceptually, there is a continuum of times for which there is a temperature. For the power generation example, there is a continuum of times for which the power is defined for each generator. The current release of YSM does not support continuous entities.

3.4 Strategies for Building and Maintaining the EIA

If there is no existing EIA, then one needs to be created; if there is an existing EIA, then it will need to be maintained as enterprise information requirements change. This chapter discusses strategies for managing these activities. §3.5 discusses some techniques that can be used by these strategies.

3.4.1 Main information gathering strategies

There are three main strategies that may be adopted in building an enterprise information aspect:

1. **Fact gathering and view synthesis**: the entities, relationships and attributes that are relevant to the enterprise activities are modelled without regard to any specific implementation or use, either present or future. This is, in effect, enterprise information modelling carried out in isolation.

2. **Modelling existing use of information**: existing data in the form of files, records, databases, etc., is studied to identify their information content. This is a form of 'reverse engineering', where the decisions of a previous design activity are undone.

3. **Identification of information to support enterprise functions**: specific enterprise functions are examined and their information usage requirements identified. These enterprise functions may be anticipated systems or they may be functions being examined in strategic planning.

In practice, these strategies are used in parallel, but the first two are usually more important than the third in initiating an enterprise information aspect.

3.4.1.1 Fact-gathering and view synthesis

Information modelling is carried out without regard to any specific implementation or use, either present or future.

There is a user view for each person (actually their role) who supports the enterprise's activities. In a similar way, within a computer system, application programs may require access to different parts of the automated database. Each of these also constitutes a user view. Each view is documented in terms of entities, relationships and attributes used.

In using this strategy, careful control must be exercised to ensure that the information modellers don't get 'too enthusiastic' and model everything that they can think of, even real-world facts that will be of no relevance to the enterprise. This is an important scoping consideration.

See §3.5.1 for discussions of heuristic techniques used with this strategy.

3.4.1.2 Modelling existing use of information

Existing information used by the enterprise is identified. This information may currently be held in the form of manual files, computer files, databases, etc. Previous design decisions on how this information is organised — this is a form of 'reverse engineering'.

This strategy has a first step in which the information aspect of existing applications is documented. This requires examination of their file documentation and existing manual files and documents.

The evidence of the existing implementation is then removed by identifying the information content in an essential representation. This representation will be mainly in terms of ERDs and attribute specifications.

These steps are discussed in §3.5.6. Normalisation techniques may also be useful here (see §3.5.8).

3.4.1.3 Identification of information to support enterprise functions

Known enterprise functions are each examined to identify the information that they need to accomplish their function. These enterprise functions may be anticipated systems or they may be functions being examined in strategic planning.

These two possibilities are discussed in §4 and §5 respectively.

Where system projects have already been initiated, their information requirements are examined. This will involve discussions with the system team about how their information requirements correspond to the use of entities and relationships that are already in the enterprise information aspect. New attributes are very often identified, but new entities are less commonly seen. Techniques related to subtyping will be especially useful, as a specific system often only 'sees' a certain subset of an entity.

3.4.2 Maintaining an existing EIA

There is an ongoing requirement to support the enterprise information aspect. Indeed, this is an open-ended activity. As company requirements change, new procedures are established and new systems are built there will be a gradual change in the information resource that is required to accomplish the enterprises goals. It is very important that construction of the enterprise information aspect is not seen as a 'once off' activity — there must be an continuing commitment to maintain it to keep it in alignment with the true information requirements.

Maintaining the EIA is not a linear activity — in other words, there is no sequential set of actions that can be carried out to a final conclusion. Rather, it is a circular, or spiral activity — the information analyst has to keep going over the same ground, gradually improving the model as time progresses. In a sense, this activity 'has no end'.

3.4.2.1 The enterprise information aspect as support for systems

Maintaining this information resource is an enterprise responsibility and when specific system projects identify information that they require in order to achieve their purpose, they 'make a request' on the service that control the maintenance activity. This service may be automated or manual. The enterprise resource manager will carry out the comparison of this required information with the existing EIA and modify it, or not, depending on aliasing, etc. Possible actions may include adding new components, resolving aliases and removing unused components. The request is then acknowledged in terms of the relevant new portions of the EIA being made available to the system modellers.

3.4.2.2 Dealing with enterprise information that is no longer required

There is also the possibility that a system no longer needs access to a component of the enterprise information aspect. In this case, it will no longer appear in the information aspect of that System Essential Model. This may mean that the enterprise no longer has any need to maintain this component as part of the EIA. This is discussed in §3.6.3.

3.4.3 Combining and organising information models

Whichever strategy is adopted, there will be a requirement that an existing information model has to be modified by combining another with it. This is discussed in §3.5.

A single large ERD should *not* be constructed. Such large diagrams may serve to impress the naïve visitor to the office, but they do little to help communication. Indeed, they completely contravene the 'lump' law [Wei75]. It is much better to organise the model into a set of well-defined views, each covering an understandable set of entities and relationships. In order to be understandable, it is suggested each view be organised around an entity or relationship of interest. This is discussed in §3.5.7.2. Note: this problem is a major concern only with ERDs — the other components of the model are naturally partitioned around the components they specify.

One problem that will have to be overcome is identification of correspondences between features on different user views, particularly where 'aliases' occur. If we build models that are 'externally oriented', this problem is much reduced. This external orientation of model building concentrates on real-world entities and events (see §3.6.1).

Text support has to be used to define each of the attributes of entities and associative entities. Any known constraints should also be captured, as available.

3.4.3.1 Checking information model supports operations

In order to check that the information model that has been captured is complete, it should be checked against the current operations of the enterprise. Any function that is currently being carried out by the enterprise should have sufficient information in the information model to support it. If this is not true, some components of the model must be missing.

This check can be carried out informally in user interviews, checking off an informal description of what is done against the draft information model.

Note: the collection of information that is never used may indicate that it should be dropped from the model (see §3.6.3).

3.4.3.2 Refining the EIA

The model, as derived from a combination of the above strategies, is only a starting point. In fact, one of the major reasons for building models is in order to improve them. Information models may be improved by improving their organisation to remove redundancy and to make them easier to understand and use. Some heuristic techniques for refining the model are discussed in §3.5.7.

3.5 Heuristics to build and refine the EIA

3.5.1 Techniques for building a new EIA

3.5.1.1 Using fact-based information modelling

Information modelling captures and abstracts the most important aspects of real-world facts used by the enterprise. These facts are not created by the analyst, but the analyst is responsible for formulating a representation and organisation of these facts.

Fact-based information analysis is a very effective information gathering and synthesis technique that may be used to build up an information model [Ken83]. This approach gathers facts and then identifies how they can be represented by the modelling tools available — entities, relationships, attributes, etc. As an example of this technique, suppose the following fact had been established during user interviews:

> Employees are assigned to departments

This is a fact-pattern with two components linked together — "employee" and "department". It is a generalisation of a whole set of real-world facts:

> John Smith is assigned to the production department

> Mark Barker is assigned to the quality assurance department.

etc., where "John Smith" and "Mark Barker" are both similar, in the sense that they are "employees". Furthermore, the pattern is elementary in the sense that it cannot be decomposed into any smaller units without loss of meaning. The atomic components are "employee" and "department".

Suppose the fact-patterns:

> Employees have names

> Employees have salaries

were also established. Again, these are elementary and involve two atomic components. However, these fact-patterns are of a different type from the preceding one. "Department"s and "employee"s are an abstraction of a role that may be played by many real-world objects and we thus model them by an entity. "Name" and "salary" correspond to properties that describe each occurrence of employee — they are therefore modelled by attributes of "employee". The synthesised picture is as below:

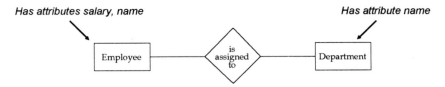

3.5.2 Fact gathering

Information modelling requires that real-world facts are:

1. **Captured**: sources for these facts include user interviews, policy statements in enterprise documents and external standards. Existing system documentation will also embody these real-world facts, although reverse engineering them may not be trivial. User interview is discussed in §3.5.3.

2. **Abstracted**: the facts are often in a very specific form, for example, "Jones signs the orders", "Jones, Smith and Marsh work in order processing" etc. . The appropriate information modelling component that corresponds to these specific facts must be identified. This is discussed in §2.

3. **Organised**: the EIM is an integrated model, although it is organised into many views. The individual model components that are identified must be placed into the model in a way that maintains model integrity and understandability. This is discussed in §3.5.6.2.

3.5.3 Interview techniques

The main aim is to establish facts, so interviews should be organised to that end.

An unstructured and informal approach is probably appropriate for initial interviews with users who have not previously been involved in such fact-gathering exercises. This will encourage confidence in the analyst's ability to communicate and not alarm the user with any new and intimidating methods. Once this initial confidence-building phase is complete, it is advised that some more structured method of capturing information is used. The following techniques are useful in identifying entities, relationships and attributes. The one chosen will depend on user reaction and training:

1. **Entity relationship diagrams may be drawn interactively during the interview**: these act as a prompt for further questions, such as 'what information do we need to store about x?', where 'x' is the name of the entity on the diagram

2. **Capture a series of fact-patterns during the interview**: these can be synthesised later and an ERD drawn from them.

3. **Use tabular representations of the information required**: this is a very low-tech approach, but works well.

3.5.3.1 Entity relationship diagrams used interactively in interview

As the interview progresses, the diagram may be extended and revised as information is agreed. The diagram also allows identification of areas that require more detailed examination — for example, each relationship will need to be discussed with regard to participation rules, etc.; each entity may be discussed in terms of what information is relevant to it (its attributes).

This technique has the benefit of precision, but sometimes has adverse user reaction to what users see as 'technical' diagrams. Whether this reaction occurs is largely conditioned by the analyst's attitude — if the analyst is not experienced in ERA modelling and thinks the approach is difficult, this approach fails; if the analyst is at home with ERA and presents the techniques in the right way, the approach succeeds without the user even being explicitly aware of the techniques.

3.5.3.2 Fact-patterns as a basis for interview

This is equivalent to drawing ERD fragments or stating that entities have attributes, etc., but has the benefit of using everyday language, rather than specific diagrams. It therefore does not require the user to be instructed in how to read ERDs. For example, rather than drawing an ERD fragment, the fact-pattern:

> A <Course> covers a <Topic>

helps clarify that there are entities called "Course" and "Topic", with a relationship between them. The information analyst is aware of the correspondence between the ERD and this statement, but the user need not be aware of it. It is, of course, the analyst's responsibility to ask the correct questions to elucidate what these entities mean and then formulate them into the correct graphic representation.

This technique again has the benefit of precision, but the analyst should be careful to explain the notation in a more informal way. The example above should be read 'Each specific course may cover a topic'.

3.5.3.3 Tables as a basis for interview

For entities, including associative entities, a tabular representation of the attributes of that entity may be useful. For example, when talking about the associative entity "Booking", the analyst might draw a table in the following format:

booking no.	Scheduled course	Customer	Delegate	Date booked

Each row of the table corresponds to one occurrence of the entity, relationship, or associative entity. In this case, the row corresponds to one occurrence of the associative entity "Booking" that refers to the entities "Scheduled course", "Customer" and "Delegate". There are two alternative identifiers — "booking number", or "scheduled course + customer + delegate". The analyst can fill in specific entries to ensure that the user does understand what the table represents and that all values will be available in the real world.

Note: references to entities in a relationship (or associative entity) are replaced by the names of those entities — the analyst is responsible for replacing these by the correct entity references in the agreed model. In this example, they would be "<Scheduled course>", "<Customer>" and "<Delegate>".

The use of tables induces positive user reaction if presented in the right way. However, it does have some risks in the sense that the links between tables are not very clearly visualised. As a consequence, when the analyst replaces the informal table by entities, with explicit links as shown by relationships, some errors may occur. It is best used as an initial investigation tool for such entities — the larger picture should be agreed with the user by means of ERDs or their equivalent. For the above example, we would establish the following ERD fragment:

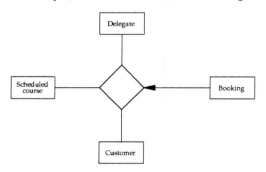

3.5.3.4 Interview medium

Pencil and paper or a whiteboard is a good discussion medium for such early discussions. On the whole, CASE tools are best avoided in the early stages of such discussions, as they tend to discourage lateral thinking. In the later stages, CASE tools are very good at capturing and storing information once identified.

Once the 'bigger picture' has been established, details can be filled in using a CASE tool. It is assumed that the analyst has entered ERDs, either during or after user interviews. These can then be used to create a skeleton model with entries for each component, but little textual detail. This information can then be filled in interactively during discussions between the user and the analyst.

If a CASE tool is not available, then the pencil and paper diagrams are revised during subsequent interviews. Textual information may conveniently be filled in on preprinted proformas for each required specification.

3.5.3.5 Scenario playing

One useful interview technique is to use 'what happens next?' types of discussion. The analyst elicits information by a sequence of questions about what happens next. These questions will usually give rise to further questions and corresponding replies. As the session proceeds, the analyst identifies facts and records them.

In recording them, they have to be identified as events (see also §5.6.2.1), entities, relationships, rules, attributes, etc. This is the responsibility of the analyst — the analyst cannot ask 'is that an entity or a relationship?'!

> **Analyst**: Tell me, after the course has been scheduled so that it is available, tell me what might happen.

> **User**: We might get bookings — hopefully we do.

> **Analyst**: (*thinks — booking sounds like an event.*) What do you mean, a booking?

> **User**: Well, customers ring up and want to book people on courses.

> **Analyst**: (*thinks — wait a minute, is this the course I heard about earlier?*) Tell me what the course is again?[1]

> **User**: Well, as I told you, the course is a specific seminar that we may run from time to time on a public basis, or for specific customers on an inhouse basis.

> **Analyst**: (*thinks — sounds like the course is run several times. Each time it runs, there may be bookings for it. If so, maybe there is an associative entity lurking in there somewhere ....*) Would it be fair to describe each time the course runs as a "Scheduled course"?

> **User**: Yes.

> **Analyst**: (*thinks — lets elucidate more about that "booking" — maybe its a relationship. If so I'd better identify what it refers to. Also, I must also identify any effect on the scheduled course when the booking event occurs — that will be a change of state of the associative entity Scheduled course.*)

and so on.

3.5.3.6 Attribution

Any values that are identified during fact-gathering will be potential values for an attribute (or state variable)[2]. This will describe an entity (which may be an associative entity). Attribution is the process of identifying which entity they 'belong' to.

To be correctly attributed, both the following must be true:

1. Each occurrence of the entity must have a single value for the attribute,

2. It must be possible to change the value of the attribute for one occurrence of the entity without affecting the values for other occurrences.

Note: at certain times, an occurrence of the entity may not have a value for the attribute. Only at certain times in the life of an occurrence of the entity will the value be available. If the value is present at *any* point in the life-cycle, then it is an attribute.

See also 'normalisation' in §3.5.8. Note: the normalisation techniques are particularly geared towards ensuring that there is no redundancy in stored attributes. Because YSM allows temporary and derived attributes, the above criteria are the ones that should be applied to determine whether a quality is an attribute.

---

1 Note: playing 'dumb' is a useful ploy in interviews, as long as it is not overdone.

2 To simplify, the remainder of this section refers to attributes, but the same requirements also apply to state variables.

3.5.3.7 Distinguishing between attributes and state variables

There are many similarities between attributes and state variables. However, a state variable has the specific characteristics:

- Its values are literal — the names of the states. They cannot be compared with any attributes or state variables belonging to other entities.

- A state variable must always have a value at any point in the life-cycle. This means that when an occurrence of the entity is created, it must have a value for each (if there is more than one) state variable.

If either of the above are contravened, then the value must be that of an attribute, not a state variable. Attributes have attribute specifications. State variables have enterprise eSTDs.

3.5.3.7.1 Access to state variables

The following are the ways a state variable may be used:

- The value of a state variable for an occurrence of an entity may be compared against a literal. For example: "<Course>.availability = planned" is allowed. This checks that this specific Course is in the state "planned", as far as the state variable "availability" is concerned.

- The value of a state variable for one occurrence of an entity can be compared with the value of the same state variable for another occurrence of the same entity.

- The state of one (or more) occurrences of an entity may be changed. If there is more than one state variable, then each can be changed independently of the others. For example: "<Course>.availability := obsolete" would mark a particular occurrence of course as no longer available.

3.5.3.8 Determining whether attributes are derivable

A derived attribute is not stored. Its value can be determined using available information at any time. For example:

- the "age" of a person can be derived from the stored attribute "date of birth" and the enterprise operation "todays date( )". Both "date of birth" and "age" describe a single occurrence of the entity "Person". They are attributes; "date of birth" is a stored attribute; "age" is a derived attribute.

- "number of bookings" is a derived attribute of the entity "Public course". It has a single, well-defined value for each occurrence of "Public course". However, it is possible to determine this information from other stored information (in fact, the number of occurrences of the associative entity "Booking" that refer to this specific course.

3.5.3.8.1 Implementation choices for derived attributes

When the implementation for the information model is chosen, there is a choice between:

- deriving the value of the attribute every time it is required. This has the benefit of guaranteeing consistency, but may be very time-consuming for some attributes.

- storing the attribute. This has reduced time to access the current value of the attribute. However, any change to any of the items that it is derived from requires the attribute to be recalculated. It may be difficult to prove that this always occurs and the complexity of the implementation is increased.

As with most implementation decisions, there is a trade-off here. However, in essential modelling, this is of little concern — the main requirement is that the model capture the policy of what the information means. Conceptually, once the attribute has been declared as 'derived', it may be referenced in the same way as a 'stored' attribute. When a stored attribute is accessed, the current value is retrieved; when a derived attribute is referenced, the value is calculated, using available information. As referenced in a minispec, there is no difference.

The rule is therefore:

> *If the attribute can be calculated using available information, including values of other attributes, then it is a derived attribute; otherwise it is a stored attribute.*

Sometimes there may be two attributes (such as "date of birth" and "age") such that one could be chosen as stored and the other as derived, but the choice could be made either way. In these cases, there is usually one that is conceptually simpler, but each case will have to be treated on its merits.

3.5.3.9 Text specifications
Entities, associative entities, relationships and attributes have associated text specifications. All these must be completed before the model is regarded as complete. In particular, the analyst should pay attention to policy constraints and rules of association where these exist. See the corresponding specifications in §2.

3.5.4 Abstraction
The user will often volunteer quite specific facts. For example, "I sign the orders before they go to purchasing". The analyst is responsible for identifying whether the fact is relevant and also for replacing specific references by role references. In this case, the relevant fact might be "Each order is signed by a buyer". The analyst has identified that the user is playing the role of a "buyer" for the enterprise when he signs the order.

If the user volunteers specific values, the analyst must identify what are all the possible legal values and what each value signifies. For example, the statement:

> The Newfoundland tracking station is at latitude 54 north.

becomes:

> Each tracking station has a location, which requires a latitude. The latitude is measured in degrees in the range −90 to 90.

In this example, not only would the analyst have to spot the abstraction required, but would also (hopefully) establish that the longitude of the location is also required.

3.5.4.1 Recognising correct model components

This is not normally a separate step in information modelling, but is part of the process of fact gathering, interviewing and abstraction. See also 'scenario playing', discussed in §3.5.3.5.

In situations where unstructured text has been used to record interviews and it is then being converted into an information model[1], the following sequence is effective:

1. **Identify entities**: these are roles that real-world things can play in their interaction with the enterprise. The enterprise may need to remember values associated with each occurrence of them (the attributes of the entity) and their relationships with other entities. Entities usually have a noun-type name.

2. **Identify relationships**: relationships are associations between entities that are set up when an event occurs. An occurrence of a relationship depends on the prior existence of the entities. Relationships normally have a verb-type name that may be used with entities to construct a meaningful sentence.

3. **Decide if there are any subtypings present**: there may be subsets of an entity that are treated differently by the enterprise. If this is the case, it may be helpful to define a subtyping.

4. **Identify attributes and determine what they are dependent on**: an attribute is a named property of an entity or relationship. If a relationship has attributes, then it is modelled as an associative entity. Attributes usually have noun type names.

5. **Establish legal values for attributes**: all possible values should be discovered. This may correspond to an existing ADT (or a restriction of one), a fixed set of values, or a new ADT (see §2.8.5).

6. **Identify state variables and build eSTD**: it is difficult to build and verify these without extensive user interview — they should only be created in a very preliminary form when extracting from textual descriptions (see §3.5.5).

The main discussion of these components is in §2, particularly §2.2. Resolving aliases is discussed in §3.5.6.4. Recognising aliases is discussed in §3.6.1.

3.5.4.2 Some common problems in abstraction to the correct model component

3.5.4.2.1 Entity, relationship or associative entity?

An example of the relativist philosophy is the modelling of associative entities. Initially an entity may be discovered by the analyst. Several users may make use of this entity in different ways. The information modeller correctly models this as an entity with an entity specification and shows it as involved in relationships on ERDs.

Subsequently, while examining other areas of the enterprise's activities, it may become clear that it is originally created as a relationship between entities. In other words, it is an associative entity. It should then be converted into an associative entity in the model (and the entity specification will be replaced by an associative entity specification).

---

1 This approach should not be the method of choice. It is much better to build the ERA model up interactively, without this intermediate text form.

This does not compromise the original insight, as captured and agreed in the previous diagrams. Indeed, the associative entity is still regarded and modelled as an entity in those views. However, in the Enterprise Essential Model taken as a whole, it is an associative entity. It has an associative entity (rather than an entity) specification.

If there were no attributes of the relationship and it did not subsequently participate in any other relationships as an entity, it would be correctly modelled as a relationship.

The associative entity may be regarded as an entity or an associative entity, or relationship, depending on the point of view.

3.5.4.2.2 Entity or ADT?

Sometimes there may be some choice over whether a real-world quality should be captured as an abstract data type or an entity. Generally speaking, if the enterprise needs to keep track of distinct occurrences of this quality, then it should be regarded as an entity rather than an ADT.

One example of such a quality is 'colour'. If the enterprise needs to record the colours of different entities, then they should be regarded as attributes of the entities and colour would then be an abstract data type. In an organisation that made paints (for example), a colour might be something that was produced in a specific way and then it would be more appropriate to model colour as an entity. Each occurrence would be a distinct colour that might be manufactured.

Generalisation entities

Consider the example of a system whose purpose is to keep records for the results of different types of tests carried out in a hospital. There may be many different kinds of test — drug sensitivity tests and infection tests might be two examples. These could be regarded as subtypes of a more general entity "Test" (say). If that is the modelling approach used, then effectively we are saying:

> There are certain facts that we need to record about each occurrence of a test. Some of these items would be relevant to any tests; some would only be relevant if the occurrence was of a certain type of test.

This seems little different to what we had above for subtypes. However, suppose we wished to record such facts as:

A drug sensitivity test requires 100 ml of blood

An infection test requires 25 ml of blood

This would pose problems if subtyping is used. For example, trying to store this information in a single table that records occurrences of "Test" leads to information being stored redundantly — every occurrence of "Infection test" would have "required blood volume" with the value 25 ml.

This is more correctly modelled using a more general entity of which "Drug sensitivity test" and "Infection test" are occurrences. If these were the only type of test with which the system was concerned, then there would only be two occurrences of this more general entity, which we shall call "Test type". Attributes of the entity "Test type" would include "required blood volume for sample".

We would still have an entity "Test" whose individual occurrences could be seen to be carried out from time to time. The relationship between these two entities is an *is a* relationship, as shown below:

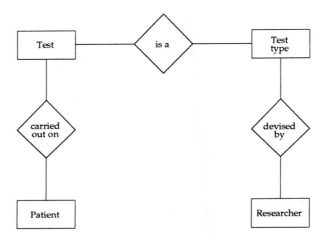

These generalisation issues can become quite abstract and difficult[1]. As a general guide, trying to think of each entity as a table helps. If the tables contain redundant information, then the model is incorrect and should be revised. See §3.5.8 for a discussion of normalisation techniques.

3.5.4.2.3 Entity or subtype?
Some of the more difficult decisions relate to generalisation — is it more appropriate to generalise using subtypings or by creating a more general entity.

Subtypes revisited
A subtype of an entity is a well-defined group of occurrences of the entity that may be regarded as an entity in its own right.

In §2.2.3.6, the following example was introduced:

---

1 Models of models have the same problem, leading to excursions into 'meta–meta land'.

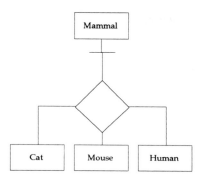

Consider the entity "Mammal". Individual occurrences of this entity correspond to Archimedes, John McEnroe, Tom (a well-known cat), Jerry (a mouse) ... .

One well-defined group is the entity "Human" containing many individual occurrences (two of which were explicitly mentioned above). The entity "Human" is a subtype of the entity "Mammal". Other subtypes of "Mammal" are "Cat", "Mouse".

An occurrence of the subtype is always an occurrence of the supertype (for example an occurrence of "Cat" is automatically an occurrence of a "Mammal"). As a consequence, any property that relates to the supertype is automatically a property of the subtype. The subtype may be said to 'inherit' the properties of the supertype (see §2.2.3.8 for a further discussion of these common properties).

A convenient mental picture is to imagine a table for the supertype. Rows of this table correspond to occurrences of the supertype. Each row is also an occurrence of a subtype. The table therefore has some columns that do not always have values, depending on which subtype a specific row is an occurrence of.

For example, the table below shows that different types of plane have different attributes (no claims for accuracy!). The entity "Plane type" has the subtypes "Passenger aircraft type", "Fighter aircraft type", "Bomber aircraft type". (Note: the entity is a generalisation entity. Occurrences are types of planes, not specific physical aircraft.)

Plane type	Number of engines	Maximum range	Number of air-air missiles	Maximum bomb load	Maximum number of passengers
Boeing 747-J3	4	10 000 km			540
Tornado-B1	2	3 000 km	6		
B1	4	8 000 km		25 000 kg	
Gulfstar	2	3 000 km			25

3.5.5 Establishing entity life-cycles

Each entity may have one or more state variables. A state variable describes a characteristic way the entity changes over time. Not all entities have state variables. Only entities for which there are different rules for the way the enterprise deals with the entity, depending on some quality 'owned' by the entity have state variables.

Some entities have more than one state variable. For example, the entity "Employee" might have the state variables "employment status" and "marital status".

If different systems use different 'status type' information, then it may be convenient to use more than one state variable. Alternatively, a single, more complex eSTD can be constructed, using a single state variable.

3.5.5.1 Identifying events that affect an entity

To identify the events, the following questions are often productive:

- **Is there more than one possible creation event?** Different real-world events may occur which cause the enterprise to create an occurrence of the entity, possibly with different attributes.

- **Are there any events that cause values to be assigned to attributes after the occurrence of the entity has been created?** Looking at each attribute of the entity and trying to decide whether it will always have a value and if it might change is a good strategy.

- **Are there any events that cause the entity to participate in a relationship?** Note: if the event creates an occurrence of the relationship, but does not require an access to the entity, this event will not appear on the eSTD.

- **Are there any related events to entities that participate in relationships with this entity?** Events often affect a group of entities and the relationships between them. Note: the use of an entity–event table is a good way of visualising the entities and events in such a grouping. It should be pointed out that any new event that is found for this entity ought to be checked back against other entities to see if it causes any access to them. Entity–event tables are discussed in §2.4.

- **Are there any known functions that utilise this entity?** If so, the real-world events that this function responds to are likely to be ones that require an access to this entity. This is a particularly useful strategy if carried out in association with a specific system project. See also §5.

3.5.5.2 Identifying states of the entity

Having identified a first-cut version of the events that affect the entity, the next step is to identify the states. For each state and event, one of the following must be true:

- **The event is always dealt with in the same way**: in this case no action is required and the state is left as it is.

- **The event is always ignored:** again, no problem and the state is left as it is.

- **The event is sometimes dealt with in one way and sometimes in another.** Note: this may be because of participation rules in a relationship or other 'memory' of what has happened to that occurrence in the past. In this case the state must be split into as many states as there are different actions when the event occurs, so that there is a well-defined action for each event in each state.

3.5.5.3 Splitting an entity life-cycle into more than one life-cycle

If there seem to be two totally independent sets of behaviour that occur in the real-world entity, then it is probably advisable to split the entity life-cycles into separate life cycles, one for each of these behaviour patterns. Each is given a state variable, with a descriptive name that describes that pattern of behaviour.

3.5.5.4 Presentation tools for entity life-cycles

For modelling entity life-cycles, enterprise eSTDs are used. These have the benefit of precision, but may suffer from the 'Emperor's new clothes' problem — because the user does not want to appear stupid, there may be a tacit agreement to anything the analyst suggests, rather than a true validation of it being the correct sequence of events. Some users will be more familiar or receptive to the use of these diagrams than others, so the analyst should choose the approach used accordingly. The eSTDs must be carefully explained and talked though. Any tool may be made to seem more user-friendly by talking through it, rather than just presenting or drawing it in silence. See also 'scenario playing' in §3.5.3.5.

It is important to examine all combinations of states and events. The tabular form provides a good discipline in this respect — every cell in the table should either contain an entry or be verified to be truly 'empty'. A standard way of indicating that the cell has been verified (such as a "—") can be used to indicate that there is no transition, rather than no consideration has been given to the possibility (a blank cell).

3.5.5.5 Identifying accesses

The accesses shown on the eSTD are those that the system needs to make to the entity when the event occurs. For each event, the access and the state of the entity occurrence after the event should be established.

The eSTD (or its equivalent) should be verified by walkthroughs and presentations.

Note: all states must be reachable by a finite sequence of transitions — if this is not the case, then there are missing transitions.

3.5.5.6 Distinction between entity life-cycle and system behaviour

When building entity life-cycles of hardware used by the enterprise, it is important to build the life-cycle of equipment use by the system, not the internal states of that equipment[1]. For example, in an enterprise building satellite systems, momentum wheel actuators may be used to rotate and stabilise the satellites. As a piece of hardware a momentum wheel is either "spinning" or "not spinning" and the events that affect it are "increase rotation rate" and

---

1 Entity life-cycles for real-world hardware are particularly important where the enterprise is involved in controlling that equipment.

"decrease rotation rate". The correct life-cycle to draw for an enterprise using them would show states that are specific to how they are used, for example "locked on star" and "rotating to target".

3.5.6 Techniques for building a current information aspect

3.5.6.1 Documenting existing information

When establishing the current information usage of a system (or systems) or the enterprise as a whole, most of the information model will be found by examining data stored and used by systems. This section mainly concentrates on identifying the 'static' portion (i.e. entities, relationships and attributes and not state variables) of the current information aspect.

Participation rules, constraints in general, legal values, etc., will be found by examining operator manuals and existing program code. Information may currently be held in many formats and using many media. In order to abstract the underlying information components, this will need to be converted into a standard form that is more tractable than the diverse implementations currently being used.

(See also §5.3.2.2.1.)

3.5.6.1.1 Computer files

Each file will consist of records, with other constructs either explicitly or implicitly being used to associate elements of one file with those of another. Such links include set (or similar constructs) in network databases, foreign keys, etc. Each of these files should be converted into the standard YSM model components — entities, relationships and attributes. Generally speaking:

- **Files** correspond to entities (or associative entities).

- **Foreign keys, pointers, sets, etc**. correspond to relationships.

- **Fields (or their equivalent)** correspond to attributes.

Note: computer files include not only those files held on databases and on conventional storage devices as files, but also data tables in main memory, or hard-coded in programs. One factor that might be easily overlooked is the significance of file order. The position in a file may act as a hidden attribute or link. For example, in a sequential file, the position in the file might be the identifier. References to a specific record might then be defined by a number, giving the position in this file. This sequence number should be treated as if it were another field in the record in these cases.

3.5.6.1.2 Manual records

Paper and card records should be dealt with in the same way. For record cards in a standard format there will be various fields of information on the card. In the case of paper records, pre-printed sequence numbers may also have some significance. These numbers may be used to identify (or sequence) the occurrence of the document. Information content of these forms is thus not just 'the written items'.

3.5.6.1.3 *Implicit information*

Care should be taken to include all information in the model, even information that is not written down. This is particularly a danger with manual operations, as many 'files' may be in peoples heads. For example, the way salesmen deal with customers may often involve implicit memory of what has happened in the past, an assessment of how important that potential customer will be in the future, and so on.

3.5.6.2 Organising the current information aspect

Note: the names of entities, relationships and attributes should be chosen taking into account the rules on uniqueness rules. Names of the components should be converted to names of real-world significance, rather than implementation names, as soon as possible.

There are usually too many files to construct the current information aspect in one step, so some organisation strategy is required. After building the information aspect, it may be possible to group files into groups of entities and relationships.[1] These would be sensible groups to start with. Unfortunately, prior to the information model being established, this will not be so obvious. Two alternatives are:

1. Convert each file to entities, relationships, and attributes. Then synthesise the model by joining together these many small fragments. Considerable redundancy and aliasing will have to be removed and then the model should be refined. After refinement, a set of views may be defined as ERDs.

2. Identify preliminary groups of related files and then convert each group into a view. After constructing the views, 'overlaps' between different views and aliases are identified and then resolved. The views are then repartitioned to form more cohesive views.

Of the two approaches, the second is likely to be most productive, as it tries to use a high-level organisational technique to control complexity — each unit can then be thought of in isolation. This is therefore the recommended approach with YSM. Warning: it is unlikely that the preliminary allocation to groups will correspond exactly to the final views identified, and the analyst will have to be prepared to carry out repartitioning as described.

3.5.6.3 Entity relationship diagram construction

For each group of files an ERD is then drawn up, with associated attribute definitions. The overriding principles are that the sum total information in the group of files is captured in an understandable form. The exact sequence of operations carried out to achieve this is not critical, but the following is a suggested sequence:

1. Draw an entity for each file. Initially this has the same name as the file, but this should be replaced as soon as possible by a name that reflects its real-world meaning.

---

1 This assumes that the original design was according to some strategy that put related information together. This is usually the case, competent designers probably came up with designs that were fairly well normalised, even before the theory was well-understood. Occasionally, really arbitrary data designs are encountered, with many entities and attributes combined in a set of overlapping files. The techniques described below should cope with both situations.

2. For each way of navigating from one file record to a record in of another, draw a relationship icon linking the two entities. (Specific mechanisms of this type include use of foreign keys[1], pointers, 'sets' in Codasyl network databases.). In effect, this step reverses a preceding implementation decision to 'post' a relationship into one of the entities involved in that relationship. This design technique will be discussed in later releases of YSM.

3. If a table uses a foreign key in the identifier, then it must correspond to a relationship or associative entity. In that case, it is likely that the other components of the identifier are also be foreign keys. (There may be alternative identifiers with keys that contain no foreign keys.) If there are no dependent attributes, the table is modelled as a relationship; if there are dependent attributes, it is modelled as an associative entity. The foreign keys are each replaced by a reference to an entity and documented (see the associative entity specification and relationship specification sections of §2).

4. Any entity that has been identified as being referred to by a foreign key (see above) must already be in the model; if not it is added. If the field corresponds to a time or date, an entity may or may not be required. If the time or date is used for other purposes (scheduling etc.) it is advisable to model it as an entity; if it is used only as a record of when something occurred, it should be modelled as a attribute.

5. If the significance of a relationship (meaning) can be determined immediately, this should be captured as the name of the relationship. Sometimes this may have to be deferred until later. The relationship may also be 'merged' with other relationships. This is discussed below.

6. Where higher-order relationships have been posted into the table, there will be several foreign keys for one relationship in the table. Removing these as relationships may lead to the higher-order relationship being replaced by several binary relationships. This is best avoided by checking that if there are several foreign keys in a table they could either be:

   • be given values separately: there is probably more than one real-world event involved and they should be regarded as distinct relationships.

   • be given values at the same time: where a pair of relationships "a" and "b" can be identified as being 'synchronous' in the sense that there is a real-world event that always gives rise to an occurrence of relationships "<P> a <Q>" and "<X> b <Y>" at the same time, they should be merged to form a higher-order relationship. (In this specific example it would involve the four entities "P", "Q", "X" and "Y".)

7. All components (fields, etc.) of each file should be documented as attributes of or state variables for the entities.

---

1 A foreign key is a field that is the identifier for another file. It is an implementation mechanism by which relationships can be represented in many data architectures. Design using these mechanisms will be covered in later releases of YSM.

8. Any attribute (or state variable) that does not seem to depend on the entity, but, rather, on a relationship that has been pulled out of the entity, should be reassigned to be an attribute of that relationship. This relationship becomes an associative entity. In effect, this step reverses a preceding implementation decision to 'post' an associative entity into an entity.

9. All attributes left in each entity should be dependent on the identifier of the entity. Semantic guidelines such as whether it can be a property of the entity in the table are the main guide here. However, normalisation may be helpful (see §3.5.8).

3.5.6.4 Resolving aliasesialiases
A common problem that will occur is that what appear to be different model components are really the same thing, called by different names. These different names may be used by di
fferent departments, designers of different systems, etc. Recognising these aliases is sometimes obvious and sometimes very difficult. The fact that they are aliases may only be realised after a very long time. This is discussed further in §3.6.1.

The information model constructed initially may have considerable redundancy in this respect. Duplicate relationships and entities will be found with different names. This will be true, whichever strategy has been adopted. Redundancy may also occur when models are merged. This redundancy should be controlled by identifying aliases and removing them. Any references to the removed components are replaced by references to the remaining component.

3.5.7 Refining information models

3.5.7.1 Naming criteria
As a general principle, names of components should have a real-world, rather than implementation bias. This will ease user review and understanding. It will also enable dependencies to be established, by clarifying what the information means, rather than how it has been physically organised in the past.

Specific criteria for names are given in the sections on specific modelling tools of §2. It is important to realise that the 'best' name is a subjective matter and can only be established by consensus and the use of review techniques (e.g. walkthroughs). The best source for possible names is the everyday speech of users and managers, *not* data processing staff!

3.5.7.2 View definition
The main modelling tool is the ERD, but this must be correctly organised. If the views chosen do not help in understanding separate 'areas' of the enterprise information aspect, then they may need to be repartitioned. This, in effect means 'moving some components to different views', 'joining views', 'splitting views', so that each view is more coherent and useful. No components are lost; they are just moved to different views.

3.5.7.2.1 Partitioning complex information structures into views
Where any view is becoming too large to review as one unit, it should be partitioned into smaller views. Repartitioning may also be required. In practice, this is carried out iteratively:

1. **Check each view for size, comprehensibility**: the view should show one related set of components (be cohesive). If too complex or not a cohesive group then split into two or more views. When splitting views, some repeats are allowed — entities may be seen on several views; relationships should only appear on one view; associative entities should appear on one view as a relationship. On other views they appear as entities.

2. **Check overlap between views:** if there is overlap between views, other than that specifically allowed above, then join them into a larger view.

3.5.8 Normalisation

3.5.8.1 General principles

The general concept of normalisation is that there are certain desirable criteria of a structural nature that the information model should conform to. These criteria have been given arbitrary numbers and if the model component conforms to the criteria, it is said to be successively in first, second, third … normal form.

It is not quite true to say that normalisation is merely structural as many of the criteria involve semantics, but it is certainly more concerned with syntax than meaning, as compared with entity-relationship modelling. However, normalisation is a useful technique, particularly in organising a model of existing stored data or for checking that the semantic techniques advocated by YSM have been correctly applied.

Under no circumstances should normalisation be mechanically applied without semantic considerations.

Note: modelling stored data does not give a complete information aspect. It also includes temporary information (data flows) and derivable information.

3.5.8.2 Tabular representation of data

To carry out normalisation, imagine a 'table of values'. This table may be a transcription of a file structure (either manual or computer). Alternatively, for a proposed entity, associative entity or relationship, the occurrences of that model component should be set out as rows in the table. The columns correspond to one of:

- **Attributes**: the entry in the cell corresponds to the value of that attribute for the occurrence of the entity which is that row of the table.

- **State variables**: the entry in the column corresponds to the current status of the variable for the occurrence of the entity corresponding to the row.

- **Relationship references**: the entry in the column corresponds to a way of identifying a row of another table. This type of column is usually called a foreign key in describing tables. YSM replaces foreign keys by the equivalent relationship, documented accordingly. In documenting existing files it may not be immediately apparent that they *are* foreign keys.[1]

---

1 The use of foreign keys in design will be covered in later releases of YSM.

Normalisation then applies tests to such a table, whatever its origin. Normalisation is usually applied to stored data only. Its primary aim is in avoiding duplication of stored data, with consequent problems in maintaining integrity. The treatment below can be applied to all stored attributes. However, YSM allows non-stored attributes to be derived from other attributes. Normalisation should not be used as an argument to say that these are not attributes.

3.5.8.3 First normal form

A table is said to be in first normal form if the table is rectangular, with a single value in each column and a single occurrence on each row. The following table is not in first normal form:

and should be converted to:

The values in a column must not contain repeats. However, the use of compound ADTs *is*

COURSE NO (identifier)	LOCATION	DATES
00137	London	10 April, 11 April, 12 April, 13 April, 14 April
00138	Paris	23 April
00139	Utrecht	28 April

allowed in a column. Thus, it would be allowable to have a column which contained an

COURSE NO (identifier)	LOCATION	START DATE	LENGTH
00137	London	10 April	5 days
00138	Paris	23 April	1 days
00139	Utrecht	28 April	1 days

address, even though all address data items might contain components. Compound ADTs are discussed in §2.10.5.

Generally speaking, anyone who is used to ERA modelling would immediately have converted any existing files to first normal form, so the above check is likely to be implicit rather than explicit.

3.5.8.4 Second normal form

A table is in second normal form if the value of each column requires the value of each of the columns that constitute the identifier to be known. This is termed functional dependency.

3.5.8.4.1 Functional dependency

A column a is said to be functionally dependent on columns b, c, … if knowing the value of each of the columns b, c, … necessarily implies a particular value for column a. There may be any number of columns in the list b, c, … . This list is called the determinant for a.

3.5.8.4.2 Checking for second normal form

A table is in second normal form if the value of each column is functionally dependent on the identifier (or identifiers, if there is more than one).

The value in each column may be an attribute or entity reference. The following table has some attributes that depend only on part of the identifier — for example, "course length" depends only on the specific course that is being run and not the specific date and location on which it is being presented.

(Identifier)				
COURSE	LOCATION	START DATE	COURSE LENGTH	NUMBER OF BOOKINGS
SBS	London	10 April	5 days	7
JCN	Paris	23 April	1 day	4
JCN	Utrecht	28 April	1 day	6
JCN	Washington	28 April	1 day	8

To convert this to second normal form, it must be split into two tables with an implied link between them. One contains all the above columns, with the exception of "course length", and the other contains "course" and "course length".

Checking for second normal form is a useful technique and should be understood by any information modeller. When applying it, the entity relationship fragments should be built up to show the links between the tables. These will become the relationships after refinement, naming, etc. For example, in the above case, examination of the table should lead to the following entity relationship fragment, with entity specifications for the two entities:

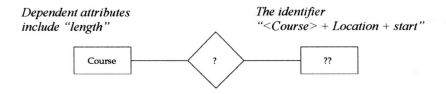

*Dependent attributes
include "length"*

*The identifier
"<Course> + Location + start"*

Note: the reference is part of the identifier here, so this must be an associative entity or relationship.

3.5.8.5 Third normal form

A table is said to be in third normal form if all attributes that are not part of the identifier depend only on the values of the identifiers. The example given below contravenes this if each consultant has a standard rate at which he is charged out.

(Identifier) CONTRACT NUMBER	CUSTOMER	CONSULTANT	START DATE	CONSULTANCY RATE
001567	SBS	Phil	19 April	320
001568	JCN	Tim	25 April	320
001569	JCN	Alex	22 April	320
001570	JCN	John	28 April	410

The "consultancy rate" is dependent on "consultant". In effect, an entity — "Consultant" and a relationship to the "consulting assignment" has been 'hidden' in the table.

CONTRACT NUMBER	CUSTOMER	CONSULTANT	START DATE	CONSULTANCY RATE

A well-known saying sums up third normal form (TNF) in:

"The value must depend on the key, the whole key and nothing but the key."

(In YSM terms, the 'key' is the 'identifier'.)

As with second normal form, the application of this criterion effectively splits an entity and relationship out of the table. This should again be documented by means of an entity-relationship fragment. The above example leads to the following fragment being identified:

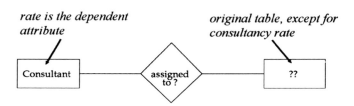

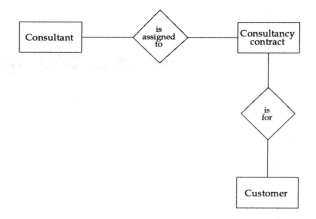

Note: the customer is also a foreign key. Replacing it by the equivalent relationship leads to:
This illustrates one of the dangers of applying normalisation as a purely mechanical technique,

without due consideration for the meaning of the information model. In fact, a better model of this would identify "Consultancy contract" as an associative entity. This would have been spotted, even with the mechanical approach, if the fact that "customer + consultant + start date" was an alternative identifier, but this might not be immediately established by the analyst.

3.5.8.6 Boyce–Codd normal form

The 'third normal form' discussed above is the original formulation, but it does not deal with multiple identifiers, particularly where some or all of the identifiers are compound. In this more general case, an amended criterion is used. This is 'Boyce–Codd normal form' (or BCNF for short). Using the definition of functional dependency given in §3.5.8.4.1:

> A table is in BCNF if and only if *every* determinant is an identifier.

An example of a table in BCNF with two identifiers is shown below:

The check for third or BCNF is useful and should be understood by any information modeller. Note: normalisation should not be applied without careful semantic checks. The above is

(Identifier)	(Identifier)			dependent attributes	
COURSE NO	COURSE	INSTRUCTOR	START DATE	DATE SCHEDULED	NUMBER OF BOOKINGS
137	SBS	Mike	10 April	5 Jan	7
138	JCN	John	23 April	7 Jan	4
139	JCN	Julian	28 April	8 Jan	6
140	JCN	Sylvia	28 April	3 Feb	8

readily identified as an associative entity ("Scheduled course" perhaps?) that refers to "Course", "Location" and "Day"). However, it contains some derivable information — "bookings" is the number of occurrences of the relationship that refers to this scheduled course. The corrected version of this part of the information model is shown below:

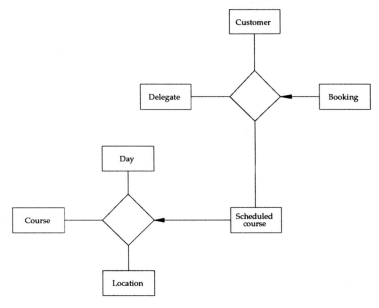

3.5.8.7 Fourth and fifth normal form

The criteria for testing these are adhered to are not explicitly used within YSM. Rather, it is advised that the meaning of information in tables is carefully examined and any relationships carefully modelled. In particular, if two separate relationships are established, they must be shown as distinct links. For example, the table:

(Identifier)		
COURSE	INSTRUCTOR	HANDOUT
SBS	Mike	notation card
SBS	Paul	SBS notes
SBS	Paul	notation card
SBS	Mike	SBS notes
JCN	John	notation card
JCN	John	JCN notes
JCN	Julian	notation card
JCN	Julian	JCN notes

which shows courses, instructors who can teach those courses and 'handouts' that are relevant to the course can be modelled by:

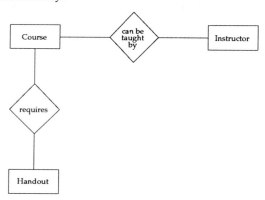

This shows that the table is really a combination of two distinct relationships into one table. If a table fails the fourth normal test criterion it is because there are several relationships within it, as in this example. Semantic, rather than normalisation techniques are the best way of avoiding these problems.

3.6 Maintaining the enterprise information aspect

Maintenance of the enterprise information aspect is an open-ended activity. The model will need to be changed to take account of the changing information requirements of the enterprise.

Some of these new requirements will be a continuation of the enterprise modelling activities discussed in §3.5.1 and some will be to support new systems as they are developed.

The maintenance activities can be divided into the following categories:

1. **Additions of new items**: these are either:

   - not identified as part of the enterprise information, although they were, in fact used. This omission now needs to be remedied.

   - New enterprise information requirements. consequent to a change in the activities of the enterprise or the environment in which it operates.

   In either case, the EEM must be extended to include the new item or items. One problem that arises is that what is potentially a new item, is not in fact so, but an existing item referred to under another name. This is referred to as 'aliasing'.

2. **Deletion of items**: over a period of time, the operation of the enterprise may change, so that certain information items are no longer required. This may lead to them being removed from the enterprise information aspect.

3.6.1 Techniques for recognising aliases

This is not a trivial problem because in a large enterprise many different names are likely to be used for the same thing. If care is not taken, a model component may be duplicated with different names. As this is totally against the spirit of building a single information model, it is very important to identify and remove these aliases. This is a general problem and there is no rigorous way of identifying aliases, but the following are some suggestions. The 'similarity' of two components is the main clue to aliases. For example:

- If two entities participate in the same relationships, or have very similar attribute lists, then they may well be aliases, even though their names are different.

- Two relationships are similar if they have the same order. If they have the same order *and* refer to the same entities, then they may be aliases. Even if they seem to refer to different entities, they may really be the same entities, with aliases.

- If two attributes of an entity have the same ADT, then they could well be aliases. If they have one or two values that are different, then this clue is less strong, and so on.

3.6.2 Techniques for adding new requirements to an existing model

These requests may be submitted manually or electronically. They may even be semi-automatic in sophisticated CASE environments that have facilities for building system models and a supported enterprise information resource manager. The discussion here will assume that all these operations and decisions are visible and controlled by the information analyst. Future automated tools may hide some aspects of this, but the principle will still be the same.

3.6.2.1 Additions to EIA

If the additional information usage requirements is not a subset of the available information, as documented in the EIA, the EIA should be updated so that it becomes a subset.

Comparing information usage with the EIA may be conveniently carried out by 'laying' the system information views 'on top' of the enterprise diagrams. For example, all system ERDs must correspond to part of an enterprise ERD containing all entities and relationships used by the system.[1] Conceptually, we may show the system's information use as part of this large ERD, with some components highlighted. For example:

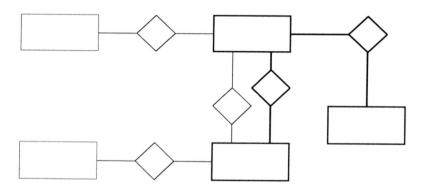

In a similar way system entity life-cycles must be a subset of the enterprise life-cycles. Attributes used must be included in those that have been specified as being available.

3.6.2.2 Modifying the existing model

If a component of the system information aspect cannot be identified in the existing EIA, even after checking for aliases, then it must be added. Where aliasing is suspected, the recommended procedure is to go back and talk to the users and subject-matter experts — guessing that 'they are probably aliases, we will assume they are' is *not* acceptable. This is discussed under each component below.

3.6.2.3 Resolving aliases

A common problem that will occur is that what appear to be different model components are really the same thing, called by a different name. These different names may be used by different departments, designers of different systems, etc. Recognising these aliases is sometimes easy and sometimes very difficult. The fact that they are aliases may only be realised after a very long time. This is discussed further in §3.6.1.

---

1 It is not recommended that this single diagram be drawn — it would contravene all modelling guidelines on complexity.

The information model constructed initially may have considerable redundancy in this respect. Duplicate relationships and entities will be found with different names. This will be true whichever strategy has been adopted. Redundancy may also occur when models are merged. This redundancy should be controlled by identifying aliases and removing them. Any references to the removed components are replaced by references to the remaining component.

3.6.2.4 Adding new entities
Checking for aliases may sometimes be quite time consuming. Good clues for aliases include similarity of names and participation in relationships with the same or similar names. If an entity is to be added to the EIA, then it must be added to at least one Enterprise ERD and be provided with an entity specification.

3.6.2.5 Adding new relationships
A good clue to two relationships being aliases of each other is that they involve the same entities. However, this is only a clue — there are many cases of where there is more than one relationship between a pair of entities. For example "<Man> is married to <Woman>", "<Man> met <Woman>", "<Man> is the partner of <Woman> (at tennis, say)" … .

For higher-order relationships, the clue, of course, is stronger.

3.6.2.6 Converting relationships into associative entities
If a system needs to store attributes for a relationship that does not currently have any attributes in the EIA, then this relationship must be converted into an associative entity. This implies that its existing relationship specification is converted into an associative entity specification. This should present few problems — the only differences between the two specifications are additions for attributes and possible additional identifiers. These specifications are covered in §2.

3.6.2.7 Adding new attributes
If an entity is described as having a specific attribute in any system entity specification, then the enterprise entity specification must show this attribute. If it does not, then the attribute should be added to the enterprise entity specification, with a corresponding attribute specification. If the attribute maps to a new abstract data type, then proceed as under adding new abstract data types (§3.6.2.9).

Care should be taken to ensure that it really is a new attribute. A common problem is 'aliasing' — the attribute may already be defined for that entity. A clue to this will be the abstract data type or value list of the attribute.

Another problem may be incorrect attribution — the attribute may have been ascribed to the wrong entity. This should be checked by ensuring that each occurrence of the entity leads to one well-defined value for the attribute. If this is not so, then the purported attribute is not really an attribute of this entity.

3.6.2.8 Adding new subtypes and subtypings
If a group of entities is found to have some property that is common to that group, but not to other occurrences of the entity, then they should be considered to be a candidate subtype. If that entity has already been subtyped, check all subtypes to determine whether it is a new subtype. Note: two subtypes of a subtyping may not overlap — if the new group overlaps an existing group, then a new subtyping must be created.

3.6.2.9 Adding new abstract data types

If an attribute maps to a new set of values, then it may either be specified as having an explicit value list, or a new abstract data type may be created. Generally speaking, if this set of values is likely to be re-used, or corresponds to an identifiable quality that may be possessed by many entities, it is better to create a new abstract data type. The definition of an attribute as an existing ADT or as a value list is covered in §2.8.3.6. See also guidelines for abstract data type specifications in §2.10.7.

Note: abstract data type specifications are part of the Enterprise Resource Library.

3.6.3 Removing unused information components

If no existing or planned system makes use of an EIA component, that component may be a candidate to be dropped from the model. This may occur because of 'over enthusiastic' information modelling going beyond the scope of the enterprise activities, or it may occur because a system project has made a request for provision of a component and that requirement no longer exists.

Dropping such unused components automatically is not advisable without careful consideration. Generally speaking, the effort required to keep it in the model is considerably less than that expended in identifying that it exists and what it means. It would also mean that strategic information modelling, in advance of system information requirements analysis, would become rather vacuous.

As a consequence, therefore, it is advised that such components are retained in the information model, provided they do correspond to real-world information that might be used by the enterprise at some point in the future. Planning an implementation to store such information is, of course, not justified in this situation. Later releases of YSM will discuss this in more detail.

3.6.3.1 Retiring unused attributes

If no system makes use of an attribute, it is not required by the enterprise and can be removed from the model. The usage of an attribute can be determined from the system entity (and associative entity) specifications. However, this is an extreme view and does not take account of the possibility that the attribute will not be important in future. Attributes should thus not normally be removed, unless they have either been replaced by an alternative attribute (this corresponds to a name change) or the attribute no longer has any meaning in the real world.

If an attribute is removed from an associative entity it may revert back to being a relationship rather than an associative entity. This will occur if there are no remaining attributes and the associative entity does not participate in any relationships.

3.6.3.2 Retiring unused relationships

If no system makes use of a relationship, the relationship is not required by the enterprise and can be removed from the model. However, this is an extreme view and does not take account of the possibility that the relationship will not be important in future. Relationships are therefore not removed, unless they have either been replaced by an alternative relationship (this corresponds to a name change) or the relationship no longer has any meaning in the real world.

3.6.3.3 Retiring unused entities

If no system makes use of an entity, it is not required by the enterprise and can be removed from the model. However, this is an extreme view and does not take account of the possibility that the entity will not be important in future. Entities are therefore not usually removed from the model, unless they have either been replaced by an alternative entity (this corresponds to a name change) or the entity no longer has any meaning in the real world.

If an entity is removed, all its attributes and any relationships in which it participates must also be removed.

If an entity is renamed, all its attributes and any relationships in which it participates in must be transferred to the new entity.

3.6.3.4 Retiring unused associative entities

If no system makes use of an associative entity, it is not required by the enterprise and can be removed from the model. However, this is an extreme view and does not take account of the possibility that the associative entity will not be important in future, either as a relationship or as an entity. Associative entities are therefore not usually removed, unless they have either been replaced by an alternative associative entity (this corresponds to a name change) or the associative entity no longer has any meaning in the real world.

It may sometimes be appropriate to convert the associative entity to a relationship, thus retaining this part of the associative entity for possible future use. Remodelling the associative entity as an entity which is not an associative entity is really equivalent to creating an entity with the same attributes, rather than converting the associative entity.

If an associative entity is removed from the model, all its attributes and any relationships in which it participates must also be removed.

If an associative entity is renamed, all its attributes and any relationships in which it participates must be transferred to the new entity.

3.6.3.5 Retiring unused subtypings

If no system distinguishes between different subtypes of an entity, it is not required by the enterprise and can be removed from the model. However, this is an extreme view and does not take account of the possibility that this distinction will not be important in future. Subtypings may be removed if they have either been replaced by an alternative subtyping (this corresponds to a name change) or the distinction is no longer of any significance.

3.6.4 Effects on the Enterprise Resource Library

3.6.4.1 Retiring unused abstract data types

If no attribute uses a specific abstract data type, it is not required by the enterprise and can be removed from the ERL. This is an extreme view and does not take account of the possibility that attributes might be declared to use this ADT at some future time. ADTs should thus not normally be removed, unless there is confidence that they will *never* be used for attributes in the future.

3.6.4.2 Retiring unused enterprise operations

A system makes use of an enterprise operation if any of its minispecs include a reference to that operation. Strictly speaking, if no system makes use of an enterprise operation, it is not required by the enterprise and can be removed from the ERL. However, this is a very extreme view and does not take account of the possibility that this operation will not be used at any time in the future.

3.7 Rules

3.7.1 Identification of model components

1. Each ERD must have a unique name.

2. The name of an entity is its identifier and must therefore be distinct from the names of all other entities and associative entities.

3. A relationship is uniquely identified by the entities it associates and its name. Relationships between other entities may have the same names), but no other relationship between the same group of entities may have the same name.

4. The subtyping name must be unique within the supertype entity, but the same subtyping name may be used for other supertypes.

5. If an entity is declared to be a subtype of more than one entity, then it must not inherit two (or more) attributes of the same name, except where these are inherited from a common 'ancestor'. See §2.5.3.2.3 for further discussion of this.

3.7.2 Specification of model components

1. Each entity must have a corresponding entity specification.

2. Each relationship must have a corresponding relationship specification.

3. Each associative entity must have a corresponding associative entity specification.

4. Each subtyping must have a corresponding specification.

5. Each attribute must have a corresponding attribute specification.

6. Each state variable must have a corresponding entity life-cycle (usually an entity-state transition diagram).

3.7.3 Visibility of model components

1. Each entity must appear in at least one ERD.

2. Each relationship must appear on exactly one ERD.

3. Each associative entity may appear in several ERDs as an entity, but must appear in exactly one ERD with its role as a relationship clearly shown as links to the entities it associates.

4. Each subtyping recognised by the enterprise must appear in at least one ERD. It may appear in more than one ERD where the number of subtypes is too large to conveniently show on a single ERD.

5. All subtypes of a subtyping must be shown to be part of that subtyping on at least one ERD.

3.7.3.1 Events and information

1. Each event shown in an eSTD must appear in the entity–event table as an event that causes an access to that entity.

2. Each events shown in the entity–event table must have a corresponding specification.

3.7.3.2 Function–entity table/ERDs and entity–event table

1. Any entity, relationship, or associative entity shown in the function–entity table must appear in at least on ERD.

2. Any access shown for an entity, relationship or associative entity in the function–entity table must have at least one event in the entity–event table that causes that access to the same entity, relationship or associative entity.

3.8 Guidelines

3.8.1 Entity relationship diagram selection

1. Each ERD should show a set of related facts. This is difficult to quantify, but the view should be capable of supporting a discussion of one area of the enterprise activities.

2. The ERD should be capable of being given a simple, functional name which honestly records all types of fact that are shown on the view.

3.8.2 Model interpretation

1. Entities should correspond to a class of real-world 'things'.

2. Both relationships and associative entities should be dependent on the entities to which they refer. In other words, an occurrence of them should not be capable of existing without the prior existence of occurrences of each of the entities to which they refer.

3. Relationships and associative entities should reference entities that interact together at a specific point in time. If they do not all interact at the same point in time, then new entities should be introduced so that the different interactions may be related. Note: these will probably be associative entities.

4. If an entity has some attributes which are optional, or is involved in relationships with constraints that restrict the participation, then it may be helpful to categorise the entity into subtypes.

5. If several entities have similar relationships or attributes of the same abstract data type, then it may be helpful to create a supertype entity of which they are subtypes.

6. If an entity is used in different ways by different people and each group uses different criteria for using that entity, then it may be helpful to subtype the entity.

7. If two (or more) entities are treated in similar ways, then it may be helpful to create a new entity that they are both subtypes of.

8. If two (or more) entities have similar attributes, then it may be helpful to create a new entity that they are both subtypes of. It is not always obvious what similar attributes means, but they must (at the very least) be of the same abstract data type).

However suspicious the similarity between two components is, the analyst should *never* assume they are the same without checking with subject matter experts. They must have the same meaning, or semantics. The above similarity checks are mainly of a structural or syntactic nature. To be considered the same, two components must have the same syntax *and* semantics.

3.8.3 Model component names

1. Names of pure entities should correspond to a role that real-world objects can play in interactions with the enterprise. In general, this will correspond to a noun phrase (often a common noun).

2. The name of the associative entity should be a noun phrase that reflects the name that is commonly used when referring to occurrences of this association. For example, the term a "booking" would be used in everyday language by staff in an organisation that ran public courses. If the name is suggestive of real-world tangible objects, then it probably better to model this as an entity, rather than an associative entity.

3. All relationships should be named in the active voice, rather than the passive voice. Thus, for the relationship that links one employee to his/her superior, the name "reports to" is preferable to "is reported to".

4. Names of all relationships should allow someone with subject-matter knowledge to interpret what the relationship means without extensive explanations or reference to the relationship specification.

5. For binary relationships, a fact-pattern constructed from the relationship name should give understandable and verifiable statements about the real world when occurrences of the entities are filled in.

6. Names of higher-order relationships should allow a simplified version of the relationship to be constructed. For example, "is married to" is still an acceptable name for the relationship that links the "Man", "Woman" and "Location" entities (rather using a name derived from the full form of the fact-pattern:"<Man> was married to <Woman> at <Location>", for example "…was married to … at …").

4: The Relationship Between Systems and the Enterprise

Table of Contents

4.1 Introduction

4.1.1 The system and its environment

A system is a collection of functions and information, organised to a specific purpose or purposes. It may already exist, or there may be plans to construct it.

A system fits into a larger environment containing organisations, people and devices with which the system interacts.

The boundary between the system and the environment defines the scope of the system, or 'how big' the system is. Choice of this boundary is referred to as 'defining the scope of the system'. Some discussion of how an initial choice of system boundary might be made is given in §4.2.

4.1.2 The enterprise as a collection of systems

The enterprise consists of many systems. From another point of view, it is itself a system, containing functions and information, organised to achieve the 'mission' of the enterprise.

In fact, the distinction between the enterprise and a system is a relative one. The enterprise is a system, within that system there may be smaller systems, within those systems, there may be sub-systems etc. There are many principles that apply to all systems, whether enterprise or smaller, right down to the smallest components. One of these, for example, is that they can always be modelled 'from the outside in', breaking down the internal requirements in terms of the external demands.[1]

4.1.2.1 System life cycles

Several system projects will run in parallel, and all will run within the enterprise. The diagram below illustrates how several system projects begin and end at different times within the enterprise. The enterprise modelling effort does not end, but continues in parallel with system projects.

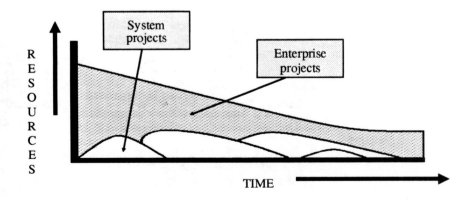

---

1 Others will be covered in later releases of YSM.

The enterprise models must support these activities simultaneously, often while they are still under construction. This requires careful management and awareness of the uses system projects make of enterprise models.

4.1.3 Enterprise and system models

Enterprise and system models serve different needs, but overlap in many areas. This overlap is beneficial to the enterprise and system projects, but must be controlled and exploited to its full benefit.

4.1.4 Enterprise support for System Essential Models

The Enterprise Essential Model (EEM) defines the sum total of information about the enterprise. Each system has an System Essential Model (SEM) that defines parts of the enterprise. Maintenance of the Enterprise Essential Model requires each system modelling team to ensure that the model they build is consistent with the Enterprise Essential Model, as shown below:

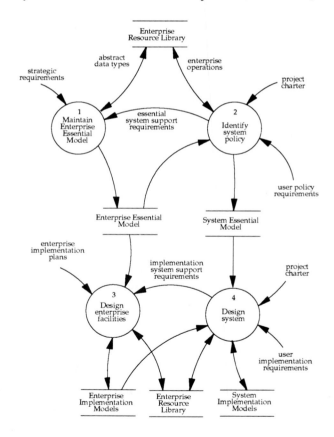

The Enterprise Essential Model is discussed in more detail in §3.

4.1.5 The Enterprise Resource Library

The Enterprise Resource Library defines all resources that are available to the enterprise. Use of the Enterprise Resource Library has the following benefits:

- **Re-use of functional logic and specifications**: operation specifications allows a function to be defined once and then used in many systems. Potential inconsistencies in the policy used in the various system can be avoided.

- **Abstraction of data representation:** abstract data types can be specified even though no attribute or other data element has yet been declared to be that ADT. Enterprise operations may be specified even though no system function has been identified as using them.

- **Implementation of shared resources**: this can proceed without duplicate work. These resources includes databases, object libraries, subroutines, comms software.[1]

This release of YSM covers essential models. The only components of the ERL that are covered are abstract data type (see §2.10) and operation specifications (see §2.17). Both ADTs and enterprise operations may be defined in anticipation of future use. They are an available resource that may be used in building the Enterprise Essential Model and System Essential Models.

The relationship between the Enterprise Resource Library and the System and Enterprise Essential Model is shown below:

---

1 Modelling re-usable software and hardware will be covered in later releases of YSM.

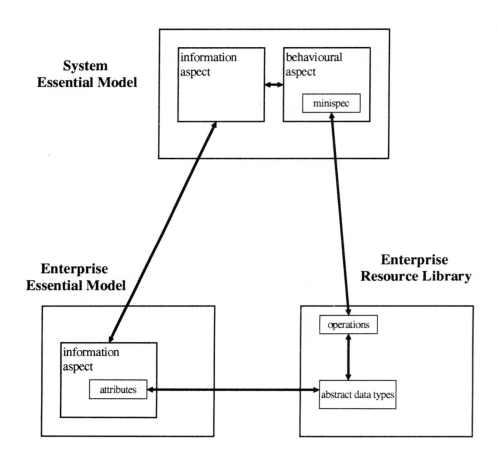

4.2 System Project Initiation

Deciding which systems to build and when is an important strategic problem. Goals of the company are examined and decisions taken on which system boundaries to choose and when projects should be initiated. This will be covered in later releases of YSM, but the treatment below shows how some of the tools described in section §2 may be used. It concentrates on the sharing of information between systems and the allocation of events to the scope of systems.

4.2.1 Enterprise models and strategic planning

System projects are initiated to benefit an enterprise. A whole discipline of business known as strategic planning exists to aid in the proper identification and initiation of system projects. Strategic planning is not covered in this release of YSM. However, these enterprise models will be key inputs to the decisions required to initiate system projects. The Enterprise Essential Model can aid in the strategic planning of the enterprise to decide which systems will be the most cost-effective to begin and the appropriate distribution of resources. Cost-benefit and feasibility studies can be made to ensure the timely initiation of a system project.

4.2.2 Project charters

A project charter is a definition of the protocol for investigation and possible construction of a system. It may include preliminary identification of the system's scope, in terms of events, functions, and entities, relationships within the system's responsibility.

4.2.3 Using the function–entity table

The function–entity table (see §2.19) is very useful in strategic planning. It shows the relationships between functions carried out and the information required to support those functions.

Function–entity tables may be used to study dependency between enterprise economic units or several possible systems. These studies are strategic planning activities.

For example, suppose three possible systems are under consideration. Each of these has had some preliminary allocation of functions. These functions would be high-level functional responsibilities, of the type that might be shown in a statement of purpose. The inter-relationship between the systems in terms of shared information can be visualised using the function–entity table.

Specifically, if system "A" is defined as creating an entity (or relationship) and system "B" as reading an attribute of the entity or matching the relationship, then system "B" is dependent on system "A". "B" will not work if "A" has not been built. This implies a dependency relationship between the systems. This must be taken into account when choosing which projects to initiate and when.

If B depends on A and C depends on B (and possibly A, too), there is not much problem. System A should be built first, then system B, and finally, system C. Unfortunately, things are not always that simple. It is more likely that B depends on A, C depends on B and A depends on C. When this occurs there are two choices:

- Develop several systems in parallel;

- Reorganise the proposed system boundaries and re-examine the dependency.

It is for this second choice that the function–entity table is useful in visualising the relationship between functions.

4.2.3.1 Ownership of information

This concept is quite useful, though not formalised in this release of YSM. Information may be loosely categorised into the following types:

- **private**: used entirely within a system. It cannot be accessed from other systems;

- **owned**: owned by one system, which is responsible for its integrity. Other systems may access the data, but do not own it.

- **shared**: used from within several boundaries. The information is truly shared between systems and no one system owns it.

Private information has little impact on system dependencies and can be ignored for the purpose of understanding the relationship between systems.

A system is defined to own a unit of information if all of the following are true:

- the system is responsible for creating and (usually) deleting occurrences of the item;

- if the item is an entity, the system is responsible for assigning values to attributes and changing the state of state variables;

- if the item is an entity, no other system assigns values to attributes or changes the state of state variables;

- no other system creates, or deletes occurrences of the item.

Any information item that is not owned or private is shared.

Although private items have no dependency implications, both owned and shared items have. An item that is owned by one system and used by another creates a dependency between the two system. This can be visualised using a function–entity table. The example below shows three proposed systems and information model components C_1, C_2, C_3 .... Within each system, possible functions F_1, F_2, F_3, have been allocated. Any item that is created, updated or deleted is shown shaded (state variables have been omitted from this table to simplify it). These items are potentially owned (if no other system has a shaded cell in this row), shared (if two systems have a shaded cell in this row), private (if no other cell in this row has an entry).

information component	SYSTEMS AND INCLUDED FUNCTIONS									
	System A			System B				System C		
	F_1	F_2	F_3	F_4	F_5	F_6	F_7	F_8	F_9	F_{10}
C_1	c	r	d		r					
C_2		c,r,u,d			r			r		
C_3	r,u,d	c				r				
C_4				r	c	d	u		r	
C_5								c,u	d	r
C_6				r	c	d				r
C_7	c,u		r	d	r		r	r		
C_8				r	c	d		r		r
C_9								r	c,u	u,d

It should be clear that if f_4 is reallocated from system B to system A, then the systems would have a simple dependency, so that A would be built before B, which should be built before C. The table can also be clarified by changing it to a more diagonal form to clarify the relationship between the systems:

information component	SYSTEMS AND INCLUDED FUNCTIONS									
	System A				System B			System C		
	F_1	F_2	F_3	F_4	F_5	F_6	F_7	F_8	F_9	F_{10}
C_1	c	r	d		r					
C_2		c,r,u,d			r			r		
C_3	r,u,d	c				r				
C_7	c,u		r	d	r		r	r		
C_8				r	c	d		r		r
C_4				r	c	d	u		r	
C_9								r	c,u	u,d
C_5								c,u	d	r

This diagonal form is the preferred form for function–entity tables. It helps to show the dependency between systems.

It may be difficult to avoid choosing system boundaries that prevented more than one system having the responsibility of creating occurrences of an entity. For example, in a service company, there will be many enterprise activities that cause a new occurrence of "Customer" to be identified.

The dependency of one system on another may also require examination of individual attributes or state variables. Although many systems might access the same entity, there might be groups of attributes or state variables that are only accessed by certain systems. These also create dependencies between the systems. This is outside the scope of this release of YSM.

4.2.4 Use of the entity–event table

The entity–event table (see §2.4) is also very useful in strategic planning. It can be used to carry out a high-level investigation into the feasibility of a proposed system. The events, entities and relationships would then be those that are believed to be within the charter of the proposed system. The effect of the events on the proposed information views of the system can be investigated for possible inconsistencies.

For example, if an event is one that leads to the creation of a relationship and that event is allocated to a system, then the system should have access to all components needed to verify the legality of that relationship. This will require match access to the entities involved in the relationship. The system will have, at least, to share the relationship and preferably own it.

If another event causes deletion of one of the entities involved in the relationship, then the system to which this event is allocated must be able to delete occurrences of the relationship. It is clearly better if these two events are allocated to the same systems. The relationship can then be totally owned by that system, although other systems can check for occurrences of the relationship.

As for the function–entity table, the situation is often more complex. However, the entity–event table is a good tool for visualising the relationship between events and information model components. It is therefore useful in helping to define the scope of systems in terms of which events they deal with.

4.2.4.1 Techniques for organising the table

Entities and relationships are should be shown together (or nearby) in the table if the same event causes an access to them. This is discussed in more detail in §2.4.4.

4.2.5 Using the function–entity and entity–event tables together

Using the two tables in conjunction allows iterative examination of choosing different system charters, in terms of the included functions, events and information components.

4.3 Building the System Essential Model

4.3.1 When the system is the enterprise

In some situations, the scope covered by a system is the whole enterprise. An example might be a software house working on one specific contract to support a client. The economic enterprise to the software house is precisely the system.

Then the scope of the enterprise models and system models are the same. They can be developed in parallel, even by the same personnel. However, all models should still be built.

It is strongly advised that the enterprise retain the distinction between enterprise activities and system project activities wherever possible. As many people have found to their cost, very few completely independent systems are built.

4.3.2 Building the Enterprise Essential Model first

Enterprise models can be built independently of any system project activity. Following the heuristics in the enterprise modelling chapters, views can be built for areas of the enterprise regardless of any potential system. These models will be independent of specific system project needs. The initial building of enterprise models becomes a project in itself.

This allows a more unbiased look at the enterprise, and frees the modellers to concentrate on areas of interest, rather then immediate need to support a system. The result may be a more neutral, cohesive set of models than other approaches to enterprise modelling.

Once the enterprise models are in place, the system projects have all this information already available to them. Building system model becomes a matter of sorting out which pieces of the enterprise model are of interest, and re-organising them into specialised views for the system. It will relieve the system project workload, as much of the work that was traditionally done within a system project to capture requirements and design systems has already done .

There is a drawback to building the enterprise models first. Depending on the size of the enterprise, building these models can take a lot of time and resources. It forces the enterprise to choose one of the following options:

- Suspend all system projects that might use these models while they are being built. This is rarely economically feasible.

- Allow the system projects to continue as before and deepen the problems of alias data and poor data integrity.

- Support both the independent and system project coordinated building of models, which will require more careful coordination and more resources.

4.3.3 Requests for enterprise model maintenance

In cases where the enterprise models are not complete, or they are currently being created, the system project may have to request that the enterprise modelling to support their area has a higher priority than independent enterprise modelling.

The system is constrained to only use information provided in those models. If, as often happens, the system project discovers some information not contained in those enterprise models, such as an event, or a required attribute of a particular entity, this request should immediately go to the enterprise modellers. They will approve the addition (or state why not) and add it to the models.

If, because of limited resources or other reasons, this does not occur, the enterprise modelling group is no longer supporting the system projects that are running. If it does not support other system projects, the enterprise modelling project is likely to fail. System projects will be forced to finish and will abandon attempts to synchronize with longer-term enterprise needs. The system will not be consistent with the enterprise definitions. In all likelihood, this means it will not be consistent with other or future systems as well. All benefits that may have been gleaned from enterprise modelling will be lost.

When the system project discovers or requires information that is considered to be under the realm of enterprise modelling, the enterprise modelling must place that work at a higher priority than independent work. Available resources should be applied to the system project request immediately in order to keep the enterprise models sufficiently complete to support the system projects underway.

4.3.3.1 Aliasing
Sometimes system investigations identify components that are already in the EEM, but under another name. These aliases should be removed by renaming the System Essential Model components.

4.3.4 Developing the Enterprise and System models in parallel

Enterprise and system models capture similar information. The major difference is their viewpoint: enterprise models capture information considering the entire enterprise; system models focus on the specific need for that information in one system.

If coordinated carefully, the necessary study of an area of the enterprise can be done once and result in both the enterprise and the system views. To achieve the best results, the enterprise view should be constructed first, with the system views created as an immediate subset, almost a copy, so the system may continue. The section on specific model support will consider how to work in parallel for each of the models.

All the system models rely on pre-existing enterprise models. In reality, no enterprise can take the time to stop all system project activity until it finishes the complete set of enterprise models. The enterprise can build enough of the enterprise models to suggest which system projects should continue, or begin. Then, provided careful coordination and management exist, it can build the enterprise and system models in the same general time span.

To build system and enterprise models in parallel, the resources needed, time and budget for building the required pieces of the enterprise models must coexist with the needed resources, time, and budget for completing a system project. If not, the system project will suffer as it must get needed information from the enterprise models before it can continue.

4.3.4.1 Overlap of system requirements

Most systems will only use part of the enterprise models, that is, they need only certain subsets of the enterprise resources that match the areas within their scope. Fewer resources will be needed for building the enterprise models if the concurrent system projects cover the same area of the enterprise.

4.3.5 The relationship between the enterprise and system information aspects

The enterprise information aspect is a static model describing each entity within the enterprise, its attributes, and the relationships between entities. This aspect also shows which events occur that will cause the creation, use and deletion of the entities.

The system information aspect uses subsets of this aspect, just those pieces of information, entities, relationships, and events that are within the system scope. These views must be a consistent subset of what is already in the Enterprise Essential Model.

The enterprise model feeds information to the system models on the system projects request. The system models may show new areas of enterprise information or resources that the enterprise models did not include. They request maintenance to the enterprise models, and take the new needed information from them when the enterprise models are complete.

4.3.5.1 Extracting the system information aspect from an existing enterprise information aspect

If the enterprise information aspect is built before the system information aspect, the system modellers follow these first steps in building the System Essential Model:

1. Build the statement of purpose.

2. Build the system information aspect (if not already done in strategic planning) by selecting those items from the enterprise information aspect that they will use.

3. Continue to build the rest of the components of the System Essential Model, using and ensuring consistency with the information aspect.

As the modellers continue to define the model in more detail, they may identify items that the system's information aspect does not include. For example, they may identify a new attribute of an entity when defining data flows. The system information aspect must be updated to reflect this attribute usage.

The system project will check the Enterprise Essential Model to see if that attribute has already been identified. If it is, it will be added to the system entity specification; the Enterprise Essential Model will not be modified. If it is not already in the model, however, the system modellers will request maintenance to the Enterprise Essential Model to add that attribute.

When the Enterprise Essential Model has been changed, the information aspect of the system can be changed, and will pass the consistency checking criteria of the System Essential Model.

4.3.5.2 Building the information aspects in parallel

It is not recommended that both enterprise and system modellers try to identify entities, relationships and attributes in parallel. (This is sometimes called the 'firstest with the mostest' technique, but is unhealthy competition.) Rather than speed the system project, it extends it as it creates aliases that take longer to correct than doing the work once in the first place. It is also very frustrating for subject matter experts to have to learn two vocabularies, one for each team of modellers.

4.3.5.3 When the system is the enterprise

When the system is the enterprise, or the system project has more available resources than the enterprise modelling, the same individual can take on both enterprise and system modelling roles. This is permitted as long as the criteria for ensuring the enterprise models remain unbiased are upheld, and the individual has the necessary skills to build either model.

4.3.6 Consistency between the Enterprise and System Essential Models

1. For any system entity, there must be a corresponding enterprise entity or associative entity.

2. For any system associative entity, there must be a corresponding enterprise associative entity, referencing the same entity occurrences with the same participation rules.

3. For any system relationship, there must be a corresponding enterprise relationship or associative entity, referencing the same entity occurrences with the same participation rules.

4. For any attribute seen in a system entity specification, there must be an attribute specification in the Enterprise Essential Model. The attributes must be for the same entities.

5. For any access relationship seen in a system entity–event table, there must be an identical access seen on an enterprise entity–event table.

6. Any condition seen on a system eSTD must be seen on the corresponding enterprise eSTD. This condition must be on a transition between the identical origin and destination states, and the action on that transition must be identical.

7. Each system event must have a corresponding enterprise event specification (it also has a system event specification, as part of the System Essential Model).

8. Each function shown in a system function–entity table must have a corresponding function in an enterprise function–entity table.

5: The System Essential Model

Table of Contents

5.1 Introduction

5.1.1 Role of the System Essential Model

A System Essential Model is a representation of a system's underlying policy. A system must carry out this policy no matter what implementation is chosen. The System Essential Model is a statement of requirements for a system.

The model is focused on the business, real-world, or subject-matter side of the system. This allows careful study and error detection without digression into a discussion of computer-technology related issues.

The System Essential Model is part of the Enterprise Essential Model. Views that are part of the System Essential Model are described with a 'system', rather than 'enterprise' prefix. In all cases, the prefix 'system' may be dropped when no confusion would result. (This chapter retains the prefix for minimum ambiguity, at the expense of being rather pedantic.)

5.1.1.1 Uses
The System Essential Model serves to aid:

- Validation of requirements with customers or users: as the model is free of issues unrelated to the subject matter itself, end users can decide if the model is an accurate statement of requirements.

- Verification against other documents containing requirements: the model structure permits checking against written documents containing system's requirements to verify that these have been covered.

- Verification against existing systems: the model structure permits tracing the model components to parts of existing systems to ensure completeness and consistency.

- Communication of requirements to all who must be familiar with the policy behind a system's functions: designers, new customers, managers, etc. can view the system's requirements either at a detailed level or in a more general overview. They can concentrate on just one area of the requirements without being forced to understand them all. Detailed requirements may be checked with full rigour.

5.1.1.2 Freedom from technological bias
The System Essential Model documents a system by suppressing any aspects of the implementation of the processing or storage requirements.

For example, a System Essential Model of a pay-roll system would show "Employee number" and "hours worked" as system inputs. It would be incorrect to show "Time card" — the use of a paper card to contain this information is a specific implementation. The model permits capture of the requirement to input "employee number" and "hours worked". This allows concentration on that information, rather than the medium carrying the information.

The System Essential Model gives the underlying policy that may be carried out by a variety of implementations, old and new. One organisation may run pay-roll on paper, another may computerise it. Each system would have the same essential requirements.

5.1.1.3 Benefits of the System Essential Model

Some of the benefits that result from the System Essential Model being used to capture essential policy, with no assumed implementation include:

- portability to different implementations;

- freedom to design the best system to meet the requirements, using any allowed technology;

- the System Essential Model is made smaller by suppressing technology. This allows it to be understood without technical details making the model larger and more difficult to review;

- the model is in a form that can be understood by subject-matter experts, who are not confronted with irrelevant technical details about the implementation.

5.1.1.4 The 'perfect technology' concept

To help suppress implementation details, the System Essential Model may be visualised as running in perfect technology. This ideal processing environment is defined as running on one conceptual processor, with:

- infinite, non-volatile storage capacity;

- zero instruction time;

- the ability to run any number of processes simultaneously;

- no errors — it never 'goes down';

- all I/O is carried out in zero time, using a conceptual, rather than any specific technology.

Although no such technology exists, imagining the model to be running in this environment helps suppress any effects that would result from any real implementations. This aids the concentration on policy, rather than implementation.

5.1.2 Assumptions of the System Essential Model

There is an assumed architecture for the System Essential Model. This is:

- data processing carried out by data processes. Many data processes can be active at the same time. Data processes can be time–discrete or time–continuous;

- an entity-relationship style stored data model, with zero access time and infinite storage capacity;

- control and sequencing is carried out by state machines. Each state machine may communicated with other state machines using event flows. State machines may activate (enable/disable and trigger) other state machines and data processes.

These are the fundamental building blocks out of which the System Essential Model is constructed.

5.1.3 Constraints on the System Essential Model

Although the System Essential Model assumes perfect technology inside the system, it does not assume perfect technology in the system's environment. The System Essential Model is responsible for dealing with these errors when they occur and protecting the integrity of the system.

The System Essential Model must respect any limitation in the environment that cannot be changed. Only features within the system scope may be changed. For example, it would be wrong to model a device that generates a signal saying 'whether it is on or not' as sending the 'speed of its motor'.

Although the System Essential Model is not targeted to a specific implementation, there must be *some* technology available that *could* carry out the requirements. For example, a system that billed customers without knowing their purchases would not be possible.

5.1.3.1 Support for ad-hoc processing

If there is a requirement that a system support ad-hoc enquiries, the enquiries should be excluded from the system scope. The response to the enquiry, whether it is a response to an incoming request, or the collection of stored information, is, however, the responsibility of the system.

Where there are known number of enquiries, whose form is pre-determined, the system carries out a response for each. These are not really ad-hoc. There will be an incoming data flow, a system response and an outgoing data flow.

Where the responsibility of the system is to collect the information and store it for ad-hoc enquiries, this information should be shown as a 'write-only' store on the context diagram. See §2.12.3.8.1 for further discussion of how to model this.

5.1.4 Viable systems

Some systems may be conceptually possible, but very expensive to implement using the existing, available technology. Although it is possible to build and verify System Essential Models of such systems, it is the responsibility of the analyst to make the customer aware of cost implications.

Sometimes the scope of the system is fixed and 'non-negotiable'. For most systems, however, the choice of system scope is a balance between cost and benefits. Of two systems, one with a larger scope, that includes the other system, the one with the larger scope is usually (but not always) more expensive than the other. There are many 'trade-off' decisions that have to made.

The System Essential Model is a significant investment, so the analyst should try to avoid building a System Essential Model for which there is no justification. To do this, estimates of the likely cost of an implementation should be communicated to the customer at early review meetings to refine the scope of the system. Full details of the cost estimates are outside the scope of this manual, but some discussion of 'sizing' of the System Essential Model is given in §5.9.

5.1.5 The System Essential Model and the system life cycle

The System Essential Model serves as both documentation of user requirements for the system and also organisation of these requirements for reference by people building or maintaining the system.

As the requirements for a system are the most critical part of systems development, the System Essential Model is very important in the life cycle of a system. This model should always be the first system model to be built.[1] This ensures the least iteration. Requirements are the basis of the design; as requirements change, designs will change dramatically. Designs may change, without affecting essential requirements.

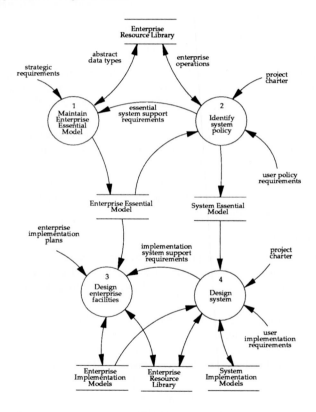

The stage of building the System Essential Model is often referred to as the 'analysis', or 'system's requirements' phase.

---

1 Some 'radical' methods do not make this demand and 'reverse engineering' might not go back as far as this model. However, for the present, YSM demands that this model *is* built.

5.2 Structure of the System Essential Model

5.2.1 System Essential Model views

The System Essential Model is too large to be understood in one sitting as a whole. Because of this, different views of the model are used. Each view shows part of the model that is used in a particular review or requirements gathering session. Each type of view is designed for evaluation and review of a specific area of concern.

The System Essential Model requires many views to be defined to complete the model. Some view types are 'single instance', for example, the statement of purpose; some view types have many instances, for example, the minispec. The views inter-relate and together produce a cohesive specification of the system's requirements.

5.2.1.1 Views used in different technical support environments

In an automated environment, all views are useful — each presents a different concern, without overloading the view with unrelated information. The automated tool is responsible for maintaining the integrity of the views. As views are created, information that has already been entered is available to help build the view. This allows a highly iterative, 'spiral' approach to model building to be used.

In a pencil and paper environment, demanding that all views are built would overload the modeller — the same information would have to be captured redundantly on many views. In pencil and paper environments, therefore, the method allows some of the views to be regarded as optional. They may be created if insight is gained from them. They do not have to be maintained as part of the delivered System Essential Model.

5.2.1.2 Modelling function, time and information

The three major viewpoints of a system are function, time, and information. Some of the views in the System Essential Model relate to one of these viewpoints. Thus the event list is concerned with the time viewpoint, the entity relationship diagram is concerned with the information viewpoint, etc. Other views relate to several viewpoints. For example, the entity–event table relates to information and time.

The diagram below shows the views used in the System Essential Model positioned according to their approximate positioning in a 'function–information–time' triangle. Higher-level views are shown at the top, more detailed views lower down.

Note: For ease of reference, these diagrams show the views in terms of the tool recommended for them. Alternative presentations are possible (for example, a tabular form of a state transition diagram is allowed).

Map of System Essential Model views

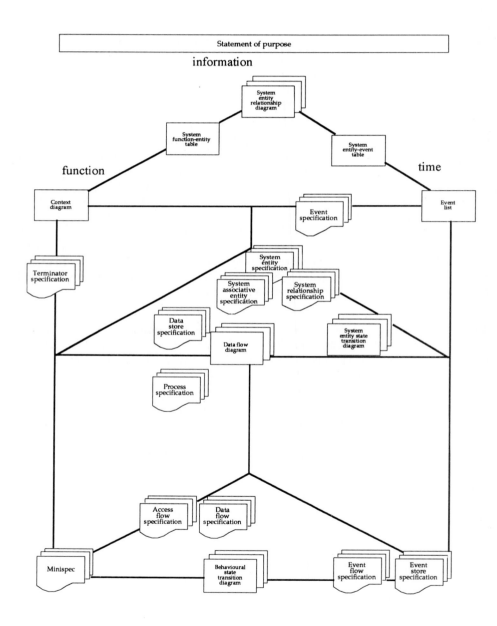

5.2.2 System Essential Model aspects

An aspect is a group of views into a model, selected to highlight one particular type of concern. (The 'view' and 'aspect' concepts are further discussed in §1.2.2.1.)

The standard aspects of the System Essential Model are shown below:

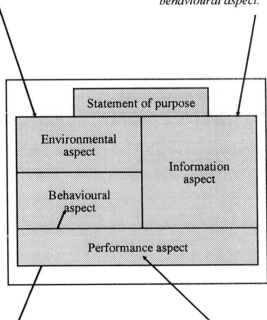

This aspect of the model highlights the system's scope and reasons for existence. It is constructed before the behavioural aspect.

This aspect focuses on the systems use and modification of information. In addition, the entity-event table is used to link this aspect to the environmental aspect; the function-entity table is used to link the information aspect to the behavioural aspect.

This aspect elaborates the functional and dynamic behaviour of the system. It is the system's response to the demands of the environment, as specified in the environmental aspect.

This aspect highlights such issues as response times, mean time between failures, amount of processing/unit time, etc.

5.2.2.1 Information aspect

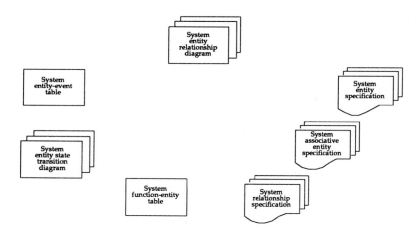

The information aspects main role is to 'link' the System Essential Model to the enterprise information aspect. The information aspect includes:

- **System entity relationship diagrams**: to declare the enterprise entities and relationships used by the system;

- **System entity state transition diagrams (system eSTDs)**: to define changes of state of entities;

- **System entity specifications**: to define the system's use of entities;

- **System relationship specifications**: to define the system's use of relationships;

- **System associative entity specifications**: to define the system's use of associative entities;

- **System entity–event table:**[1] shows the access to entities and relationships when events occur. This table also acts as a link to the environmental aspect;

- **System function–entity table:**[2] this shows which system functions access which parts of the information aspect. It acts as a link to the behavioural aspect.

Specifications such as the attribute and subtyping specifications, which are part of the Enterprise Essential Model, are also used to check the System Essential Model

---

1 Optional view, depending on support environment.

2 Optional view, depending on support environment.

5.2.2.2 Environmental aspect

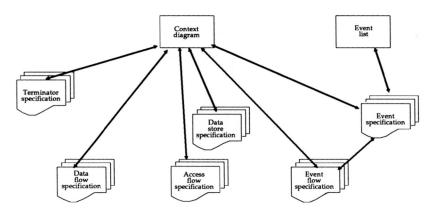

This aspect includes:

- **The context diagram**: to model the system interfaces. this may be a single diagram, or broken into several partial context diagrams;

- **Terminator specifications**: one for each terminator;

- **Data flow specifications**: for each data flow shown on the context diagram;

- **Event flow specifications**: for each event flow shown on the context diagram;

- **Data store specifications**: for each store seen on the context diagram. These define which entities and relationships are 'inside' stores;

- **Access flow specifications**: for each named access flows seen on the context diagram;

- **Event list**: defining the events that the system must respond to;

- **Event specification**: each event has a specification, either in frame or table format.

5.2.2.3 Behavioural aspect

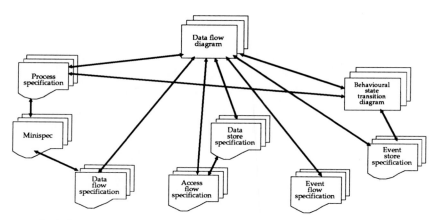

This aspect includes:

- **Data flow diagrams**: the context diagram and all its 'child' DFDs;

- **Process specifications**: one for each process in the system;.

- **Behavioural state transition diagrams**: to specify changes of system behaviour. (As an alternative, this may be shown in tabular form);

- **Minispecs**: for each function used by the system. There will be at least one data process that uses this function;

- **Data flow specifications**: for each data flow shown on DFDs;

- **Event store specifications**: for each event store;.

- **Event flow specifications**: for each event store shown on DFDs;

- **Data store specifications**: to define which entities and relationships are 'inside' stores shown in DFDs;

- **Access flow specifications**: to define the access to stored data by each process.

5.2.2.4 Performance aspect

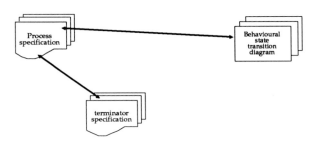

The performance aspect is used to capture requirements about system performance. It is an important input to 'sizing' calculations that are carried out in allocating to processors. It is concerned with the amount of 'work' that the system has to do and the reliability of the system and subsystems.

This aspect includes:

- **Process Specifications**: each process has a required response time, mean time between failures (MTBF) and mean time to repair (MTTR). The required processing (number of 'instructions') for each activation is also defined. Continuous processes have a discrete sample frequency that they must exceed. Where a process is multi-instance, the number of instances (or maximum number, where this is variable) of the process must be given;

- **Behavioural state transition specifications**: each change of state must have a maximum time for the transition to be completed after the condition becomes true;

- **Terminator specifications**: each terminator is single or multi-instance.

Information from the Enterprise Essential Model is also used. In particular, each event has expected frequencies and each entity, relationship and associative entity has an 'expected number' of occurrences. In effect, the System Essential Model 'inherits' this information from the Enterprise Essential Model.

This aspect defines how much the system is allowed to deviate from 'perfect technology'. The System Essential Model may be checked for logical completeness and consistency without it. However, it is required information in the sense that any implementation model which does not satisfy these requirements does not meet the users requirements.

5.3 Strategies for Building the System Essential Model

5.3.1 Sources for finding essential policy

The main sources for the System Essential Model are:

- **charter of system project**: this defines in general terms what the scope of the system should be. The charter is used to derive the statement of purpose and also to determine the way the project is run;

- **subject–matter experts**: the required policy for the system can only be ascertained by discussion with people who can state what this policy should be. Depending on the type of system, these subject–matter experts might be any of end-users, managers, business analysts, engineers. Policy is elicited from these experts through interview (see §5.6.2.1 for an example showing the identification of events);

- **existing systems**: it is often the case that a project is mandated to build a replacement to an existing system. In these cases, examining the existing system may be profitable (see §5.3.2.2);

- **the Enterprise Essential Model**: where this exists, even in part, it is an obvious source of system policy. The system is part of the enterprise and the System Essential Model is therefore a restricted subset of the Enterprise Essential Model.

5.3.2 Deciding on the best source for the system requirements

A system's requirements can exist in the real world in many sources: existing systems (automated or manual), existing documentation, papers, people's minds, etc.

If there is an existing system, the project must determine if it is beneficial to build models of existing requirements and add any new policy, or to glean requirements and information from these systems and build them directly into one model of the new system's requirements.

5.3.2.1 Modelling the new system directly

If there is going to be a dramatic change in a system, both in its essence and its implementation, building any model of the existing system may not be worthwhile. Much of it may change, and the modeller may be biased against change.

It is best to ignore an existing system if it is not understood by the project participants, is poorly documented, or is known to be full of erroneous policy.

Sometimes, such as when developing a new piece of research equipment, there is no existing system.

5.3.2.2 Modelling existing systems

Modellers may build a System Essential Model of the existing system first, before moving on to a new system's model. The modellers use the modelling tools to document a specific area of study, possibly including manual and automated systems, as it exists at present.

There are two ways of modelling existing systems:

1. Modelling the existing implementation, and then 'distilling' the essentials of the system.

2. Modelling the essence of the existing system directly.

5.3.2.2.1 Modelling the existing implementation

DeMarco [DeM78] recommended building a 'Current Physical Model' (similar to the Processor Implementation Model) of the existing system. This model, organised around the technology, includes all functions existing to support the implementation. After the Current Physical Model has been completed, the modellers remove the implementation characteristics of the system from the model, and reorganise it to reflect the essential functions only. At this point, they add, change, or delete essential policy as required for the new system.

This approach can still be used, although, because YSM now recognised more than one implementation model, there are several options on which model to build. A full 'reverse' engineering procedure would require that the Code Implementation Model, Human Implementation Model and (possibly) Hardware Implementation Model be built. The Processor and Software Implementation Models can be derived from these by a process of re-organisation.

This technique has the advantage of easy traceability to existing systems; the processes found in the implementation models will match to real people and computers. It is very thorough; it documents all details, including details of the current implementation.

That thoroughness incurs a high cost because the (current)implementation models may be very large. Further, they may be irrelevant after they are reduced to the System Essential Model by removing technology effects. After building it with great time and effort it is then discarded.

When studying the existing implementation at length, the modeller may find it difficult to start ignoring the old implementation when reducing the model to its essence. Sometimes the analyst will never uncover the underlying policy.

Because of these limitations and the great cost involved, a full set of models of the current implementation should only be built if either of the following is true:

1. Traceability back to the current technology is required.

2. One or more of the (current)implementation models has a use in its own right.

Usually, this is when the current system is not going to be completely replaced but only slightly modified. Maintainers may need the (current)implementation model to locate where the policy specified in the System Essential Model is located.

5.3.2.2.2 Modelling the essence of an existing system

Building a (current)implementation model is rarely recommended, but building a System Essential Model of the existing system is often the best course. The system under development quite often has very little change in essential policy from the existing system. For example, if a manual pay-roll system is being automated, the policy will remain the same, although the technology changes. For an existing automated system, moving it from batch sequential files to an on-line database is unlikely to affect it's underlying policy.

When the underlying policy is unlikely to change, and especially if the existing system is well documented, it may pay to build a System Essential Model of the system as it exists before considering desired policy changes. The modeller should model the existing system, ensure the model's quality, and then add, change or delete essential policy to reflect the new system's requirements.

This technique has the advantage of traceability to the existing system, discovery of unknown errors and problems, and provision of a base model to provide a place to restart if new requirements have been incorrectly added.The modeller should 'scratch out' a diagram of an existing implementation whenever it helps in building the System Essential Model. These diagrams do not become a formal part of the model.

5.3.2.3 Deciding which strategy to use

The following algorithm shows whether it is worthwhile building the (current)implementation or (current)System Essential Model:

5.3.3 Heuristics for constructing the System Essential Model

5.3.3.1 Definition of a heuristic

A heuristic is a means of discovering a solution to a problem. While there is no guarantee that using this technique leads to the *correct* (or best) solution, it is a pragmatic way of tackling the problem to obtain *a* solution. The result should be reviewed and improved to obtain a better

> *If there is an existing system which is well-understood, with the correct policy which is largely to be retained:*
>
>   *1.   If much of the implementation is to be retained:*
>
>   *1.1.1. Build (current)implementation model.*
>
>   *1.1.2. Remove technology features to give (current)Essential Model.*
>
>   *1.1.3. Re-organise (current)Essential Model to an event-partitioned format.*
>
>   *1.1.4. Add new policy to give (new)Essential Model.*
>
>   *1.1.5. Map onto (new)implementation models, using any relevant technology features from (current)implementation model.*
>
>   *else*
>
>   *1.2.1. Build (current)Essential Model.*
>
>   *1.2.2. Re-organise (current)Essential Model to an event-partitioned format.*
>
>   *1.2.3. Add new policy to give (new)Essential Model.*
>
>   *1.2.4. Map onto (new)implementation models.*
>
>   *end*
>
> *else*
>
>   *2.1. Build (new)Essential Model.*
>
>   *2.2. Map onto (new)implementation models.*
>
> *end*

solution. Heuristics provide a quick and cost-effective way of approaching complex problems that would be impossible or very expensive to solve with no guidance.

5.3.3.2 Reasons for heuristics

All of the model-building activity requires iteration and input from subject-matter experts. The rules thus require human input — the reason for modelling something in one way rather than another cannot be written down in a rule that would *always* apply.

On automated tools, there are two types of rules that can be applied to models. Those that can be applied without manual intervention are referred to as 'closed' rules. An example would be constructing the parent process of a DFD by suppressing all inter-process flows on the lower-level diagram. Other operations require manual intervention, because it is not possible to write down a closed rule to carry out the operation. These 'open' rules correspond to heuristics.

Heuristic techniques are those designed to give help in these more 'open' situations.

5.3.3.2.1 Heuristics in context

As mentioned in §1.4.6, YSM does not force any one set of procedures on the system modeller. Rather, model structure and modelling tool syntax and semantics are defined. A set of techniques (some of which are heuristics) are suggested as helps in building the models.

There are rules about the structure and completeness of models, the way the modelling tools are used etc., but in other areas, YSM still leaves a degree of choice for the system modeller.

5.3.4 Construction sequence for the System Essential Model

Standard procedures have been developed for building System Essential Models. The most common are outlined here. They are recommendations, not required methods of proceeding. There is no one strategy for what to model, nor any single way to build the model. Any means that can provide all needed components of the required quality are acceptable.

5.3.4.1 Parallel approach to building the System Essential Model

This approach 'flits from view to view, gaining some insight, before going on to the next'. It is very iterative.

It is not a cost-effective strategy without automated support, because of the large amount of reworking that has to be carried out.

5.3.4.2 Step by step approach

Rather than a parallel approach, a more 'step by step' approach to constructing the model may be adopted. Each view is completed before moving on to the next. There is less reworking of views than in the parallel approach.

There is no single sequence in which the views of the System Essential Model must be built. However, for the sake of planning, there is a standard order. This order can be changed if a 'short cut', or easier sequence of building the views is identified.

5.3.4.3 Parallel and sequential approaches as two extremes

The parallel and sequential approaches may be described as being at two extremes of a spectrum of approaches. All required views can be built in parallel, or they can be built in sequence. In reality, neither of these are practical options.

True parallelism, where the views are built at the same time, is not the natural way human minds work — 'one thing at a time' is a more natural way of thinking.

Neither is trying to build the views in a strict sequence realistic — no one view of the model is 'done' until the entire model is done. As in any modelling process, the building is iterative — a view is constructed, on the basis of this another view is built, based on this new view, the first view is changed, etc.

The approach adopted will depend on the technical support, modelling expertise and previous use of YSM. With the correct automated support, a parallel approach is suggested. Any part of the requirement can be captured in a suitable view as soon as it becomes available. The system modeller can switch from one view to another as soon as a view is complete or becomes difficult to proceed with.

5.3.4.3.1 Sequence of building aspects
Where a sequential approach is adopted, the following algorithm describes the order in which aspects should be built:

---

1. Draft the statement of purpose.

*2.   If the enterprise information aspect seems to be fairly complete for
          this project area:*

> *2.1. Build the system information aspect.*

else

> *2.2. In parallel:*

>> *2.2.1. Build the environmental aspect.*

>> *2.2.2. Build the system information aspect.*

>> *end*

> *end*

3. Build the behavioural aspect.

(The enterprise information aspect is kept in alignment with the
system information aspect *)*

Note: Where the term 'in parallel' is used, the best technique is to work in a very iterative fashion. If a more sequential approach is used, the order does not matter.

5.3.5 Reviewing the System Essential Model

Most review sessions should cover only part of the model. When reviewing in detail, for example, one DFD and its associated textual specifications are as much as usually can be reviewed effectively in a session. The views are chosen so that one or two views can be reviewed in isolation from the rest of the model. If the standard structure of the model is adhered to, the system modeller can be confident of the overall quality of the model.

5.4 Building the Statement of Purpose

Before detailed work can begin on specifying the system's requirements, the modellers must be certain that they know what 'system' encompasses. This is modelled using a statement of purpose (see §2.11).

5.4.1 Use of the system charter

Somewhere there should be a statement of system charter. This may be an output of strategic planning, feasibility studies, or steering committee reports. This charter acts as the initiative for building a system. It implies specific system responsibilities, which must now be more clearly defined. The charter may also cover the following points:

- intended technology;
- required delivery time scale;
- personnel requirements;
- project management guidelines;
- quality assurance guidelines.

Although each of these will be important for successful project conclusion, they are over and above the purpose of the System Essential Model, which is to define the policy that the system supports. They must therefore be 'filtered out' of the charter and excluded from the statement of purpose. The charter is broader and less 'focussed' than the statement of purpose, which specifies the functional responsibilities of the system.

The modellers should find the essential policy components of the charter and outline these as system responsibilities. Any specific exclusions should also be added to the statement of purpose. (As modelling proceeds, additional exclusions may be added if ambiguity about whether the system *is* responsible for those areas is removed.)

5.4.2 Checking the statement of purpose for consistency

The statement of purpose cannot be checked in a 'mechanical' way. It just lists functional responsibilities of the system, together with specific exclusions. This does not allow any rigid rules to be defined that must be satisfied by the statement of purpose. However, it should be reviewed by subject matter experts to try to check that all responsibilities seem to contribute to the same goal.[1]

5.4.3 The statement of purpose as early definition of system scope

The statement of purpose should be presented to major customers and project management for early approval. This is very important, as errors in the implied scope of the system can lead to much wasted work. Such errors should be caught as early as possible.

---

1 This is a 'cohesion' type of property.

5.4.4 Revision of the statement of purpose

Detailed study of the system requirements may imply additions, deletions, or changes to the functional responsibilities of the system. As soon as these are identified, they should be 'fed back into' the statement of purpose and re-agreed with the customer and project management.

5.5 Building the Information Aspect

As described in §5.2.2.1, the information aspect defines the information usage of the system. It may be thought of as 'a window' into the enterprise information aspect, as shown below:

Different systems see different parts of the enterprise information aspect (EIA). The EIA serves

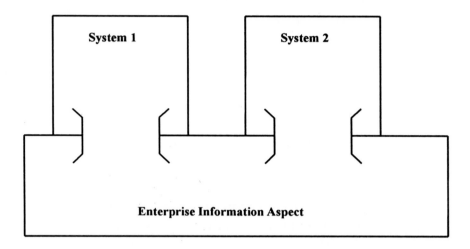

to tie them together through shared information.

One of the most important strategic techniques for defining system charters involves the 'fit' between the information aspects of potential systems. This is outside the scope of this version of the reference, which will assume one of the two following cases:

- The enterprise information aspect will be built in parallel with the System Essential Model. This is described in §4.3.5.2.

- The enterprise information aspect exists and is expected to cover most of the information used by the system. This 'filtering' technique is described in this section.

5.5.1 Sequence for building the information aspect

The following is an algorithm for deciding the order in which the views of the information aspect should be built:

> *1. Build a system function-entity table.*
>
> *2. Build system ERDs.*
>
> *3. Start system entity and relationship specifications.*
>
> *4. If there are enterprise eSTDs built:*
>
> > *4.1.1. Build system eSTDs.*
> >
> > *4.1.2. Build an entity-event table for the system.*
>
> *else*
>
> > *4.2.1. Build an entity-event table for the system.*
> >
> > *4.2.2. Build system eSTDs.*
>
> *end*

5.5.2 Building the system function–entity table

This gives a view of the information used by each of the system functions (see §2.19). At the lowest-level, it can be used as a view of the entities and relationships accessed by one data process; at a higher-level, it may be used as an overview of the system's use of entities and relationships.

Created initially, it gives a preliminary view of the system functions and their expected use of information. After the behavioural aspect has been completed, the function–entity table may be made to correspond to the functions shown in the levelled set of DFDs. Each data process or process group is then treated as a function in the table. Used in this way, the function–entity table may be used as a high-level view of the system's use of information.

The use of the function–entity table depends on the technical support environment (see §5.5.2.6). If the view is not built, then the comments about building this view should be applied to the system ERD instead (see §5.5.3).

5.5.2.1 Extracting from the enterprise function–entity tables

If an enterprise function–entity table exists, it should be examined for any functions that are included within the scope of the system, as defined by the statement of purpose. It may be that the enterprise function–entity table was used in defining the system's charter. If so, one or more functions shown in the enterprise table were chosen as being the responsibility of the system. The system function–entity table can be derived by a 'restriction', as defined below.

5.5.2.1.1 Restricting the enterprise function–entity table to the system scope

The system function–entity table has a column for each function that the system is responsible for. A row is added for each access that is seen in this column of the enterprise function–entity table. The entry in the cell of the system table is copied from the corresponding cell in the enterprise table. In the example below, part of the enterprise function–entity table is 'restricted' to a system whose responsibility is for three of the functions seen in the enterprise table.

5.5.2.1.2 Mismatches between the statement of purpose and the enterprise function–entity table

	FUNCTION				
	f_1	f_2	f_3	f_4	f_5
entity$_1$		m		r,u	
entity$_2$	c,r		u,r		
entity$_3$		m			
relationship$_1$		c			m
relationship$_2$	u,m				

	FUNCTION		
	f_2	f_4	f_5
entity$_1$	m	r,u	
entity$_3$	m		
relationship$_1$	c		m

Enterprise function-entity table

System function-entity table

Sometimes there is no exact correspondence between a preliminary list of system functions (as identified from the statement of purpose) and the functions shown in the enterprise function–entity table. In this case, one of the following will bring them into alignment:

- If a system function can be reworded to bring it into alignment with an enterprise function, it should be.

- If an enterprise function can be broken into several functions, one of which corresponds with a system function, then the enterprise function should be split in this way. This corresponds to adding new columns to the enterprise function–entity table.[1]

- If the system function can be split into several functions, one of which is an existing enterprise function, then it should be. The remaining functions should also be 'matched up'.

---

1 These columns are related to the original column, which remains as it was. There is now a link of the type 'this function includes the following functions' between the enterprise functions. This hierarchical relationship between enterprise functions is not covered by this release of YSM.

Functions should be added to the enterprise function–entity table if a system function cannot be made to match any existing enterprise functions using the above techniques.

5.5.2.2 Identifying system functions from the statement of purpose

The statement of purpose is a useful source for finding high-level functions. Any function that seems to be within the scope of the system (as derived from the statement of purpose), will have a column in the table. Each of these functions should be examined carefully, to try and identify what entities and relationships they might need to achieve their purpose. These will become the rows of the table.

Although the statement of purpose is a useful source for finding system functions, the statement of purpose should *not* be reworded to make it into a function list. The statement of purpose is not a list of high-level functions — if it were it would be entirely redundant after the System Essential Model is complete. No automatic check between the statement of purpose and system functions should be attempted.

5.5.2.3 Establishing entities and relationships used by a function

If there is an existing enterprise function–entity table, the information can be extracted from that table. If there is no existing enterprise function–entity table, then the entities and relationships required to support the system functions must be identified 'from scratch'.

To determine which entities and relationships are required to support functions, consider general statements about the functions. Potential entities and relationships should be identified in these statements (see §3.5.1.1 for some ideas on this). These should be the rows added to the system function-event table.

If these are new entities and relationships, they will need to be added to the enterprise information aspect (see §4.3.3).

5.5.2.4 Finding additional entries from the enterprise information aspect

Once the system function–entity table has some entries, the enterprise ERDs are good sources for other entries in the table.

5.5.2.4.1 Checking whether relationships are used for the system

All enterprise ERDs that show an entity that is already known to be accessed by the system should be examined. Each relationship that the entity participates in should be carefully checked to decide whether they are significant to the system. If they are, they should be added to the system function–entity table.

5.5.2.4.2 Checking entities involved in relationships are 'visible'

If a relationship (or associative entity) appears in a system function–entity table, then all entities that participate in that relationship must be accessed by the system. They must therefore be shown in the table. However, where the entity is a supertype, the system may only need to deal with some of it's subtypes. If this is the case, only these (and not the supertype) appear.

5.5.2.4.3 Checking subtype visibility

If the enterprise information aspect has any subtypings for an entity shown in the system function–entity table, these subtypes *may* be relevant to the system. If the system uses any relationships that only refer to the subtype, or attributes that are specific to the subtype, then the subtype should be added to the system function–entity table.

5.5.2.5 Deciding whether the system deals with the supertype, subtypes, or both

The following rules may be applied to decide whether to show the supertype and/or the subtypes as being used by the system:

1. If the system only deals with certain subtypes and *never* deals with the more general supertype, only the specific subtypes (and not the supertype) appear in the function–entity table.

2. If the system *ever* deals with the supertype without 'caring' which subtype it is, then the subtype should also appear in the system function–entity table.

3. If the system *never* distinguishes between the subtypes, then only the supertype should be shown.

5.5.2.6 Effect of technical support environment on system function–entity tables

For 'pencil and paper' environments, the system function–entity table should not be regarded as part of the System Essential Model, but merely as an optional working view. This is because of the redundancy between this view and other views in the System Essential Model, with the risks entailed by possible divergence of duplicated requirements.

For 'automated' support environments, the table is a required view into the model. However, it is not required that the system modeller builds this view. It may be generated from other information by the support tool. (For example, it can be partly derived from DFDs and data store specifications. It can be fully derived using the references to the entities and relationships in minispecs.)

5.5.3 Building system ERDs

The system ERD describes the entities and relationships used by the system (see §2.2.4).

System ERDs may be extracted from enterprise ERDs, using the function–entity table. If the function–entity table has not been built, then the comments in §5.5.2 about identifying entities and relationships that the system uses still hold. However, they are now used to decide which ones to include on the system ERD (or ERDs).

The simplest situation is where a single ERD is used. For example, the pair of diagrams below show how an enterprise ERD is restricted to give a single system ERD.

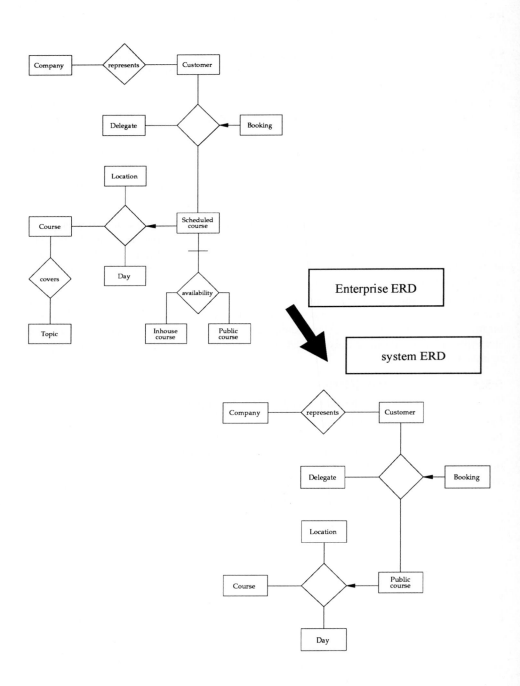

5.5.3.1 Multiple system ERDs

For large systems, it may be helpful to use several system ERDs. If any ERD is too complex (see §2.2.6.3), it should be split into several views, each showing a sub-functional area. The way this is interpreted depends on the technical support environment.

5.5.3.2 Effect of technical support environment on system ERDs

For 'pencil and paper' environments, the system ERDs show a specific relationship in only one of the system ERDs. This is to minimise redundancy, with the risks entailed by possible divergence of duplicated information on several views.

For 'automated' support environments, a system ERD could be shown for any functional area. Thus a system ERD could be shown to support one data process, or a process group. The entities and relationships appearing in this view would be all those needed to support that function. Redundancy between the views is controlled by the support tool.

5.5.3.2.1 Choosing which ERDs to show the relationships on in pencil and paper environments

In pencil and paper support environments, it is recommended that a relationship is only shown in one system ERD. This also applies to associative entities in their role as a relationship. There is no restriction about repeating entities on several views (nor is there any special notation to show this). Schematically, we might have three system ERDs, as shown below:

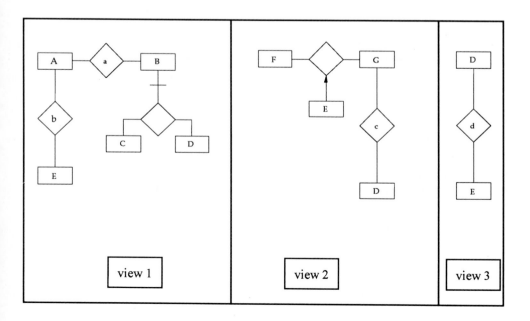

To decide on which ERD to show a relationship on, the following algorithm may be used:

If the system creates the relationship:

    1. If only one system function creates the relationship:

            1.1. show the relationship in the ERD corresponding to this function.

      else

            1.2. show the relationship in any of the ERDs corresponding to one of these functions.

      end

elsif the system deletes the relationship:

    2. If one system function deletes the relationship:

            2.1. show the relationship in the ERD corresponding to this function.

      else

            2.2. show the relationship in any one of the ERDs corresponding to these functions.

      end

else

    3. show the relationship appears in any ERD corresponding to a function that uses a relationship (a "match").

end

(Recall that there is no problem about repeating relationships on several ERDs in an automated support environment.)

5.5.4 Building the system entity–event table

Entity–event tables are used to model the events that are significant to an entity (see §2.4).

This modelling tool may be used for high-level strategic planning and definition of system charters (compare function–entity table, §5.5.2). As for that tool, however, this section restricts itself to building the entity–event table for one system

5.5.4.1 Creating the entity–event table

An initial set of sources for information in this table include:

- Enterprise entity–event table (if it exists), is an obvious source. It should be 'filtered', taking into account the system's purpose.

- System ERD. Each entity and relationship should be examined and an attempt made to identify any event that would cause the system to access it.

- System function–entity table. For each function, events should be discovered that cause some part of the function (or all) to be carried out. The accesses shown in the function–entity table are allocated to each event identified for this function and shown in the entity–event table.

- System event list (or enterprise event list, if it exists and the system event list has not been built yet). Each event should be considered carefully. Taking into account the system's purpose, the system modeller should try to decide which entities and/or relationships that the system need to access when the event occurs. These should be added to the table.

All events that system deals with must be included in the enterprise event list. Care must be taken that it really is different from all previously discovered event. If a system event is the same as an enterprise event, except for wording, then the system event should be renamed (any other references to the event must also be updated).

If no correspondence between an event in the row and an event in the system event list can be found, the event needs to be added to the system event list. (If the system event list has not been built yet, this step can be deferred — the event list, when it is created, can be derived from the entries in this table.)

5.5.4.2 Creating the entity–event table from system eSTDs

If system eSTDs have already been built (see §5.5.5), parts of the entity–event table can be derived from them.

5.5.4.3 Consistency between the entity–event table and entity and relationship specifications

When the system entity and relationship specifications have been built, the access actions in the table should be the same as those given in the specifications. If there are any inconsistencies, these must be resolved.

5.5.4.4 Effect of technical support environment on system entity–event tables

For 'pencil and paper' environments, the system entity–event table should not be regarded as part of the delivered System Essential Model, but as a temporary, working document. This is due to the redundancy between this view and other views into the System Essential Model and the risks entailed by possible divergence of duplicated information in multiple views.

For 'automated' support environments, the entity–event table is a required view, with the support tool controlling the visibility of the components in several 'overlapping' views. A sophisticated support tool might be able to generate it from the event specification and minispecs. This would mean that this view could be generated at any time to provide an overview of how the system's functions (as defined thus far) access the information aspect when a given event occurs. Alternatively, if this view is built early, it provide a guidelines for many of the other views into the System Essential Model (for example, writing minispecs for responses to these events).

5.5.5 Building system eSTDs

These are partial view of the enterprise eSTD (see §2.3.5).

If enterprise eSTDs exist, then the system eSTDs can be obtained by deleting all events and accesses that are not within the system's scope (as described in §2.3.5). If enterprise eSTDs do not exist, then they should be built, as described in §3.5.5.

5.5.5.1 Entity state transition diagrams required

Any entity (including associative entities) used by the system may have 0, 1, or more system eSTDs. There may never be more system eSTDs than enterprise eSTDs for a particular entity, but there may be less. Sometimes, although an entity is used by a system, the values of a given state variable are irrelevant to the system.

Potential entities are those shown in the system ERD(s), entity–event and function–entity tables. Any, or all of these may be used to identify the entities.

5.5.5.2 Building one system eSTD

For each entity, there may be several enterprise eSTDs (building these is discussed in §3.5.5). Each of these is for a specific state variable. If the system distinguishes between different values of this, in terms of treating the occurrence of the entity differently, then there must be a system eSTD for this entity and state variable.

This is derived from the enterprise eSTD, with only certain events selected. These are the events shown in the system event list (and entity–event table). If either of these definitions of events already exist, constructing the system eSTD is automatic — the other events and transitions are just 'deleted' from the enterprise eSTD. For example, the enterprise eSTD below is restricted to give a system eSTD:

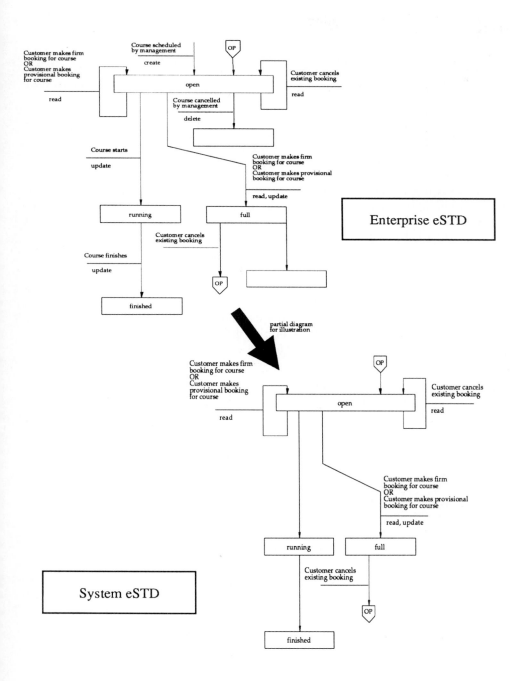

If the system events have not been identified yet, these should be discovered as in §5.6.4.

The access actions may be added by referring to the system entity specifications.

5.5.5.3 Effect of technical support environment on system eSTDs

For 'pencil and paper' environments, the system eSTD should be regarded as an optional part of the delivered System Essential Model. This is because of the redundancy between this view and other views into the System Essential Model.

For 'automated' support environments, an eSTD view is created for each entity used by the system. The support tool controls the visibility of the events and accesses in several 'overlapping' views. A sophisticated support tool would allow automatic generation of the system eSTD from the enterprise eSTD.

5.5.6 Building system entity and relationship specifications

Specifications must be provided for each entity, relationship or associative entity used by the system. These define the use of that item by the system. Each entity and relationship shown in the system function–entity table, entity–event table and system ERD(s) is thus specified.

5.5.6.1 Distinction between enterprise and system entity and relationship specifications

System entity and relationship specifications should be carefully distinguished from enterprise entity and relationship specifications. An enterprise entity specification truly specifies the entity; a system entity specification defines which parts of that enterprise specification are 'visible' to the system. When the system is the enterprise, then there is only one entity specification, which serves both purposes, because there is nothing of interest about the entity other than those things needed by the system.

Note that an associative entity may be used by the system as:

- an associative entity: in this case, it has an enterprise associative entity specification and a system associative entity specification;

- an entity (but not as a relationship): in this case, it has an enterprise associative entity specification and a system entity specification;

- a relationship (but not as an entity): in this case, it has an enterprise associative entity specification and a system relationship specification.

5.5.6.2 Starting and updating system entity and relationship specifications

Initially, each of these specifications may be incomplete, or incorrect. However, they should be created and used as an aid in building the other views of the System Essential Model, particularly minispecs, which will refer to specific occurrences of these entities and relationships, and values of attributes.

As the model progresses, more detail can be added to these specifications. Errors and inconsistencies will also need to be resolved. Some of these changes will imply updates to the enterprise information aspect. This is discussed in §4.3.3.

5.5.6.3 System entity specifications

These are described in §2.5.4.

5.5.6.4 System relationship specifications
These are described in §2.6.3.

5.5.6.5 System associative entity specifications
These are described in §2.7.4.

5.6 Building the Environmental Aspect

As described in §5.2.2.2, the environmental aspect describes the 'fit' of the system with the environment. In a sense, it is a description of the system 'from the outside'.

The environmental aspect is usually started before the behavioural aspect, but may be built in parallel with the information aspect.

5.6.1 Deciding whether to build the context diagram or event list first

Both the context diagram and event list/specification define the scope of a system — each supports the other. The event list and event specifications highlight the dynamics of the system; the context diagram highlights the system inputs and outputs. Depending upon the type of system, beginning with one will make it easier to build the other.

Whether the event list or context diagram is started first, the two should be built iteratively, cross-checking, as described in §5.6.5. Whenever it becomes difficult to continue, switching to the other view is advised. The best way to build them is 'in parallel', repeatedly switching from one to the other.

It is of little import which is started first, but the following guidelines have been suggested. They may be used if helpful. They are not mandatory.

1. If the system is information based, handling queries and generating reports, the context diagram should be started first. This is because the reports are probably already defined, to some extent.

2. If the system is responsible for collecting information when particular events occur, but does not generate many reports, then the event list is easier to build first. This is particularly true if the information aspect has been started (the entity–event table and system eSTDs are prime sources of events).

3. If the system is a control system, operating machinery, interpreting sensors, etc., the context diagram should be started first. If the system changes its behaviour over time, it is *not* cost-effective to complete the context diagram without starting the event list — parallelism is definitely recommended.

4. If the system is based on an existing system, the context diagram is the easiest to start with. It is created by summarising the inputs and outputs from the existing system.

5. If top-down functional decomposition is to be used, the context diagram is started first. The events are discovered after low-level functions have been elaborated. This is discussed in §5.7.4.4.1.

5.6.2 Interview with subject matter experts

The main input to the environmental aspect is interview with subject matter experts. For each functional area that the system is concerned with, one or more experts should be identified and interviewed. During the interviews, information relating to events, system inputs and outputs, terminators and events will be volunteered in a fairly unstructured manner. It is the analysts responsibility to organise these into the two main environmental views and prompt the subject expert for other information, as well as seeking clarification of ambiguities.

5.6.2.1 Scenario playing

The example given in §3.5.3.5 might be revised to highlight discovery of events:

> **Analyst**: After the course has been scheduled so that it is available, what might happen?

> **User**: We might get bookings — hopefully we will.

> **Analyst**: *(thinks — booking sounds like an event.) What do you mean, a booking ?*

> **User**: Well, customers ring up and want to book people on courses.

> **Analyst:** *(thinks — I think that is an event "Customer wants to attend a <Scheduled course>".)* Tell me, what happens when the customer wants to attend the course?

> **User**: Well, it depends on whether the course is full or not ….

> **Analyst**: *(thinks — I know that the associative entity "Booking" relates to one "Delegate" wishing to attend a "Course". Let's try to talk through its life-cycle.)* For a given booking, what happens next?

> **User**: Nothing, until it is time to send out the joining instructions, which we do about a week before the course. We also send a list of delegates to the course instructor. When the course starts, we ….

> **Analyst**: Could I just interrupt you there, please. *(thinks — there seem to be several events here. Let's check them out one at a time. I also get the feeling that there must be an event that leads to the instructor being allocated to teach the course.)* Are the joining instructions always sent out at the same time as the delegate list?

and so on.

Note that, the interview lead to other information being gathered, not just the events. For example, in the above, flows such as "joining instructions" and "delegate list"; system functions such as "Send joining instructions" … would also be identified. The analyst is not allowed to say "Tell me the events that affect the system"!

5.6.3 Building the context diagram

5.6.3.1 Finding terminators
The devices, departments, people, agencies or other systems that will be giving inputs or outputs to the system should be identified as terminators.

There are several sources for identifying terminators, including:

5.6.3.1.1 From the statement of purpose
Each responsibility should be examined to try to identify terminators that are needed to accomplish this responsibility or from which the responsibility derives. For example, if the responsibility of the system includes "Maintain an up to date record of course bookings", then the terminator "Customer" would be identified as the person initiating a booking and "Manager" as someone who wanted to know about bookings.

5.6.3.1.2 From event list and/or specifications
If an event list exists, the subjects of the events are likely to be terminators. For example, the event:

> Customer makes deposit

suggests that "Customer" will be a terminator on the context diagram. Of the two 'nouns' present in this, the "Customer" is identified as the signalling agent (active) and the "deposit" as part of the information being signalled (passive). This approach is greatly aided by using the information aspect (see §5.6.3.1.3) in parallel.

5.6.3.1.3 From the system ERD
Each fact that the system needs to know about must be communicated to the system by some agent. This agent is the terminator. For example, suppose:

> <Delegate> is booked to attend <Scheduled course>

is relevant to the system. How does the system know about this fact? By user interview, the system modeller might find out that the customer phones in a booking, or sends in a standard form requesting a place. This would identify the "Customer" as the terminator.

If the system needs to record who communicated the fact to the system, then the model would have used:

> <Delegate> is booked to attend <Scheduled course> by <Customer>

which would have made identifying the terminator event more obvious. Certainly, any entity is a potential terminator, but not all terminators are entities (see the first version of this example).

5.6.3.1.4 From system inputs and outputs
Whenever a flow is known to cross the systems boundary, the flow must be generated or be used by a terminator. Any known context flows should be examined to identify terminators. Sources for information about this are existing documents or interview with subject-matter experts.

5.6.3.2 Finding context flows
There are several sources for identifying flows, including:

5.6.3.2.1 From terminators
For each terminator, system inputs can be identified as data and event flows that are generated by that terminator. Only those items that are required by the system should be included — not all of the potential outputs of the terminator will be relevant to the system.

For each terminator, system outputs can be identified as data and event flows that are required (used) by that terminator. Only those items that the system is responsible for generating should be included — other systems, devices may generate some of these inputs to the terminator.

For terminators that are people, they (or a representative) should be interviewed. Terminators that are outside organisations cannot be interviewed as such — a representative must be identified to state the required interface to them. For hardware interfaces, project engineers, or specifications for existing devices are the main source of information.

5.6.3.2.2 From the event specifications
If the event specification exists, the entries for data flow input, data flow output, event flow input and event flow output correspond to flows on the context diagram. (If an event table is used, then there will be equivalent entries for system inputs and outputs.)

5.6.3.2.3 From the behavioural aspect
The context diagram acts as the top-level DFD in the behavioural aspect. In the rare cases when the bahavioural aspect is built before the context diagram, it may be derived as a summary of all DFDs. (This is only really conceivable when reverse engineering current systems.)

5.6.3.3 Finding context stores
Context stores are used to show stores that are interfaces between the system and its environment.

Showing all stores that are shared information between the system and other systems within the enterprise may lead to very complex context diagrams. This is true, even if the only ones shown on the diagram are those created by the system and used by other systems (or vice-versa). In these cases, showing the interface between the systems as stores on the context diagram is optional.

A better tool for modelling the shared information between systems is the enterprise function–entity table (see 2.19). The use of this tool to define the best scope for different systems is a strategic activity and outside the scope of this release of YSM.

Where the terminator is:

- a device that 'owns' a store accessed by the system (for example, registers, or memory that are part of the device)

- an organisation outside the enterprise (for example, another company, or government department)

the store *must* be shown, as the function–entity table only shows functions within the enterprise.

5.6.3.3.1 From enterprise function–entity tables
If an entity or relationship is used by several systems, as defined in the enterprise function–entity table, it may be shown on the context diagram as a store, with a corresponding name. This is optional, as discussed above.

5.6.3.3.2 From the system entity–event table
If no enterprise function–entity table exists, or the system is the enterprise, the system entity–event table may be used to identify stores.

If the entity–event table shows any entity or relationship whose access is "match" or "read", but not "create", it must be information created by a terminator for use by the system.

If the entity–event table shows any entity or relationship whose access is "create", "update", "delete", but not "read", or "match", it must be information created by the system for use by a terminator.

In either case, a store may be shown on the context diagram (but see discussion in 5.6.3.3). In other cases, the system creates and uses the entity or relationship. The information may still be shared between systems, but it is not conventional to show the store on the context diagram.

5.6.3.3.3 From terminators
Some information may be created by a terminator and stored in a form that allows the system to access this information, whenever it is required. If the terminator is another organisation or system, this might correspond to part of a database, machine readable file, printed tables of recommendations used by the system, etc. If the terminator is a device, there might be 'registers', or other storage within that device that may be read by the system, whenever it needs to.

Alternatively, the system might create information that is held in a form that may be accessed by terminators, as and when they require that information. In either case, this should be modelled as a store on the context diagram.

5.6.3.3.4 From the event specifications
If the events have been partially specified, the entries for data store input and data store output correspond to potential stores on the context diagram. However, some of these items will be entirely 'within' the system and therefore should not be shown on the context diagram. Some of these may be 'shared' between the system and its environment, or other systems — these are shown on the context diagram.

The event specification is not the best source for this information. Use of the enterprise function–entity table and interviews about terminators and their interfaces is more natural. However, event specifications may provide supplementary techniques.

5.6.3.4 Checking stores really are interfaces
If the store is entirely used by the system, and by no other function or device, it should not appear on the context diagram.

Note: The difference between a store and a flow on the context diagram is discussed in §2.12.6.5.

5.6.3.5 Finding access flows

If the system creates, deletes, or updates any item in the store, an access flow 'into' the store from the system should be shown.

If the system reads or carries out a match of an item in the store, an access flow 'from' the store to the system should be shown.

Optionally, the flow may be named and specified as described in §2.21.

5.6.3.6 Organising the context diagram

It is important that the context diagram is easy to review. This requires some care in organisation. Techniques to help make the diagram more understandable include:

- Repeating terminators on the same diagram: each may be 'tagged' with a "$*$".

- Splitting the context diagram into several smaller diagrams: each of these is referred to as a partial context diagram.

- Packaging data and event flows: where there are a number of flows that are related, they may be packaged together as a group or multiple data flow 2.18.3.3). This group is given a name to reflect its meaning. Note that data flows should only be packaged together if they are generated by the same terminator, at the same time, or used by the same terminator and generated by the system at the same time.
  Similar rules apply to grouping event flows.

If there are a large number of stores on the context diagram, there are two solutions that may be adopted:

1. Do not show stores that are an interface between this system and other systems. These may be modelled by use of the enterprise function–entity table.

2. Group related stores together into a single store and give it a name that represents the group of entities and relationships that it 'contains'. Guidelines for choosing and naming stores are given in §2.20.5.

5.6.3.7 Adding text specifications for components of the context diagram

5.6.3.7.1 Specifying terminators

Each terminator must be specified by describing its meaning. The number of instances of the terminator and how they are distinguished are defined as part of the performance aspect (see §5.9.3).

5.6.3.7.2 Specifying data and event flows

All flows must be specified, including any that are defined as being part of a group or multiple data flow.

5.6.3.7.3 Specifying stores

Each store should be specified as described in §2.20. Stores are a 'window' from a DFD (of which the context diagram is a specific example) into the information aspect. See §5.7.3.6 for a discussion of this.

Stores on the context diagram may be specified when the context diagram is built. Alternatively, when the context diagram is built, stores may be named to describe their expected contents and not specified. When the behavioural and information aspects are constructed they will be specified as containing one or more entities (and possibly relationships).

Access flows may be named, but this is optional. Specification of access flows is deferred until the behavioural aspect is built.

5.6.4 Building the event list

The event list gives all the events that a system must respond to (see §2.22.1.1). Each event has a number and description.

Each event has a full system specification (see §2.22).

An alternative view, the event table, may be used when several events need to be considered with partial specification for each (see §2.22.1.2). The event table is particularly useful in initial collation of information about events. A lists of events, with inputs and outputs for each event is easily understood by the system modeller and end-user. Additionally, the stimulus may be designated in some way.

There are several strategies that may be adopted in establishing the events to which the system must respond. Each of these has its own merits and will be easier or harder to use in a particular system. The system modeller should look at all these strategies; one of them is likely to be more 'productive' than the others, but use of all of these techniques is complementary.

5.6.4.1 Using the entity–event table as a source of events

If an entity–event table for the system exists, the event list is just the consolidated list of all events shown in that table.

If an event table is to be used, then there will be a row of this table for each event.

Each event must have a full specification. This may be in a frame format or table, as long as the information is captured. However, it is not usual to complete this until part of the behavioural aspect has been completed. As and when information becomes known, it is added to the specification of each event.

5.6.4.2 Using the context diagram as a source of events

If the context diagram exists (even in part), it is a good source for identifying events.

5.6.4.2.1 System outputs

Every output on the context diagram will be output as a result of a response to one or more events. Each output should be considered and the event (or events) that cause it to be produced added to the event list, if not already present.

If an event table is being built, new events require rows to be added. If a separate event specification is being constructed for each event, each new event will require a specification.

If the output is a data or event flow, the output name should be added to the system output entry.

5.6.4.2.2 System inputs

Every input on the context diagram is used to either (or both):

- detect the event,
- as a required input to the response to the event.

Discrete inputs that *always* (as a matter of policy) occur at the same time as an existing event do not require a new event. If a discrete input can *ever* occur without any of the known events occurring, then there must be an event that has not been identified. This event should be given a distinct name and added to the event list.

Continuous flows may have events associated with a change in their values. For event flows, this might be when the flow becomes "TRUE", or "FALSE"; for data flows, there might be one or more events associated with changes of value.

If the input is used to detect the occurrence of the event, then this is added to the 'stimulus' entry of the event table (if used) and the corresponding entry in the event specification. If the input is not used to detect the occurrence of the event, but is used in determining the response, then it is added to the 'input' column of the event table and the corresponding entry in the event specification.

Note: Some inputs may be associated with several events.

5.6.4.3 Finding events through interview
The primary source for identifying system events is through user interview. Events are 'things that happen' in the environment, to which the system responds in some way. During interview, the modeller should try to identify these occurrences and elucidate the system response.

5.6.4.4 Grouping events
Sometimes an event may be identified, which proves to be a whole set of similar, related events. In such cases, it is convenient to place these into a group and number them accordingly. For example, the event:

5	<Customer> books place on <Scheduled course> for <Delegate>

might be broken down into the three events:

5.1	<Customer> makes provisional booking on <Scheduled course> for <Delegate>.
5.2	<Customer> makes firm booking on <Scheduled course> for <Delegate>.
5.3	<Customer> converts status of existing <Booking> from provisional to firm.

Note: The above events would be best studied by drawing up an eSTD for the associative entity "Booking" — two of its states ("provisional" and "firm") are referred to here.

5.6.4.5 Finding events from a levelled set of functional DFDs
This is described in §5.7.4.4.1. See also the discussion about using the event-partitioning and functional decomposition in parallel in §5.7.2.4.

Much of the specification of the event may also be derived from a functional decomposition (see §5.7.4.4.2).

5.6.4.6 Completing entries for system inputs, system outputs and stimulus
The event specification should not be completed until the behavioural aspect is built (see §5.2.2.3). It is helpful to add the entries for stimulus, system inputs and system outputs before building the behavioural aspect. Building a first-cut event–response DFD (see §5.7.3.3) is made much easier if these entries are identified.

These entries may be identified by examining the context diagram.

The stimulus is a system input flow that is used to detect that the event has occurred. The name of the flow is entered, or the word "time". Any flow entered must be an input from the context diagram (it may be a contained flow within a multiple flow). If the entry is "time", then the event is temporal (a "time" data flow is *not* shown on the context diagram).

The system inputs are any flows required to make the response to the event. These are also seen as system inputs on the context process.

System outputs are any flow produced as part of the response to the process. These are also seen as system outputs on the context diagram.

5.6.4.7 Completing the event specifications

5.6.4.7.1 System inputs, outputs and the stimulus

The context diagram may be used to help identify events, as discussed in §5.6.4.2. As events are identified in this way, the system input, system output and stimulus may be added to the event specification or event table. Whichever way the event list is constructed, when it is complete it should be reviewed to complete these entries. It is recommended that these entries are completed before commencing the behavioural aspect.

Stimulus

For each event in the event list, the system will either use time or the occurrence of a flow to detect the occurrence of the event. If time is used, the word "time" is entered in the stimulus entry. If a flow is used, the name of this data or event flow should be entered. Any data flow used must also appear in the context diagram or be part of a flow on the context diagram.

System inputs

Each event should be reviewed to identify any system inputs that are needed by the system to carry out the response to the event.

System outputs

Any outputs that are produced by the system as part of its response to the event should be added to the event specification. If the flow does not appear on the context diagram (if the context diagram exists), then it should be added to it.

5.6.4.7.2 Entries deferred until the behavioural aspect

Other items in the event specifications are added after cross-checking against the context diagram and environmental aspect, but before starting on the behavioural aspect.

Event–response processes

The event–response entry gives the name (or names) of any data processes that are a response to that event. When the System Essential Model is complete, this entry will contain exactly those names. Before the behavioural aspect has been built, however, these names will not be available. Two approaches may be used:

- leave this entry blank until the behavioural aspect has been completed,
- write in a general name for the transformation of data carried out by the system when the event occurs. This will be replaced by the exact names of the data processes when they are identified.

Any data flows identified as a system input associated with an event, will be used by these event–response process (or processes). This includes stimulus flows.

Finding event–response processes is discussed in §5.7.3.3.4.

Event–detection processes
Many event–response processes are able to detect and respond to the event. They are entered as event–response processes only (see above). Some events require separate data processes to detect that they have occurred. This process, if it is different from the response process, is entered here.

It is not usual to complete this entry until the behavioural aspect is built (see §2.22.5.9).

Use in control processes
Some events cause changes in behaviour of the system. This entry gives the condition that the event is part of and the control process in which it occurs. It is not usual to fill in this entry until the behavioural aspect is built. See §5.7.3.4.3.

5.6.5 Cross-checking the context diagram and event specifications

The context diagram gives insight into what the events are. Conversely, when an event is fully specified, system inputs and outputs are implied — these must appear on the context diagram. This interaction has been described in the above two sections.

5.6.5.1 Annotating the context diagram to show events

One useful way of checking that the context diagram deals with events in the event list is to annotate the context diagram with event numbers. Any flow that is associated with an event as an input (possibly the stimulus) or output is annotated with the number of that event. After all events have been thus annotated on the context diagram, all flows should have at least one number on them. Some flows may be associated with more that one event and thus have more than one numbered 'tag'. Any flows that are not numbered are an indication that either the flow is not really within the system's scope, or an event has been omitted from the event list.

An example is shown in §5.6.6.3.

Note that this annotation of the diagram should not be regarded as a required format for the context diagram; it is a useful temporary expedient for cross-checking the context diagram and event list. In pencil and paper support environments, this should be carried out by adding these tags to a hard-copy for review. With an automated support environment, this tagging could be carried out automatically by a support tool, whenever desired (it might even be 'popped-up' on the context diagram).

5.6.6 Example of building an environmental aspect

This example shows how an environmental aspect may be developed for a system that is relatively 'stand-alone'. (Its interface to other systems is discussed in §5.6.6.1.3.)

A corresponding example for developing a statement of purpose and context diagram for a system from enterprise modelling will be given in a later release of the manual.

The example system is required to deal with electronic pictures of the earth, taken by a camera on a satellite. These pictures may then be viewed by researchers (geologists, or agronomists, perhaps). The statement of purpose for this system is:

SYSTEM: Satellite Camera Control System

GENERAL DESCRIPTION :
The purpose of the system is to take and display pictures of land resources from a satellite. The satellite will contain an automatic electronic camera that will (when activated) collect light until the exposure is complete. The image may then be read out before sending a signal to start another exposure.

RESPONSIBILITIES :
1. Accepting requests for photographs from accredited researchers and scheduling a suitable time for these photographs to be taken.
2. Controlling operation of the camera and storing pictures.
3. Using the satellite attitude control instrumentation to point the camera at the target.
4. Displaying pictures on request.
5. Keeping an up to date record of the satellite's orbit so scheduling can be achieved. To do this an on-board accelerometer is used, together with ground tracking information.

SPECIFIC EXCLUSIONS :
1. Changing the satellite orbit.
2. Tracking the satellite. This function is carried out by ground tracking stations.

5.6.6.1 The context diagram

5.6.6.1.1 Finding the terminators
Some of the terminators can be identified without much effort. For example, "Researcher", and "Camera" are specifically mentioned in the statement of purpose. However, some care is required in defining terminators for embedded systems. Because the system may contain devices and people, in addition to digital and (possibly) analog processors, a device may be within the system boundary or an external device, depending on the definition of the system boundary.

In this case, suppose that discussions with the customer established that designing and building the camera was not part of the system's charter. This implies that the camera is outside the scope of the systems and should be modelled as a terminator.

The statement of purpose also indicates that the satellite is tracked from the ground to determine its orbit. This tracking might be carried out by part of the same organisation, or it might be input from tracking stations belonging to other organisations. In either case, the terminator is "Ground tracking station".

Other terminators are not so immediately obvious. For example, the statement of purpose refers to 'Using the satellite control instrumentation to point the camera at the target'. The term 'using' is a clue that one or more terminators are used by the system. It is the analyst's job to determine exactly what these terminators are. The discussion might proceed:

> **Analyst:** How is the camera 'pointed at the target'?

> **Project engineer:** Well, there are two ways we achieve this. We mount the camera on a moveable platform. This gives us about 30 to 40 degrees rotation away from the nominal orientation. For larger rotations, we use the momentum wheels.

> **Analyst:** What are the momentum wheels?

> **Project Engineer:** There are three wheels, mounted at right angles. Each spins at a constant rate to keep the satellite at a fixed orientation. If the spin rate is increased, the satellite rotates in the opposite direction.

> **Analyst:** *(thinks — sounds like moments of inertia, Newton's third law and so on. I did that a long time ago. Hope I don't have to look it up again!)* So what would you do to rotate to a new position?

> **Project Engineer:** The momentum wheels have a servo control of their own and will maintain the wheels at a constant rate that can be changed by means of the interface. There are three analog signals that are used to define the required spin rate. To rotate to a new position, these values are changed until the measured angles are the required ones. The spin rates are then returned to their original position.

> **Analyst:** *(thinks — measured angle! I'd better clarify that. It sounds like we need to measure an angle)* How do you measure the angle?

> **Project Engineer:** All angles are measured relative to an inertial platform that is stabilised by gyroscopes. Unfortunately, over times greater than about an hour, there are problems with 'drift' and we have to stabilise the platform and the angular readings. We do this by locking onto a reference star (we use Canopus, by the way) and the earth's horizon.

and so on. The analyst tries to increase knowledge about what the terminators are. At the same time, the interfaces to each of these may be elaborated. This is recommended in practice. It helps to define the nature of the terminators.

Potential terminators identified in the above discussion are: "Camera platform", "Momentum wheels", "Servo control" (for momentum wheels), "Inertial platform", "Gyroscopes" and "Attitude measurement system" (for measuring angles).

Not all of these are truly terminators. Some are merely intermediate 'handlers' (see §2.12.6.4) and should not be shown as terminators. The servos are of this type. They allow the system to rotate the satellite by outputting a required rotation rate to the momentum wheels. The inertial platform and associated gyroscopes provide a reference against which angles can be measured. The system needs attitude angle. The transducers that provide this are mounted on the platform, but the system does not care. These terminators are therefore called "Attitude measurement system".

Further discussion identified that there was a telescope with a detector that gave a signal to detect the star. This is called the "Star tracker telescope". Regarding the star as the terminator is a temptation, but it is incorrect. The system input is not 'light from the star', but the angle between the satellites orientation and the direction of the star.

For detecting the alignment (or otherwise) with the earth's horizon, the terminator "Horizon scanner" is used.

5.6.6.1.2 *Finding context data and event flows*

People
In this example, discussion with an end-user (the researcher) or someone who knows the end-user requirements will identify the following system inputs: "picture taking request", containing all information to define when the picture should be taken; "picture display request", containing all information to identify a picture that the researcher wants to see.

"Picture display" is output to the researcher. It contains all the information content of the picture. It does not make any commitment to display hardware, formatting, user interface, etc. Those concerns are addressed in the Processor Implementation Model.

Outside organisations
There is only one outside organisation — the "Ground tracking stations". They provide information about the satellite's orbit. It is important to determine the exact nature of this information. It might be information about the current position and velocity of the satellite, or it might be the current orbit parameters. This determines whether calculation of the orbit parameters is inside or outside the scope of the system. The example assumes that the ground tracking stations supply orbit parameters as the discrete data flow "orbit update".

Devices
For this system, the interface to the device terminators is critical. Errors in modelling this interface will have a serious effect on the System Essential Model. It is important to establish the form of this interface early in the modelling process.

1. The accelerometer was determined to provide continuous analog signals, representing the accelerations in each of three axes. This continuous data flow is called "acceleration".

2. The attitude measurement system provides three continuous analog signals, representing the angles. This continuous data flow is called "attitude".

3. The outputs to the momentum wheels are three continuous analog values. This continuous data flow is called "rotation speed".

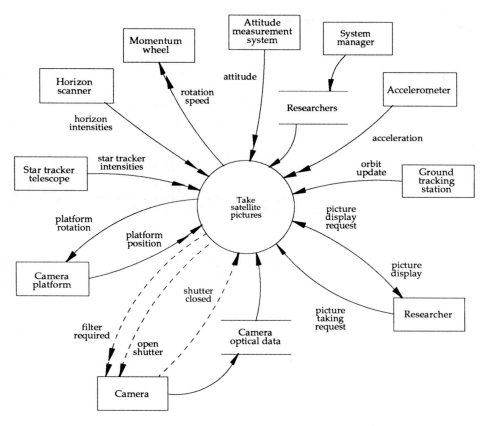

4. The star tracker telescope has a beam splitter that divides the incoming light into four sectors. If a star is exactly at the 'centre' of its axis, the intensities are equal; for other situations, the inequalities in the intensities allow the pointing error to be corrected. The star tracker telescope will provide four analog signals, representing the amount of light falling into each of four quadrants. This continuous data flow is called "star tracker intensities".

5. The horizon scanner could be implemented in several ways. For simplicity, suppose four points, equidistant around the horizon are used. This could have a beam splitter, like the star tracker telescope. The continuous data flow "horizon intensities" is used to model this.

6. The camera platform is mounted on a 'gimbal mounting', which gives rotation in each of two directions. The rotation is achieved by a pair of 'stepper motors', with their own controller microprocessors. For each, the system must output a given number of steps, together with a sense (forward or back). Once this is output, the microprocessors 'count out the steps' — no further intervention by the system is required. To model this, a discrete data flow "platform rotation" is used.

7. The position of the camera platform at any one time is sensed by transducers, returning two continuous analog signals. These are modelled with the continuous data flow "platform position".

8. The camera requires a transient digital signal to operate the shutter, This is modelled, using a discrete event flow "open shutter". There is a filter in front of the camera. It is operated by a solenoid, so that as long as current is switched to the coil, the filter is held in front of the lens. If the current is removed, the filter is removed from the optical path. To model this, a continuous event flow "filter required" is used.

9. When the exposure is complete, the camera generates a transient digital signal. This is modelled using a discrete event flow "shutter closed".

5.6.6.1.3 Stores

The camera includes an electronic store that can be read at any time. This is modelled by a store "Camera optical data".

The system keeps a record of which researcher requested which picture. This allows the system to display pictures that were requested by a specific researcher only. (It may also allow more general access.). It is also possible that only accredited researchers can use the system. Keeping track of who are allowed researchers *could* have been included within the system's scope. For the sake of example, however, it has been excluded. This means that the entity "Researcher" is accessed by the system, but not created. This interface is modelled by the store "Researchers". The "System manager" is responsible for creating new occurrences of "Researcher" and changing their access privileges.

5.6.6.1.4 Context diagram

The final version of the context diagram is shown below:

5.6.6.1.5 Terminator specifications

Some of the terminator specifications are given below:

TERMINATOR : Camera
MEANING : Provides electronic image.
NUMBER OF COPIES : 1

TERMINATOR : Momentum wheel

MEANING : Used to rotate the satellite by changing its spin rate.

NUMBER OF INSTANCES : 3

IDENTIFIER : <Wheel>.number

TERMINATOR : Researcher

MEANING : A potential user of the system.

NUMBER OF INSTANCES : variable

MAXIMUM NUMBER : 100

IDENTIFIER : <Researcher>.id

5.6.6.1.6 Data flow specification
An example of a data flow specification is given below:

DATA FLOW : acceleration

MEANING : Measurements from linear accelerometer. These are caused by the action of non-gravitational forces on the satellite.

STRUCTURE : multiple

CONTAINED FLOWS : x_acceleration, y_acceleration, z_acceleration

5.6.6.1.7 Event flow specification
An example of an event flow specification is given below:

EVENT FLOW : filter required
MEANING : While TRUE, the filter is held in front of the lens of the camera.
PERSISTENCE : continuous
STRUCTURE : element

5.6.6.1.8 Store specification
The store specification for "Researchers" is given below:

DATA STORE : Researchers
ENTITIES : Researcher
RELATIONSHIPS :
STORES INCLUDED :

5.6.6.2 Event table

The development of an event table might proceed as follows.

Interview would naturally proceed with any of the end-users of the system. In this case, the only end-user obviously identified are the "Researchers". Discussion with them might identify two events: "Researcher wants picture to be taken" and "Researcher wants to see picture".

In some cases, the end-users may not be available for interview, but there may be someone who can represent their interests. This is particularly important when the terminator is an organisation or individual outside the enterprise.

It is also important to identify interface requirements for any devices that the system interfaces with. Obviously, these cannot be interviewed, but an engineer (or equivalent) should serve as a 'stand in'. In this case, there are some obvious devices, as already seen in the context diagram. Part of the control of the satellite involves orientation, both to take pictures and also to re-align on a reference star. This needs to be done to avoid 'drift' of angular measurements. Discussion of this rotation might elucidate events such as "Time to rotate satellite for picture", "Satellite reaches correct orientation for picture", "Satellite oriented to reference star, as measured by attitude measurement system", "Satellite sights reference star" etc.

Some of these events are shown in an event table below:

Event	Stimulus	System inputs	System outputs
1. Researcher wants picture to be taken.	picture taking request		
2. Researcher wants to see picture.	picture display request		picture display
3. Time to rotate satellite for picture.	time	attitude	rotation speed
4. Satellite reaches correct orientation for picture.	attitude		
5. Reference star sighted.	star tracker intensities	horizon intensities	rotation speed
6. Satellite oriented to reference star, as measured by attitude measurement system.	attitude	star tracker intensities, attitude	rotation speed
7. Time to move camera platform.	time	platform position	platform rotation
8. Platform in position.	platform position		
9. Time to start taking pictures (target in range or beginning of 'time slot').	time		filter required, open shutter
10. End of time slot for taking pictures.	time	attitude	rotation speed
11. Satellite inserted in orbit.		acceleration	
12. Ground tracking station gives updated orbit parameters.	orbit update		
13. Shutter closes.	shutter closed		

5.6.6.3 Cross-check against the context diagram

As described in §5.6.5, the context diagram and event list need to be cross-checked. This can be done in several ways, one of which 'tags' the context diagram, as shown below for a few of the events:

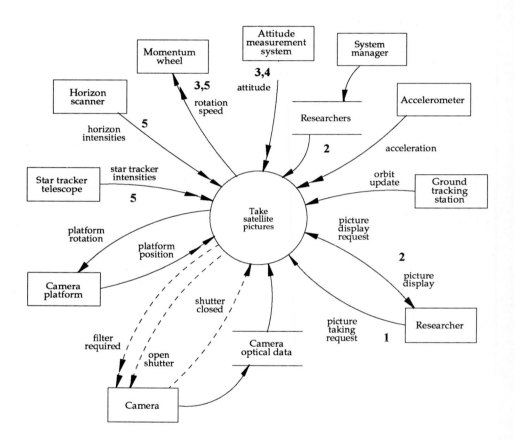

5.7 Building the Behavioural Aspect

5.7.1 Event-partitioned structure

However it is built, the behavioural aspect should be organised to have an event-partitioned structure. This organises functions and control according to the events that the system has to respond to.

5.7.1.1 Event-response DFD

5.7.1.1.1 For data processes only

For a system that has no changes of behaviour, there are no control processes and then the event-partitioned structure has the following generic form:

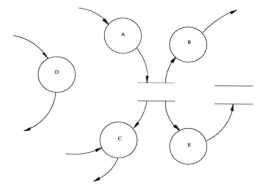

More complex systems of this type may have more data processes and individual data processes may have more inputs and outputs. There may be more stores. However, no data process is directly 'linked' to any other data process by a data flow.

The types of data process shown are:

- A: this collects information from the environment and stores it when the event occurs. The stimulus for the event might be time or it might be one of the inputs to the data process.

- B: this uses stored information to produce outputs at certain times. The stimulus for a data process of this type is time.

- C: this type of process has a stimulus data flow and uses stored data to carry out its function. It produces system outputs.

- D: this type of process transforms inputs to produce outputs without any use of stored data. This type is relatively unusual, but may occur if choices of behaviour are 'hard-coded' into its minispec, rather than as stored data. In effect, there could be a table of values 'within' the minispec.

- E: this type of process modifies stored data at certain fixed times. Although unusual, these are allowed. An example would be "Recalculate customer credit rating" for a system that reset the ratings for each customer every month, depending on their previous purchases, account balance etc.

5.7.1.1.2 Showing control processes

For systems that change their behaviour over time, there are one or more control processes, each specified by a bSTD. In addition to transforming data, data processes may detect and signal events that occur. On the basis of these detected events, the control processes may prompt (enable/disable or trigger) processes when required.

On DFDs, using the standard notation, this is shown using 'dotted' icons (control processes, event flows and prompts). If these are suppressed from the diagram, the event–response DFD still has the previous structure. No data process is allowed to link directly with any other data process using a data flow. (This may be regarded as a 'filter' being applied to the DFD to remove the dynamic part of the behaviour.)

See guidelines on connecting data processes in §5.10.5.1.

Data processes are allowed to detect events and signal that they have occurred to control processes. They may not send signals to other data processes. A control process may have several data processes that signal to it. Control processes may prompt other processes, either singly, or as a group. For example, the following is an allowed event–response DFD:

Any system has the above generic event–response DFD. There may be more control or data

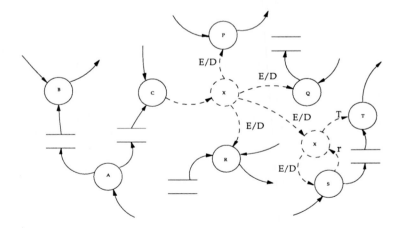

processes, but the generic format is the same.

5.7.1.1.3 Systems with constant behaviour

Some systems operate to a constant policy over time. Their behaviour does not vary. System Essential Models for such a system will have no control processes or bSTDs.

5.7.1.2 Showing terminators on the event–response DFD

Terminators may be shown on the event–response DFD. This is optional, depending on the support environment. The event–response DFD then becomes equivalent to multiple context diagrams (see §2.12.4.3).

5.7.1.3 Partial event–response DFDs

There is one conceptual event–response DFD for each system. It is of the generic format shown in §5.7.1.1.2. It shows all functions and control processes. The interface between data processes is only via stores and control processes.

If any subset of this diagram is shown on a DFD, then this DFD is referred to as *an* event–response DFD. It is part of *the* event–response DFD. It may be retained as part of the final levelled set of DFDs (in which case it has a number and name), or it may be used as a 'working view' (in which case it is un-named). (See 5.7.1.5.1 for a discussion of whether these partial DFDs are 'retained' views.)

An event–response DFD can be shown with control processes, event flows and prompts suppressed. This view highlights the functions and their use of data.

If an event–response DFD is shown with control processes and event flows then it highlights event detection and dynamics.

5.7.1.3.1 Partitioning complex processes

Some events have complex responses. These may be partitioned. For example, when a data process is examined, it may turn out that its minispec is far too large to understand and verify as a unit. The data process would then be partitioned into several smaller functions shown on a lower-level DFD.

The data process that was partitioned now becomes a process group. However, the DFD that it appears on remains an event–response DFD. The child diagram is a set of data processes that deal with the same event. These functions are synchronised by having data flows between them. This child DFD is referred to as being 'below the event level'.

5.7.1.3.2 Grouping processes

If there are a large number of event–response processes, it may be helpful to organise them into groups and provide overview diagrams showing the interfaces between these groups. The processes on these higher-level DFDs are process groups that deal with more than one event or data processes that deal with one event. The interfaces between each of these groups are stores and control processes — no direct data flow links are allowed.

Where there are control processes that enable/disable and trigger processes, it is usual to form these processes into groups.

The definition of these 'overview' DFDs is covered in §5.7.7.

5.7.1.4 Event-partitioned levelled set of DFDs

This term is used to describe a set of DFDs, with the following organisation:

1. There is one context diagram: this is a particular instance of a DFD, showing one process only.

2. Processes on a DFD are the following types: data processes, control processes and process groups.

3. There is a child DFD for each process group shown in a DFD. For example, there is one DFD that is the child diagram of the context process. The child diagram always has the same number and name as its parent process. Conventionally, the context process has the number 0, so figure 0 is the child DFD of the context process.

4. A DFD is either below the event-level or not. It is either:
   - **below the event level**: it deals with one event only; data processes and process groups may be linked by data flows;
   - **at (or above) the event level**: it deals with several events; process groups and data processes cannot be directly linked by data flows.

5. Each data process deals with a single event and has a corresponding minispec.

6. There is a bSTD for each control process (alternatively, a tabular presentation may be used).

The generic structure is similar to that shown below (see next page).

There may be more processes on the event–response DFD. There may be more overview diagrams. More (or less) of the event–response processes may be broken down into lower-level DFDs. Some systems have no control processes or bSTDs. Otherwise, the behavioural aspect of any System Essential Model includes views with this generic structure.

The behavioural aspect is referred to as being 'event-partitioned' if the above rules hold. Specifically, the responses to one event are collected together and shown as a process on a DFD. This process is referred to as the 'event–response' process. If this response is simple, the process is a data process and has a corresponding minispec. For more complex responses, there is a lower-level DFD. On the diagram showing the event–response process, the interface between this process and other processes can only be via stores.

Whatever technique is used to construct the behavioural aspect, it is suggested that the event–response DFD is constructed before grouping functions.

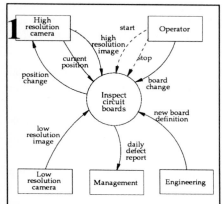

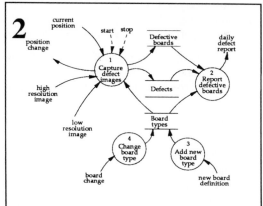

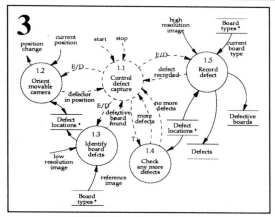

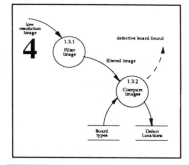

Minispec: Report defective boards

Minispec: Change board type

Minispec: Identify board defects

Minispec: Record defect

Minispec: Compare images

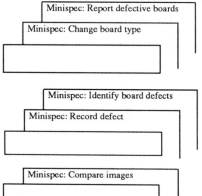

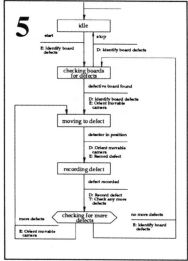

5.7.1.5 Advantages of event-partitioned organisation

There are many advantages to this organisation. Primarily, it helps separate functions into components that are activated at different times, with a simple interface between them. This aids understanding and is also a great advantage when it comes to choosing an implementation.

5.7.1.5.1 Event–response DFDs as retained views

The event–response DFD is an important view into the System Essential Model. It should be built early in constructing the behavioural aspect, because it gives good insight into other views.

Even after a set of levelled DFDs have been produced from it, it is very useful in mapping onto implementation models.

Partial event–response DFDs may also be retained views, but they are not mandatory.

In the past, this event–response DFD, showing only data processes, stores and data flows, was sometimes referred to as a 'preliminary DFD'. However, it is not discarded — it remains a useful view.[1]

Dependence on support environment

With an automated support environment, the event–response DFD is a required view into the model.

With pencil and paper support environments, the event–response DFD *may* be regarded as a temporary, working diagram. Even in these environments, it is strongly recommended that the completed event–response DFD view is retained. (This shows control processes event flows and prompts in addition.)

5.7.2 Heuristics for building the behavioural aspect

5.7.2.1 Two sequential strategies

There are two techniques often used in deriving the behavioural aspect of the System Essential Model:

1. Event-partitioning: this was developed to build an System Essential Model rapidly by organising its construction around the detection of things that happened in the environment of the system (events), and the responses the system needs to make to those stimuli. To aid in rapid organisation and definition of the model, this technique (first documented in [McM84]), requires a careful definition of the environment and all the events to which the system must respond, then focuses analysis on the response to each event. After building an event–response DFD, the system modeller then creates overview diagrams to provide a 'map' around different functional areas.

---

1 The term 'preliminary DFD' is no longer used in YSM. The diagram is no more preliminary than any other view. *All* views are produced in preliminary, first-cut versions and then refined.

2. Top-down functional decomposition (TDFD): this was developed before event-partition-ing. TDFD starts by defining a system as several high-level functional areas. The interfaces to each of those functions are considered and documented in a DFD. This diagram is revised as more detail emerges about each functional area, but it provides a starting point in controlling the complexity of the whole system. Each of these functions is then considered in turn and broken down to lower-level functions, until each lowest-level function can be verified and documented. See guidelines on this in §2.16.6.1.

5.7.2.1.1 Distinction between event-partitioning and event-partitioned System Essential Model

YSM promotes a model that is event-partitioned. Whichever strategy is adopted, the final, delivered System Essential Model is organised in a way that organises responses to a specific real-world event into a coherent part of the model, rather than distributed throughout the model. The event-partitioned model is described more fully in §5.7.1.

5.7.2.2 Comparison of event-partitioning and TDFD

5.7.2.2.1 Event-partitioning strengths
The major advantages of event-partitioning are:

- Control of complexity: the effort required in finding out the response of the system to an event is largely independent of the number of events that the system has to deal with. The total effort required is therefore approximately proportional to the number of events. Other strategies for identifying policy (such as TDFD) require effort which rises more rapidly with the number of events. As a consequence, for large systems, this strategy is the most cost-effective.

- Avoidance of 'analyst bias': it is difficult for an analyst to avoid their pre-conceived ideas about what the system should do having an effect on the model they build. This introduces 'analyst bias' into the model, which is undesirable. An analyst should be a neutral collator of the user's real requirements. Event-partitioning avoids analyst bias better than most strategies, providing the event list is carefully built and verified.

- Localisation of response to events: event-partitioning naturally leads to a model that is in event-partitioned format. Whatever strategy is adopted, it is recommended that the final model is organised in this form. If event-partitioning is used, this organisation step is short-circuited.

5.7.2.2.2 Event-partitioning weaknesses
Because of the great advantages of event-partitioning, it may sometimes be thought to have no weaknesses compared to TDFD. Although the benefits far outweigh the problems, the use of event-partitioning requires the analyst to have more technical knowledge than if TDFD is used. In addition to DFD modelling, the analyst must understand events and event-partitioning.

Successful use of event-partitioning is critically dependent on the correct identification of events. This is a problem of semantics and may lead to problems — two analysts might find very different event lists for the same system. The heuristics described in this reference manual (see §5.6.4) have largely overcome this. However, this technique remains more complex than TDFD, in the sense that more tools and techniques are used.

One type of system in which events offer little guidance are those with very few events. In a system used to calculate the neutron flux levels in a reactor for specific geometry, there is only one event — "Engineer wants to know flux levels for specific geometry". This is a classic system for which event-partitioning fails. In such systems, there is a complex information aspect, which can be modelled using entity-relationship-attribute modelling. The behavioural aspect, however, has to be modelled using TDFD.

5.7.2.2.3 TDFD strengths

TDFD allows the analyst to stay out of the detail when dealing with very large scopes, and to move the system to a more manageable size, a piece at a time, before trying detailed requirements analysis. The reduction of complexity and size is a benefit of TDFD.

TDFD also has strong management appeal. It tends to be the natural way they think about the organisation. Although tempting, this is a trap. The current functional organisation is often the result of implementation or policy decisions that were taken in the past. These company 'norms' condition the way of thinking in the company and it seems impossible that the enterprise could ever function in any other way.

This trap should be avoided at all cost. Event-partitioning, is, in any case, capable of providing high-level overviews (high-level DFDs) to give the management what they want. The need for an overview should *not* drive the method.

5.7.2.2.4 TDFD weaknesses

At times it is easier for the analyst to see the details than it is to see the general functions of a system. TDFD forces study of details to be delayed until relatively late in building the System Essential Model. There is also a risk that the organisation of processes at a high-level is implementation-oriented and the policy fragmented between lower-levels.

A more serious problem is that:

> *It is difficult to organise the system into coherent functional areas (top-level decomposition) until the system is understood; it is difficult to understand the system until it has been decomposed into smaller, functional areas.*

This is an increasingly difficult problem to overcome for larger system and is sometimes referred to as 'analysis-paralysis'. Very skilled analysts were able to overcome this problem. It is a semantic problem, like that in identifying events. Unfortunately, the techniques to overcome this problem have been sadly lacking and thus it remains very difficult to use correctly for most modellers.

5.7.2.3 Deciding whether to use TDFD or event-partitioning

If the System Essential Model is built directly, without creating either a (current)implementation model or (current)System Essential Model, it is recommended that event-partitioning is adopted. The model obtained is still decomposed into separate cohesive functions.

If the modeller can derive a fairly detailed list of events early on, direct event-partitioning is preferred to TDFD followed by reorganisation. The organisation around events is a better organisation of the essential requirements, and is useful to the designer of the system. The event level usually proves to be a helpful level of discussion with users; it is often neither too detailed nor too general.

5.7.2.4 Using event-partitioning and functional decomposition together

There are times that the modeller cannot picture the system as a whole in sufficient detail to determine what the events are. In those cases, it will help to start by carrying out a top-level functional decomposition into a small number of functional areas. Decomposition continues until each subsystem is small enough to apply event-partitioning.

Each subsystem is then examined for events. This corresponds to building the environmental aspect for each of the subsystems. As soon as this has been achieved for the subsystems, the event lists are compared, with any aliases, inconsistencies etc. being removed. The results are then merged to give a corrected environmental aspect for the whole system.

The rest of the model is then completed as before to give an event-partitioned model.

5.7.3 Using Event-partitioning

5.7.3.1 The relationship with event specifications

The event specification serves to specify the event and also the system's response to the event. Whether frame layouts or their table equivalent are used, the event specification is a useful view in building the behavioural aspect.

The event specifications may be:

- completed before building the behavioural aspect
- in parallel with building the behavioural aspect

Whichever is adopted, it advisable that the event specification is built. It helps in building the aspect and also in cross-checking the final System Essential Model. This section will assume that the event specification has been built, as described in §5.6.4.7. If the event specification is built in parallel with the behavioural aspect, the procedures in this section are carried out in parallel with those in §5.6.4.7.

5.7.3.2 Sequence for building the behavioural aspect

There are several types of view that need to be built to complete the behavioural aspect. The major view types are:

- data flow diagrams,
- behavioural state transition diagrams,
- minispecs.

There are also other views:

- process specifications,
- data and event store specifications,
- data, event and access flow specifications,

which give insight into the major views. In particular, the store specification helps to build the event–response DFD and produce minispecs.

Views such as process and event flow specifications are generally left until the end. However, they may be produced at any point in building the behavioural aspect.

The following algorithm describes the sequence in which the behavioural aspect views are built:

Note: Where the term 'in parallel' is used, the best technique is to work in a very iterative fashion. If a more sequential approach is used, the order does not matter.

Although the choice of starting with bSTDs or DFDs is not usually critical, it is best to start

1.   In parallel:

> *1.1. Produce event-response DFDs from event table or*
> > *event specifications.*
>
> *1.2. Specify stores.*
>
> *1.3. Produce bSTDs from event table or event specifications.*
>
> *end*

2.   In parallel:

> *2.1. Specify data flows.*
>
> *2.2. Add control to DFD.*
>
> *2.3. Write minispecs, partitioning complex processes.*
>
> *2.4. Organise event–response DFD into groups to give overview*
> > *diagrams.*
>
> *end*

with the one which gives most insight into the system. It will then help in building the other views. On completion of the behavioural aspect, they must all be present and checked. Deciding on a specific sequence is more a matter of efficiency of working.

The following is a simple way to 'get a feel' for whether function or dynamic viewpoint provides the most insight:

> *1. Let N_r be the number of event-response entries shown in event specifications (or the event table).*
>
> *2. Let N_c be the number of conditions shown in the event specifications. This is the expected number of changes of state. (If the condition entry has not been completed yet, a count of how many events cause a change of state may be used instead.)*
>
> *4. If $N_r > N_c$ start with the DFDs.*
>
> *5. If $N_c > N_r$ start with the bSTDs.*

Each event–response process is a data process that the system uses to respond to an event. If the numbers are roughly equal, then it probably makes little difference which views are built first. If they are very different, then it is more important.

5.7.3.3 Building the event–response DFD without control

This DFD is created from the event specifications to show only data processes, stores, data flows and event flows. It will be refined and completed to give the event–response DFD by adding control processes.

There are two types of data process shown in this DFD:

- event–detection processes: these detect that the event has occurred;

- event–response process: these carry out a response when the event occurs.

These may be combined (see §5.7.3.3.5), but the description below shows how to find them as separate processes. With experience, this step can be short-circuited and the combined processes identified at the same time. However, the separate identification approach followed by combination is easier to describe and is also open to less personal bias.

The DFD is drawn as the processes are identified.

For event–detection processes, all required inputs to detect the event are shown as inputs to the process. The only output from an event–detection process is a discrete event flow, indicating that the event has occurred.

For event–response processes, all required inputs needed to produce the response and all outputs that the response produces are shown connected to the process.

If the detection and response processes are combined, the flows are just 'added'.)

5.7.3.3.1 Use of event–response DFD to produce levelled DFDs
The event–response DFD is the primary input into the process of building levelled DFDs (see §5.7.7).

5.7.3.3.2 Finding event–detection processes
There are two types of event:

- **temporal**: the event is detected using 'time';
- **non-temporal**: the event is detected using a flow.

Temporal event detection

To detect temporal events, the system must do one of:

1. Match the time to a constant. For example:

```
time of day( ) = midnight
```

   This would take place in a data process (see §6.9.18 for definition of "time of day"). This data process would have no input on a DFD. (The 'time' input is not shown, nor is the enterprise operation. Both are 'hidden' inside the minispec.)

2. Match the time to a time in stored data. For example:

```
<Picture sequence>.start time = current time( )
```

   This could be used in a data process to detect any occurrences of this associative entity that were 'due to start now' (see §6.9.6 for definition of "current time"). On the DFD, this would be shown as a match access to the store containing the associative entity. Sometimes, the event–detection process would also change the state of this occurrence (a change access).

3. Wait a specified time after a given event. For example, when a "pedestrian request to cross" event occurs at a crossing, a system which was responsible for controlling traffic lights might signal cars to stop and then wait 3 seconds before telling the pedestrian to cross. The event "time to let pedestrians cross" is detected by:

```
SECS(state) = 3
```

   in the state machine that carried out this timing and sequencing control. (The bSTD for this example is shown in §2.24.3.4.6.) (Because this type of event is entirely handled by a state machine, it may be ignored for now.)

5.7.3.3.3 Non-temporal event detection

Non-temporal events are detected using a flow (the stimulus). The presence (or change in value) of the flow is used to detect the event. The event–detection process must therefore have this flow as an input. Where a value is checked, it could be against a fixed value ('hidden' inside the process), against a stored value (in which case a store must be read), or against another continuous flow (in which case, this flow is also an input). Sometimes additional checks must be made to be certain the event has occurred — these may require other inputs.

5.7.3.3.4 Finding event–response processes

If the system has to carry out a transformation of data in response to the event, then there will be a specific data process with this function. This is referred to as an event–response process. Event–response processes may already have been tentatively identified in building the environmental aspect (see §5.6.4.7.2). Usually, this is deferred to when the behavioural aspect is built, as described here. The event-specification is updated to reflect the event–response processes used.

If the response to the event is complex, or consists of several 'parallel' responses, then these responses are regarded as a process group that is the response to the event. This process group is elaborated in a lower-level DFD, showing the separate, independent responses.

The inputs to the event–response process are all data flows and event flows that are needed to produce the outputs. These must all be continuous, excepting the stimulus.

5.7.3.3.5 Combining event–detection and event–response processes
The above describes how, at most, there are two processes for each event. However, this can be simplified in some cases, by combining processes.

Discrete responses with stimulus
The event detection and response processes for an event can be combined into a single data process, providing all both of the following are true:

1. The event is detected by the occurrence of a discrete data flow,
2. The response is a discrete data process.

In this case, a single discrete process is shown. The data process detects *and* responds to the event. The flow is the stimulus. Combining the event–detection process and event–response process is usually done as the DFD is drawn. It is not usual to draw the two processes separately and then combine them.

If there is a change of behaviour when this event is detected, then this process must generate a discrete event flow to indicate that the event has been detected. This will be 'handled' by a control process. In this case, it is optional to separate the event–detection and event–response process. The control process can trigger the response when it receives the discrete event flow indicating that the event has occurred. If the response does not require any part of the stimulus, it is helpful to do this; if the response requires the value of the stimulus flow, then it is probably better not to separate them.

Combining an event–response process for one event with an event–detection process for another
A continuous response to one event may be combined with detection of another event, providing that the event being detected is detected by change in value of a continuous flow (data or event).

Generally speaking this is a good approach if the response process already uses that flow — simplification then results. If the flow is not used by both processes, then they are probably better kept separate.

Care must be taken if the continuous process is not permanent – the event might occur when it was not enabled. Sometimes, it can be guaranteed that the event being detected cannot occur when the continuous process is not active. Combining detection and response is allowed in these cases.

1. The data process carries out some actions (e.g. control an external device) and also detects an event that requires those actions to be terminated. This is commonly found with an associated control process, which disables the data process when the event is detected. The data process has the responsibility of 'telling' the control process that the event has occurred by means of a discrete event flow.

2. The data process carries out some actions and additionally monitors inputs to detect events. When the events occur, the process generates corresponding discrete event flows, but continues to carry out its actions. In this situation, it may be helpful to split the process into two parts — one for event detection and one for carrying out actions.

There are two cases:

3. All responses to the event are discrete data processes. In this case, they may be defined to have this temporal event as their stimulus. In effect, each of them 'wake themselves up' at this time. No event flows or control processes are required.

5.7.3.3.6 Example of developing the event–response DFD
This refers back to the satellite picture system described in §5.6.6.

For each event, the following questions were asked:

- How does the system know the event has occurred? If this requires a data process, this is an event–detection process.

- What does the system do when the event occurs? If this is a transformation of data, this is an event–response process.

For each, the inputs and outputs may also be identified from those shown on the context diagram. Recall that it is illegal to link processes that respond to different events by a direct data flow. They can only be connected via stores and/or control processes. For now, the control processes are ignored. Stores and how they are related to the ERD are discussed in §5.7.3.6.2.

The diagrams below show these DFD fragments. Detection and response processes have been separately shown. Some will be combined to simplify the model. The combination step could be carried out after drawing the detection and response fragments separately; alternatively, they could be combined as the DFD is drawn.

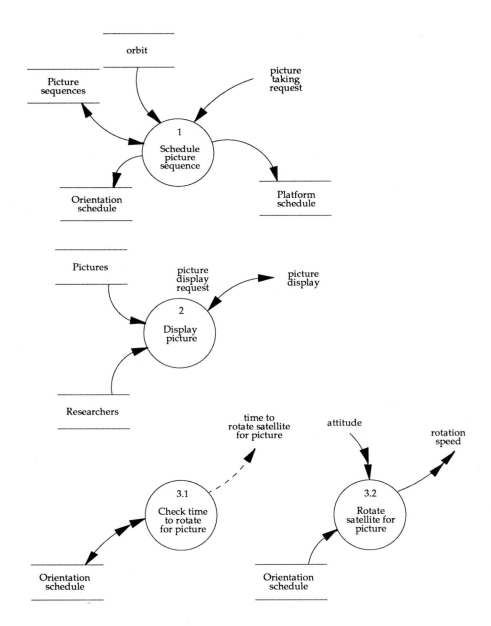

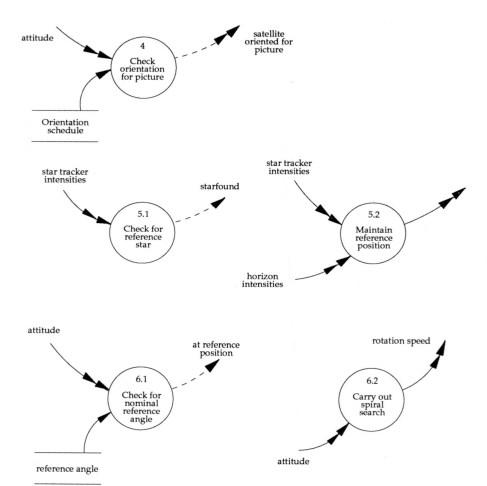

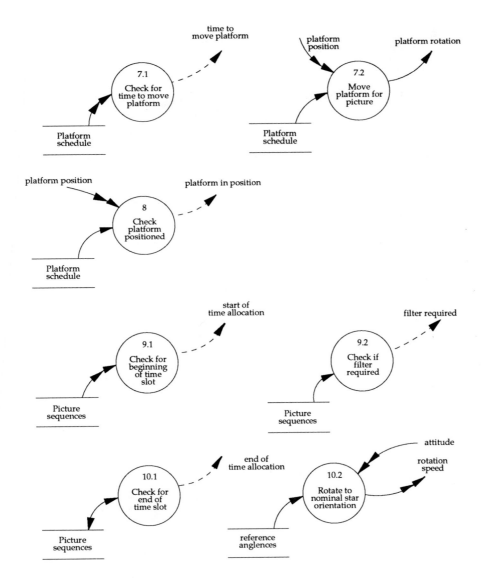

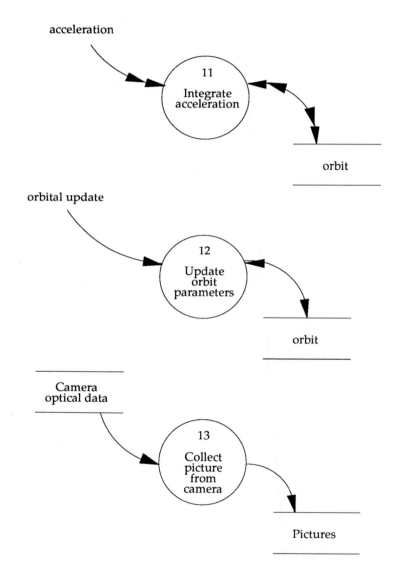

Comments

1. For event 1, a single discrete data process that detected the event (using the stimulus data flow "picture taking request") and also responded to this event.

2. For event 2, a single discrete data process that detected the event (using the stimulus data flow "picture display request") and also responded to this event.

3. The event–detection process ("Check time to rotate for picture") is permanently enabled and monitors the store for any occurrence of "Required orientation" that is due. When it detects this event, it changes the state variable "status" of that occurrence to be "due". The event–response process ("Rotate satellite for picture") reads the required orientation for the occurrence of "Required orientation" that has status = due and then rotates the satellite. As continuous outputs are required, "Rotate satellite for picture" is continuous.

4. "Check orientation for picture" is a continuous process. When it is enabled, it reads the required orientation for the occurrence of "Required orientation" that has status = due and then checks the attitude until it is equal to this angle.

5. For event 7, the detection process is rather like that for event 3. However, the response process can be discrete, because the output ("platform rotation") is discrete.

6. The event–detection processes for events 9 and 10 are different. Because event 9 must be detected at any time it occurs, "Check for beginning of time slot" is permanent. It must therefore carry out a continuous match access to check for 'due' picture sequences. Event 10 only needs to be detected (and could not be handled at any other time, anyway) after event 9 has occurred. "Check for end of time slot" is therefore enabled and disabled as required. When enabled, it can read the required time for the 'due' picture sequence. When the end of this time slot is reached, it changes the status of this picture sequence to "complete" and generates the discrete event flow "end of time allocation".

Combining event–detection process and responses
Sometimes, the response to one event may be combined with the detection of another event. In this example, there are several examples of this type.

For example "Search for reference star" and "Check for reference star" would probably never have been shown separate. Regarding the search as a response to event 6 and the finding as the detection of event 5 is very pedantic. So these could be combined into one data process "Search for reference star".

Other possible merges of the response to one event with the detection of the next include processes 3.2 and 4, 6.1 and 10.2. Note that it is not possible to combine 7.2 and 8, because 7.2 is a discrete process.

5.7.3.4 Building bSTDs

5.7.3.4.1 Determining control processes
Some of the events may cause a change in the behaviour of the system. For each of these, the entity (or entities) that is being controlled must be identified. These will be referred to as controlled entities. Ideally, each controlled entity will have its own control process. This control process should be given a name to reflect the control being carried out. This name will

typically be of the form "Control name" where "name is the name of the controlled entity. For example:

```
Control camera platform
```

for a control process that determines the correct system outputs to the device "Camera platform".

This name guideline is a good starting point, but should not be rigidly adhered to. After data processes have been grouped with the control process that activates them, a more 'honest' name may have to be chose (see for example, that in §5.7.6.2.

As soon as the control processes are determined, their names can be added to the event specification (see §5.6.4.7.2).

5.7.3.4.2 Determining the events that the control process needs to use

The starting point for building bSTDs is the events that it will have to deal with. One obvious source for these is the eSTD for the controlled entity. Note: the eSTD describes states of the real-world entity; the bSTD describes states of system behaviour (hence their names). They are different and should not be confused. However, there is a close relationship between them. In a sense, the system behaviour is prescribed in a way to cause the desired states of the entity to occur. [1]

If an event will cause a change in state of more than one controlled entity, it will be handled by each of the control processes for those entities.

5.7.3.4.3 Determining the condition

The condition is not the event. The way the condition is formed depends on what type of event is being dealt with. Primarily, it depends on how the event is detected.

Non-temporal events

If the event is a non-temporal event, there are three cases:

1. **Stimulus is discrete event flow**: the condition is this event flow;

2. **Stimulus is continuous event flow**: the condition will be a "rising" or "falling" operation on the continuous event flow;

3. **Stimulus is data flow**: the condition will be the name of a discrete event flow generated by the event–detection process.

Temporal events

If the event is a temporal event, there are three cases:

1. **Fixed time, or pre-scheduled time**: this will have been detected by an event–detection process that generates a discrete event flow when the event occurs. This will become the condition.

2. **Fixed delay after a preceding event and there are no intervening events**: the condition will be a state clock.

---

1 This relationship will become even closer in object-oriented techniques.

3. **Fixed delay after a preceding event and there may be intervening events**: the condition will be a named clock.

5.7.3.4.4 Creating the state transition framework

The following algorithm may be used to set up the 'framework' for a bSTD. This framework shows only the states and transitions, with associated conditions.

---

1. Identify the events that can affect the entity.

2. Document any sequences in which events in this group can occur, noting particularly cycles.

3. Start with an initial state (*often the 'doing nothing' state, but in fact, the choice of the first state to draw does not matter *).

4.   For each sequence of events:

    4.1.   If this event is handled by an existing state:

        4.1.1.   mark this as the "current state".

    else

        4.2.1.   add a new state.

        4.2.2.   mark this as the "current state".

    end

    4.2.   For each of the other events in the sequence:

        4.3.1.   Determine "new state" when event occurs and the system is in "current state".

        4.3.2.   If no existing state corresponds to "new state":

            Add "new state" to diagram.

        end

        4.3.3.   Add transition from "current state" to "new state".

        4.3.4.   Add event as condition for this transition.

        4.3.5.   set "current state" to "new state".

    end

  end

5. Determine the appropiate state of system behaviour when the state machine is enabled. This is the initial state.

6. Draw an initial transition to the initial state. This has no condition.

Each new state is according to the continuous behaviours occurring while the entity is in this state. When an event occurs, there may be two (or more) possible states, depending on the result of a data transformation. When this occurs, a decision point should be added, with a name describing the decision being made. It should have transitions for each possible result of the decision.

5.7.3.4.5 Compounding conditions

For conditions that are relevant to only one controlled entity, the conditions will always be simple conditions, as stated in the above algorithm.

For conditions that are relevant to more than one controlled entity, there are two cases:

1. The entities are controlled independently. In this case, the change of state in controlling one entity (and associated actions) may be carried out independently of the other entity (or entities). In this case, simple conditions may be used in each bSTD.

2. The controlling of the entities may not be carried out independently and transitions may only be made in controlling one entity if the state of the other entity (or entities) is known. In this case, compound the condition with a continuous event flow that will show whether the other entity is in a proper state for the transition to occur.

As an example of the above, we might have an event condition "x", that affects two entities "A" and "B". These might be controlled by the control processes "Control A" and "Control B". If the required transition for "Control A" depends on the state of "B" (but not vice versa) say, then the condition in "Control A" would become "x AND b_ok", where the continuous event flow "b_ok" is generated by "Control B". The condition in "Control B" would remain "x".

After this type of check has been carried out with all the conditions and the bSTDs modified, the event specifications should also be updated.

5.7.3.4.6 Adding actions

Actions in a bSTD may generate event flows, enable, disable or trigger processes, or access an event store.

Adding event flow actions to the bSTD

If more than one control process is involved, the event flow should be generated by the process controlling the entity most closely related (in meaning) to the flow.

If the system output for an event is an event flow, and that event is recognised as a condition of the bSTD, then there are two cases:

1. The bSTD can decide the appropriate action (signal, raise, or lower) at the instant the event occurs, on the basis of that event alone. If this is the case, the action should be added to the transition associated with this condition.

2. The bSTD cannot decide the appropriate action at the instant the event occurs, on the basis of that event alone, but needs to activate a data process first.

If a continuous event flow is used to reflect a combination of states of several entities, a separate bSTD should generate this flow. It will use input event flows representing each of the individual entities. (This bSTD will have to have a corresponding control process created.)

Adding process activation to the bSTD

For any condition/control process entry that also has an event–response process, determine which bSTD (if there is more than one in this entry) should activate the process. (Determining process activation is discussed in §5.7.8.1.2.)

For each transition, consider all the processes that are enabled while in the current state. If they should not be running in the destination state, disable them.

Trigger a process to make the decision as an action on the transition leading to each decision point.

Consider the initial state. If the entity is not in this state at system startup, activate any processes or send any event flows that are required to achieve this state as actions on the initial transition. Enable any event–detection processes (unless they can be permanently enabled) that may be needed in the first state.

Consider any required event–detection process. Add actions to enable this process before the event occurs unless the process can be permanently enabled. Add actions, if needed, to disable this process when the event should not be detected.

5.7.3.4.7 Example of building a bSTD

For the satellite picture system (see §5.6.6), one of the controlled devices is the momentum wheel, of which there are three instances. For one wheel, an event sequence can be identified. This would lead to three separate control processes. As a simplification, these may be combined. This corresponds to treating the momentum wheels as a single device.

For rotation, the following is a possible sequence of events: {3, 4, 10, 6, 5}. This is a cycle, in the sense that event 3 follows event 5. If the algorithm is applied, the bSTD framework below is created:

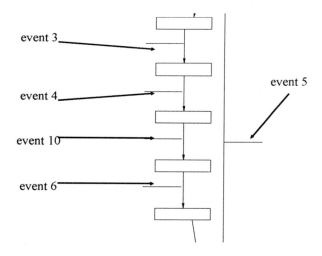

5.7.3.5 Adding control to the event–response DFD

Consider each bSTD built. Draw a control process on the DFD containing the processes controlled by that bSTD. If the bSTD controls no processes, place it on a view organised around the entity it controls.

Determine the source of the event flow for each condition that is an event flow. If the event flow is seen as a stimulus or systems input, the source is the environment. If it is seen as an output of a process, that process is the source. If it is seen with a signal, raise, or lower action on a bSTD, the control process for that bSTD is the source. If necessary, add actions to bSTDs, or existing processes to create this flow. If no existing process can (within its normal function) create the flow, add an event–detection process. Draw the flow between its source and the control process.

Determine the destination of each event flow used in a signal, raise or lower action. If the event flow is a system output, its destination is the environment. If it is an input to a control process, that is its destination. If no destination can be found, the event flow may be an alias of another; rename it consistently. Or, the flow may not be needed. Or, a condition was missed on the event table (and/or event specification).

Consider each activation on each bSTD. On the DFD, draw a prompt between its control process and the process activated. If there is no process with that name, determine if one of the processes on the diagram serves the same function. If it does, change either the bSTD or the process to make the names consistent. If it does partially, break the process into smaller subfunctions, one of which does what the control process needs, give it the name used in the bSTD, and draw in the prompt. If no process exists, and it is required, draw in the process with needed inputs and outputs.

Consider any process not activated by a control process. If it can run regardless of the state of any entity, it needs no control process to activate it. If it runs at different times, depending on the state of the entity, enable and disable actions must be added to the bSTD that monitors the states of that entity. Determine the states of the bSTD that correspond to the states of the entity (as seen in the eSTD) that the process should be enabled in. Enable and Disable actions should be added to the bSTD and corresponding prompts to the DFD to achieve this.

If more than one bSTD activates a data process determine which bSTD will 'own' the data process, and replace the activation on the other bSTD with needed signals or continuous event flows to the controlling bSTD. These will cause the needed process activation or deactivation.

Consider each discrete event flow used by a control process as an input. If this flow reflects the availability or request for an external entity, and the system does not need to respond to it before intervening events occur, change the event flow to an event store between the process that detects the event and the control process.

For any event store, add necessary initialisation to a process which also signals the event store.

5.7.3.6 Specifying data stores

Data stores are representations of entities and relationships seen on the system ERD. The data stores serve to 'tie together' the behavioural aspect and the information aspect of the System Essential Model.

The only stored information in the system is occurrences of relationships and values of attributes.

A store 'contains' all occurrences of an entity or all occurrences of a relationship. This is the simple situation and should generally be chosen in low-level DFDs where a data process accesses the store. For higher-level DFDs, where there would be a large number of stores, it is permitted to group several stores into one store, but this should only be done if they form a natural group (see §5.10.5.4).

5.7.3.6.1 When to define data stores

If entities and relationships used by the system are identified before any DFDs have been drawn, the data stores containing these can be defined. These will be the stores shown in the event–response DFDs (see §5.7.3.3).

If the DFDs are drawn before building the information aspect, stores are given temporary names, which reflect their expected content. When the information aspect is completed, these are revised to take account of the entities used.

5.7.3.6.2 Example of identifying data stores

For the satellite picture system (see §5.6.6), the system ERD is:

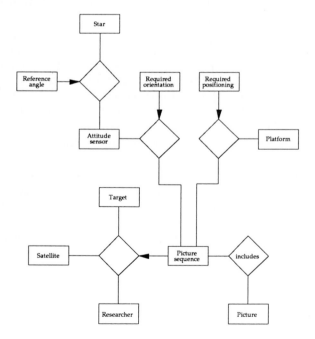

Note: This is not complete. In particular, the structure of a picture in terms of pixels has been suppressed. However, it is enough to show the general principles.

Stores

The stores used by the system are defined as:

- Pictures: this contains the entity "Picture" and the relationship "includes".

- Picture sequences: this contains the time allocation for each "Picture sequence", and a reference to the "Target" and "Researcher". Information that is fixed for a sequence of pictures (such as whether a filter is required is an attribute of the entity "Picture sequence".

- "Platform schedule": this contains the associative entity "Required positioning".

- "Orientation schedule": this contains the associative entity "Required orientation".

- "reference angle": this contains the associative entity "Reference angle".

- "orbit": this contains the attributes of the satellite. In fact, the only attributes stored are the current orbit parameters — hence the name chosen for the store.

Other entities are implicit (hard-coded) into the system.

5.7.4 Using top-down functional decomposition

5.7.4.1 Construction sequence for top-down functional decomposition
The following is an algorithm for deciding the order in which top-down functional decomposition should be used:

> *1. Create context diagram.*
>
> *2. Build behavioural aspect using TDFD.*
>
> *3. In parallel:*
>
> > *3.1. Create information aspect by examination of stored data and data flows.*
> >
> > *3.2. Identify events by deciding what caused a data process to be 'run'.*
> >
> > *3.3. Identify events by examining entities and relationships.*
>
> *4. Re-organise behavioural aspect to give event-partitioned levelling.*

5.7.4.2 Building the top layers

1. Begin identifying the major functions of the system. The statement of purpose should be a source of information. Also, check the enterprise function–entity tables (or the system function–entity tables if they exist). They should have functional groupings for any of the entities and relationships seen on the system ERD.

2. Draw a process group for each of these functions, and give it a name to describe this function.

3. Determine the needed inputs and outputs of each of these functions.

4. Group the entities and relationships accessed into data stores, and add the needed accesses to these stores from each of the functions.

5. Specify all data stores, data flows and named access flows.

5.7.4.3 Defining functions in more detail
Take each major function specified, and create lower levels for it (see §5.7.5.1). Continue until the point where the functions are sufficiently small and well-defined that they can be reorganised by the events they respond to.

5.7.4.4 Re-organising the functional decomposition

5.7.4.4.1 Identifying events
The lowest-level functions (data processes) should be examined to determine what caused them to be carried out. For each, there must be an event that occurred in the system's environment that caused that function to be needed. This event may already be in the event list, but if it is not, it should be added to the event list.

Search for the event responses among the low-level functions identified. Choose the highest-level function that completely responds to the event. If it does not exist, as the processes are grouped differently, list the lower-level responses.

5.7.4.4.2 Re-levelling the model

Finding event–detection processes
For each event, the way it is detected should be identified. It may be part of a process that has already been identified. This is allowed, but if a change in behaviour takes place when the event occurs, or several quite different types of processing occur, a separate event–detection process should be created.

Collecting together responses to one event
For a given event, all the data processes that respond to that event should be collected together and given a collective name. The function should be noted in the event response entry of the event table or event specification.

Identifying required control processes
For each event, any required change of behaviour should be determined. This may correspond to one of the identified processes 'switching' to another mode of behaviour. Any behaviour changes of this type must be provided with a control process to supervise them. Processes that have several modes of behaviour should be split into several parts, each of which has constant behaviour.

Updating the event specification
Trace the inputs to the function out to the real world. If a group of the inputs come in at the time of event occurrence and are from the same terminator, they should be grouped together and given a descriptive name. If they are the way the system knows the event has occurred, they should be placed in the stimulus entry. If the inputs are not stimuli, but are needed for the response, note them in system inputs entry. If a store has an input access by the context process, note the store in the system inputs entry.

Finding system outputs
Consider the outputs of the function. If the output goes to a terminator from the context process, it is a system output.

5.7.4.4.3 Levelling a behavioural aspect derived by TDFD
After the processes, events and entities have been identified, the behavioural aspect should be organised as described in §5.7.3.3. From here, completing the System Essential Model is as for the alternative strategy of event-partitioning.

5.7.4.5 Deriving the information aspect if TDFD is used
If top-down functional decomposition is used, then it is likely that an existing system is being examined. Even if there is no existing system, it is likely that the initial effort at modelling information will be to examine and organise existing information. This information should be:

1. Identified by examining functions for their information usage, in particular, stored data.

2. Organised to conform to a entity–relationship–attribute model.

Modelling existing information is discussed in §3.5.6. One useful technique for organising this required information is discussed in §3.5.8.

5.7.5 Specifying data processes

Minispecs can be written at any time that it seems certain that the data process is stable. Writing the minispec will clarify its interface to the rest of the system. It is a waste of time trying to write the minispec if there is no confidence that the data process has been identified as yet.

Minispecs (together with bSTDs) contain the true rigour of the behavioural aspect. All DFDs and associated specifications are redundant with the minispecs and can be derived from them.

The true test of a minispec's quality is its ability to specify all requirements with sufficient rigour that any two subject-matter experts could prepare test cases and independently achieve the same (correct) outputs for those test cases when 'playing through' the minispecs. A CASE tool may simulate this if a formal minispec grammar is used.

When 'playing through' a minispec (or simulating it using a CASE tool) requires that adequate test coverage is used. This requires careful preparation of test cases.[1]

5.7.5.1 Creating lower-level DFDs

When specifying data processes, sometimes a data process may need to be broken into smaller data processes, particularly if it is too complex (some guidelines on partitioning data processes are given in §5.10.5.3).

If the process is broken into several data processes in this way, a lower-level diagram (with the same number and name as the process being partitioned) is created. This diagram shows the child processes as separate processes, with all required input and output flows including those between processes.

Note: It is only beneath the event–response level that data flows can connect processes directly.

If this lower-level diagram is too complex, an intermediate level diagram should be created by grouping the processes seen on the diagram (as described in §5.7.7.1).

This procedure of 'breaking down' complex data processes and/or organising groups of data processes is followed until both the following conditions are true:

1. No data process is too complex.
2. NO DFD is too complex.

5.7.5.2 Choosing an appropriate style and language

It is very important to choose a style for a minispec that will be easily read and understood by those who need to review it. This style will vary from specification to specification as the audience and type of operation varies. It is not required that only one style be chosen and used for all minispecs.

Minispecs are requirements documents, not programs. While rigour is required, it need only be sufficient to guarantee consistent and correct results for all possible inputs. The use of defined

---

1 This will be covered in later releases of YSM

terms, relaxed language, graphics can all aid in the 'readability' of a minispec by non-technical personnel.

If, however, one wants to run the model as a prototype, a formal grammar is provided and can be used. It is important that, whatever specification style and language is used, all reviewers of the specification understand it. If they cannot, full reliance must be made on the prototype run, and extensive testing must occur.

5.7.5.2.1 Relaxing formal versions of minispecs

If the full rigour of the minispec cannot be understood well enough by the subject matter expert, it can be 'relaxed' and presented in a more user-friendly format. This is not the same as writing the minispec informally. It is first written formally and then presented in a more informal way at an interview or walkthrough by the analyst. The analyst is responsible for ensuring that the logic is not compromised in carrying out this relaxation. There are then (in effect) two minispecs — one is formal: this is stored as the rigorous specification; one is less formal: it is designed to be read by the user. The analyst is responsible for ensuring the correspondence between these two minispecs. The informal version may (or may not) be stored. The formal version must be.

Writing the minispec in an informal way is not acceptable. Only if it can be fully checked is the integrity of the system guaranteed. Sloppy minispecs cannot be defended by saying that "the user couldn't understand anything more complex" — it is the responsibility of the analyst to have the expertise to handle this formalism, while shielding end-users from having to know all its details.

5.7.5.2.2 Internal and external specification

YSM allows a minispec to be written as:

- **an external specification of the process:** this is useful for hiding internal details. It is also in a form that facilitates testing and QA.

- **an internal specification**: this is useful for use a blueprint in building the process. It may also be helpful for 'playing through' the logic.

Usually only one of the above is chosen. However, YSM allows both to be used, if desired.

5.7.5.3 Determining input groupings

To reduce the complexity of a minispec, it is helpful to consider input combinations where the needed transformations will be the same. For example, in most minispecs that must validate data, there are two major groupings of possible inputs; correct ones, which will undergo transformation, or incorrect ones, which will usually be discarded with an error message to the source of the input. A pre- and post-condition framework can be set up where the pre-conditions contain those comparisons required to insure correctness, and the post-conditions are the response. A general framework would be:

```
Pre-
        Data is valid
Post-
        Situation after normal response
Pre-
        Data is not valid
Post-
        Situation after error handling
```

Substitute actual conditions referring to the data and its correct values for the pre-conditions, and descriptions of the values of the output data for the post-conditions.

There may be several different valid groupings, each specified by their own set of pre-conditions.

5.7.5.4 Using the entity–event table
The entity–event table contains all the accesses that the system will make when the event occurs. If there is only one event–response process, then all these accesses will take place in that process. If there are several, the accesses may take place in several of them. This is a useful input to building the minispec.

5.7.5.5 Using eSTDs
Entity state transition diagrams can be a great help in writing minispecs. If a system eSTD exists, then it will show all events that cause change to the state of the entity. For each of the events, one of the event–response processes (there is often only one) will need to change the state of the entity. This requires a statement of the form:

```
<Scheduled course>.status := closed
```

(or the equivalent) to accomplish this. This specific example changes the occurrence of the "Scheduled course" entity to a new state.

The eSTD will also show the accesses that take place when the event occurs. These are identified by event and state, so after determining processing for each state, the minispec should be checked against the eSTD. They should be updated to be consistent.

Processing for a minispec can often be 'broken down' according to the states of the entities that it deals with. This is covered below.

5.7.5.5.1 Creating a pre- and post-conditions framework from eSTDs
If a minispec deals with only one entity that has a state variable, a pre-condition should be set up for each possible state that the entity could be in (as shown on the eSTD). For each of these, any possible transitions that could take place when the event occurs are identified from the eSTD.

If there are no possible transitions for that state and the event being handled by the process, then the post-condition may be left blank — the event is ignored.

If there are possible transitions, then the situations in which the transition *are* made must be checked for. If it is a unconditional transition, then the pre-condition is just:

```
pre- (matching)<Scheduled course>.status = open
```

(for example). In other words, if the correct occurrence of the entity is in that state, then the actions are carried out.

If the transition is only made, dependent on specific information, then other checks need to be added to the pre-condition. For example, when a "Booking" referring to a "Scheduled course" occurs, there are actually two possible transitions that could be made from the state "open" — the course might need to be closed if the last place has been taken. In this instance, the pre-condition becomes:

```
pre- (matching)<Scheduled course>.status = open
      AND <Scheduled course>.number of bookings +1
          less than
          maximum number of delegates for (matching)<Course>
```

In fact, there is another pre-condition for this state, corresponding to another transition, where the booking is rejected:

```
pre- (matching)<Scheduled course>.status = open
      AND <Scheduled course>.number of bookings + 1 =
          maximum number of delegates for (matching)<Course>
```

The above describes what happens for a single state variable of one entity. If the minispec needs to deal with more than one state variable (usually from several entities, but possibly for one entity), then a pre-condition is set up for each combination of states.

In all cases, the post-condition should be checked to see if any required accesses (as seen on the eSTD) are carried out. If the state of the entity needs to be changed (as is clearly shown on the eSTD), then this must be stated in the post-condition. For example:

```
pre- (matching)<Scheduled course>.status = open
      AND <Scheduled course>.number of bookings + 1 =
          maximum number of delegates for (matching)<Course>
post- (matching)<Booking> exists with booking date = todays date( ) ;
      <Scheduled course>.status = full
```

5.7.5.5.2 Creating a minispec from subtypes

If a minispec needs to access an entity that is a supertype, then it may be that the policy is different for each of its subtypes. A framework can be constructed for each subtype, whether pre-/post-conditions or structured text is used. There is one pre-condition (or selection) for each subtype.

If the minispec deals with several supertype entities, then each combination of subtypes should be considered carefully. It may be that there will need to be different specifications for each combinations of subtypes.

5.7.5.5.3 Using subtypes and state variables

The above two cases are readily extended to be used in combination. For example, if a minispec deals with an entity with three subtypes and an entity with four states, then there are (potentially) twelve different situations to be considered. Some of these will have the same policy and can therefore be combined in the minispec, but each should be carefully considered.

5.7.5.5.4 Use of decision tables

Rather than pre- and post-conditions, the techniques discussed above may be applied to a decision table. Indeed, if there are more than two entities involved, a decision table may be a more convenient specification tool.

5.7.5.6 Using a decision table for access specification

1. To use a decision table, set up one row for each entity accessed in response to the event. (Suppose there are two entities — there will therefore be two rows.)

2. Multiply the number of states for each entity times each other. This will be the total number of columns. (Suppose the entities have four and three states. This requires twelve columns.)

3. Fill out a column for each possible combination of states. One way of guaranteeing that all combinations are found is to do this entity by entity:

   - Take the last row, and divide the number of states for that entity into the number of columns. This gives a 'repetition factor' for that entity. (For the first entity in this example, these would be three repetitions for the first entity and four for the second.)

   - Fill in the columns by choosing the first state, and filling in 'repetition number' columns with that state name. Then take the next state, and fill in 'repetition number' of columns for that, and so on, until all columns are filled. If the result isn't exact, an arithmetical error was made in computing the number of columns or the repetition factor. (For the first entity, there are therefore three columns for the first state, three for the second ...)

   - Move up to the next entity. Its repetition factor is the repetition factor of the last entity divided by the number of states of this entity. Fill in the columns as described above. Continue moving up. If the top entity does not have 1 as its repetition number, an arithmetical error was made. (In the example, the repetition factor is one for the second entity. Its states are therefore entered in sequence: state 1, state 2, state 3, state 1, state 2, ... state 3.)

4. Refer to the eSTDs, and group the sets of actions required for each combination of states seen in a column. Add other actions needed for other outputs of the process.

5. Create a lower set of rows, one for each different combination of actions.

6. Mark which actions are to occur for each column.

See §2.16.4 for some examples of decision tables.'

5.7.6 Error handling

5.7.6.1 Adding error handling to bSTDs

Review the state transition diagram, adding any error checking or handling required, including time-outs when an event fails to occur as expected. Add additional events representing this error condition to the event table (and/or event specification), and consider all required event detection and response. Build in necessary states, transitions, and actions needed for this error handling.

5.7.6.2 Correct allocation of error handling

The control process will not always need to know about all errors. Sometimes these are handled 'locally' by a process controlled by the control process. The control process never knows about the error. For example, consider a system that played tic-tac-toe against a human player. It might have the following DFD:

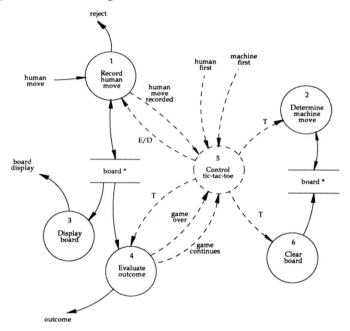

Errors in the human move inputs are not 'seen' by the control process, so no corresponding conditions or actions for this error occur in the bSTD for "Control tic-tac-toe". (Note that "Display board" is always active.)

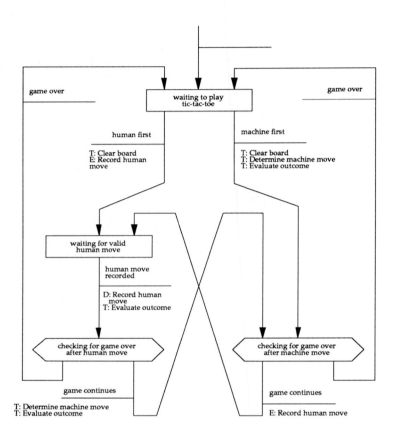

5.7.7 Grouping processes (levelling the behavioural aspect)

The event–response DFD is a major view of the system's behaviour. However, particularly for manual review it may be too large. This is dealt with by organising it into groups of processes. Each of these is shown on a partial event–response DFD. Higher-level DFDs are used to show the interfaces between the groups.

Grouping is always by event. That is, detection of an event and response to the event should always be visible in the same view.

After grouping processes that deal with one event, other groupings can be determined in various ways. The strongest reason for grouping processes is control — processes are grouped together with control processes that enable/disable or trigger them. Any processes that cannot be grouped in this way are dealt with by grouping using some weaker guideline. For systems that have no control, all process groups are formed using these weaker guidelines.

5.7.7.1 Grouping control processes and data processes

Each control process and the data processes it controls should be regarded as a group. This group is shown on a DFD. If a process is an event–response process required by more than one control process, and the response is exactly the same, a copy of the process should be added to each of the groups. These process will be enabled/disabled or triggered as required.

In addition to the controlled processes, any event–detection processes required by the control process should be added to this group.

If the event–detection processes are required by other groups, one of the two following strategies should be adopted:

1. Duplicate the event–detection process in each of the groups that needs to know the event has occurred.

2. Choose one group to detect the event and place the event–detection process in that group. The control process in that group should signal that the event has occurred to other groups by generating a discrete event flow.

Note: the occurrence of this situation means that the response to the event has been partitioned between the groups.

5.7.7.1.1 Creating groups of processes within a control group

If there are several data processes in a group that are always activated simultaneously (and in the same way, i.e. enabled, disabled or triggered) by the control process, they can be regarded as a process group. These processes are shown on a lower-level diagram, with a process group in the diagram showing the control process. The prompt is now shown to the process group, rather than to the individual components of it.

5.7.7.1.2 Partitioning control processes

If a control group seems large or complex, consider breaking the associated bSTD into two bSTDs, with event flows to communicate between them. Great care should be taken when this strategy is adopted. It means that the bSTD is also partitioned — this often makes it more, not less difficult to understand. A DFD showing one control process, with many data processes that signal to the control process and are activated by it is *not* complex. Traditional 'counting components on the DFD' limits should *not* be applied — the critical criterion is whether the bSTD can be understood or not.

If the bSTDs *can* be cleanly partitioned, allocate the data processes to one or the other of them, and form two groups (on two different DFDs). Show both process groups on a higher-level DFD, showing event flows between them.

Guidelines on partitioning bSTDs are given in §2.24.6.5.

5.7.7.2 Adding permanent data processes to existing groups

For data processes that are not controlled, consider whether they contribute to the same general purpose as any of the existing control groups. If they do, and adding the process does not cause too much complexity, they can be added to the group.

The group, as it is finally defined must be cohesive, with all processes contributing to the same general function. One good clue to whether the group is cohesive is whether it can be given a clear, honest and precise name. If it can, it is a cohesive group and is acceptable; if it cannot, the group has low cohesion and should not be chosen.

5.7.7.2.1 Dealing with non-allocated processes

For processes that do not fit any existing process groups, determine if they can be logically grouped under some higher-level functional name. Look to group processes that share the same data stores, or produce common outputs. For system with no control processes, these guidelines help in defining groups — there are no control groups.

5.7.7.3 Completing higher-level DFDs

Each group determined using any of the above techniques has a DFD. It is named according to the function carried out by the group. It may have child groups, because event–response processes were decomposed, or the group was considered too complex.

A higher-level diagram shows each of the groups identified as process groups. There is one process group on this diagram for each group identified. The name of the process group is the same as the name of the child diagram.

Consider each process group on the higher-level diagram. Draw in all input and output flows seen on its lower-level processes as coming from outside the group. Draw in all data stores with the access flows. Only those stores accessed by the group and some other process outside the group (or a terminator) should be shown on the parent figure.

After drawing in all the inputs and outputs, If any one process has too many flows around it for legibility, consider grouping some of the flows into one well-named multiple flow.

If the diagram is not too complex, no more grouping is required. If the diagram is too complex, higher-level groups should be identified on the diagram. Each of these has a DFD showing the components of the group. A new higher-level diagram is also created to show the interfaces between the groups.

In defining the groups, one or more processes may prove difficult to group together with other processes. This is acceptable — placing a process in a group with other processes to which it is only weakly related is *not* advised. Any processes of this type are copied unchanged to the parent diagram.

Note that the overview diagram will show processes that have different numbers of child processes (and/or a different 'depth' i.e. number of levels below them). This is normal. There is no requirement to 'balance the complexity' between the different child diagrams.

5.7.7.3.1 Numbering process groups and DFDs
The child diagram of the context diagram is numbered diagram 0, with processes 1, 2, 3, ... The child diagrams of each of these processes have the same name and process number as their parent process (usually — for more details see §2.14.3.9). Process numbering on these diagrams is defined in §2.14.3.20.1.

5.7.8 Completing the behavioural aspect

5.7.8.1 Filling in the process specification

5.7.8.1.1 Determining process persistence
There are two types of process:

- Continuous processes: these continuously transform their inputs to produce outputs.

- Discrete processes: these transform their inputs to produce outputs at single instants of time. Their outputs do not exist at other times.

If a process needs to produce any continuous output, it must be continuous. Event–detector processes that monitor continuous inputs to detect the occurrence of an event must also be continuous.

Other processes are likely to be discrete. If they have a stimulus, they may be enabled and disabled.

5.7.8.1.2 Determining process activation
Every process seen in the model must be checked to be sure that it has its activation properly defined.

There are several activation mechanisms, whose effects depend on whether the process is continuous or discrete.

Running continuous processes
For continuous processes, the process is either permanent, or it is enabled and disabled at specific points in time. Enabling/disabling is defined by appropriate actions on a bSTD, with corresponding prompt on the DFD. There is no entry in the process specification.

Running discrete processes
For discrete processes, there are several possibilities. The process may be triggered when required, or it may have a stimulus that causes it to run. These possibilities are the allowed values for the activation mechanism entry in the process specification. In any of these cases, the process may be shown to be "enabled" and "disabled" or "permanent", as for continuous processes.

If a discrete process is enabled and disabled, the times the process can run are limited by the control process. When the control process enables the process, the process is allowed to run. It runs whenever the stimulus or timed event occurs. When the control process disables the process, the process will not run, regardless of any arriving stimulus.

If a discrete process is triggered, it only runs at the moment it is explicitly triggered by the control process. A triggered process has no stimulus — it runs when it is triggered.

Triggering process groups is discussed in §2.14.3.23.1.

Enabling/disabling process groups is discussed in §2.14.3.17.3.

5.7.8.1.3 Checking stimulus of discrete processes

A discrete process that is not triggered has one, and only one, stimulus. The stimulus is the point in time at which the process performs its data transformation. It is either the occurrence of a temporal event, or the arrival of a discrete flow.

If a data process runs whenever an input data flow occurs, it should be modelled as having only one input discrete data flow, all of whose components will arrive at one time. The flow might be a choice of other data groups, but only one set of components will actually arrive at any one time. This flow is the stimulus for the data process.

If a data process uses one flow and then waits for another to arrive, it cannot be a single discrete data process. The first flow to occur will be lost unless it is stored by the process. This situation could be modelled by a process group with two data processes. The first flow stimulates one data process that stores the incoming data items. The second flow acts as a stimulus for a process which uses the stored value. The intermediate value must be stored; representing it as a data flow would require that both stimuli arrive at the same time, as the data flow would not persist until the second stimulus arrived.

5.7.8.2 Specifying flows

All data flows and event flows must be specified. Named access flows must also be specified.

5.7.8.2.1 Specifying elemental data flows

Each elemental data flow is specified as being an attribute of an entity, or non-attributed data item. (Note that specifying an elemental data flow as an attribute is *not* the same as saying it is stored data!)

Specifying elemental data flows as attributes

If an elemental data flow describes an entity, it is an attribute of that entity. If the flow seems to describe something, but there is no corresponding entity, careful consideration should be given to whether it is an attribute (in which case there must be a corresponding entity), or a non-attributed data item.

If the item is attributed, then there may be an existing entity. Care must be given to deal with possible aliases — the entity may not be found under the name initially allocated to it (aliasing is discussed in §3.6.1). If the entity does not exist, it must be added to the Enterprise Essential Model.

Once the entity is identified, it should be checked to see whether the attribute is already in use. If it is, the attribute should be added to the system entity specification and the data flow specified as being that attribute. The attribute may be a new attribute, in which case it must be added to the Enterprise Essential Model. Again, care is required to detect possible aliasing (see §3.6.1).

If the entity had not been previously identified as being used by the system, it will have to be added to the system ERDs. (Building the information aspect is discussed in §5.5.)

Specifying elemental data flows that are not attributes
If the elemental data flow is not an attribute, its possible values must be defined (this is referred to as typing.) If it is a numeric value that will have mathematical operations performed upon it, it must be an ADT (for example, a cardinal, integer or real). Common character data element values, such as dates, days, etc., will also be defined in terms of ADTs.

If there is no existing ADT that has the required characteristics, a new ADT may need to be specified. However, the standard YSM ADTs will cover most situations. Where the ADT has associated parameter, these need to be defined for this data item. If not all values are allowed for this specific data item, then they can be restricted to a subset of the full set of values provided by the ADT.

As an alternative (rather uncommon), the elemental data flow may be defined in terms of a value list, if it takes one of a set of values that are not used elsewhere.

Note: the most common examples of these non-attributed elemental data flows are error messages.

5.7.8.3 Specifying control processes
The usual specification tool for control processes is the bSTD. If the control process is discrete, it can be specified using a minispec with a limited set of operations (see §2.16.3.13.12). Discrete control processes are stimulated by the arrival of a fixed point in time (e.g. 'every day'), or a discrete event flow.

Note: if a continuous process does not exhibit changes of state, but determines output event flows from continuous input event flows, then it is a data process. The continuous event flows are treated as data items of type boolean within the data process. This is discussed in more detail in §2.16.3.13.12.

5.8 Gaining Insight By Comparing Aspects

5.8.1 Finding scope errors

Errors in either the enterprise information aspect or the System Essential Model can often be found by comparing the information aspect with the environmental or behavioural aspects. The consistency rules between these help to discover this type of error. These checks should be carried out as soon as possible in building the System Essential Model — this saves wasted time elaborating detail for incorrect requirements.

5.8.1.1 Example, showing failure to identify entity

For example, imagine this situation at an insurance company illustrated below:

A system ERD shows "alternate payee" as an entity. The system is responsible for creating this entity in response to the event "Member designates alternate payee". The system entity specification for "alternate payee" defines "id" to be the identifying attribute.

This event is listed on the event table, with the associated stimulus "designation data". However, the specification of "designation data" does not include any components that are attributed to "alternate payee". This is because the old system, on which this model is based, does not uniquely identify "alternate payee"s, but keeps information about them stored redundantly for each "Member".

One of two things has to occur to synchronise the views:

- the scope should be expanded to define a procedure for uniquely identifying "alternate payee". This will require that a context data flow must include information that makes it possible to identify a unique occurrence of this entity. It is likely that this would be achieved by adding an input data flow that is typed as an identifier of the entity "Alternate payee".

- the enterprise information aspect should be altered to no longer recognise "alternate payee" as an entity.

5.8.1.2 Example, showing required change to information aspect

Sometimes it is the views in the information aspect that might change. In a banking enterprise, the entity "Holding company" might be identified as created, read, and modified by a specific system. However, when the environmental aspect of that system is developed, there might be no system inputs or outputs that are attributed "Holding company".

This is a case where the necessary events for the creation either have to be added to this system's scope (its statement of purpose and context diagram may need to be modified too), or they need to be assigned to another system. The decision might be made to remove "Holding company" from the system information aspect, the justification being that the system does not require any of the information attributed to "Holding company" or the relationship "<Bank> owned by <Holding company>".

If these are removed from the system information aspect, the entity "Holding company" remains as part of the enterprise, even though this system does not need to reference any information about it.

5.8.2 Maintaining data integrity

Data integrity rules force the System Essential Model to check consistency with the views in the information aspect.

5.8.2.1 Maintaining relationship reference integrity

One rule is that any process deleting an entity must either:

- delete any relationship occurrence that references that entity occurrence;

- not carry out the deletion if any relationship occurrence referencing that entity occurrence exists .

Furthermore, if a process is deleting a relationship with mandatory participation for an entity, it must also delete the occurrence of the entity to which the relationship refers, or ensure that another occurrence of the same relationship refers to the same entity.

Checking to be certain those rules are obeyed can lead to the discovery of requirements errors.

5.8.2.1.1 An example of problems in deleting an entity

For example, consider a system that is used by a car manufacturer to keep track of dealers and their orders. Manufacturing runs will be determined on the basis of outstanding orders. Part of the System Essential Model for this is shown below:

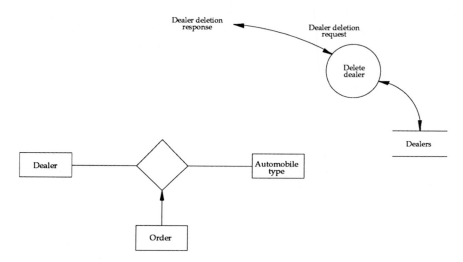

The process fails to guarantee the integrity of an automobile order — if the dealer is deleted without checking whether existing automobile orders are in place referencing that dealer, orders may be produced for which there is no dealer. Policy must be defined for whether the deletion occurs and includes deleting all orders referencing that dealer, or whether the deletion is rejected until such time as there are no orders referencing that dealer. The System Essential Model must then be changed to support the policy of the organisation. For example, it might become:

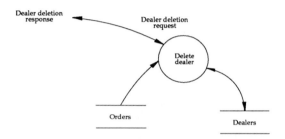

showing that the request to delete the dealer is not acted on while there are still outstanding orders (a state variable for the dealer would probably be changed to indicate that they are no longer active).

5.8.2.2 Ensuring proper updating of derived data

YSM allows attributes to be declared as 'temporary' (part of a data flow that is not stored), 'stored' and 'derived'. Problems can arise if this concept is not used correctly. In particular, if an attribute is declared to be stored, but it can be derived from other information, there is potential for inconsistencies to arise. Derivable data items should *never* be declared to be of type "stored" when they are really "derived".[1]

A derived attribute is always recalculated when it is required. In an implementation, this might be carried out by the DBMS (if it was sufficiently sophisticated!) or under the control of an application process. In a System Essential Model, the derivation is carried out automatically when the derived item is referenced.

5.8.2.2.1 Example of incorrect use of derived information

Consider the example below, taken from a banking environment:

---

[1] In implementation, it is possible that a designer may choose to store a derivable data item. This is sometimes done for efficiency reasons. However, great care must be taken in designs of this type. The trade-off between performance and complexity is carefully weighed. Essential models should never store derivable data.

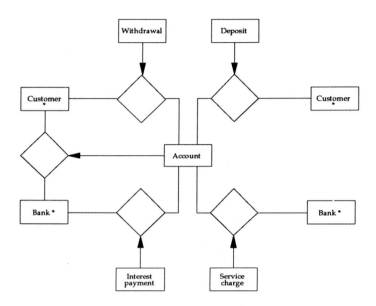

Current balance was attributed to Customer, which is correct. However, it was defined to be "stored". In fact, it is derived from withdrawals, deposits, interest payments, and service charge deductions (in this simplified example). From time to time, balance adjustments were applied to each account. The minispec dealing with this might directly update the attribute "current balance", using a values contained in "current balance adjustment".

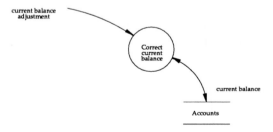

At first sight, this seems correct. However, subsequent checks of the account would give inconsistencies between the current balance, deposits, interest payments, service charges and withdrawals. To correct this problem, the current balance should be defined as being "derived" and a new associative entity "balance adjustment" created. The current balance derivation will include the sum of all deposits, less withdrawals, with the adjustment for that account added in.

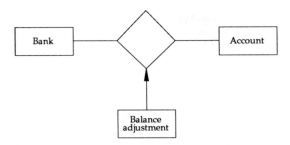

This error could have been caught when "balance adjustment" was defined as part of the input data flow composition; it should have been recognised as an attribute of an associative object, not defined as a non-attributed elemental data flow. Very few elemental data flows are non-attributed; system modellers should carefully check the entities accessed by the system before resolving that the elemental data flow is truly non-attributed. *All* stored data in the System Essential Model must be attributed.

If a elemental data flow seems to be attributed to an entity that does not yet exist, that entity should be added to enterprise information aspect.

5.9 Building the Performance Aspect

This release of YSM does not give full coverage to the performance aspect. However, additions to specifications have been made for parameters such as mean time between failures (MTBF), response time, number of occurrences (of entities), event frequencies. Future releases of YSM will provide more guidance on how to capture these parameters and then use them in 'costing' of systems and as inputs to the implementation models.

5.9.1 Estimating processing cost of the system

This is achieved by estimating the required processing for the system and then multiplying this by some factor that represents the cost of one unit of processing power.

The main parameters used in this calculation are:

- **Event frequencies**: this is given in the enterprise event specification (see §2.23);

- **Numbers of occurrences of entities and relationships**: this is given in the enterprise entity, associative entity and relationship specifications (see §2.5.2.3.8, for example);

- **Storage/occurrence of entities and relationships**: this is given in the enterprise entity, associative entity and relationship specifications (see §2.5.2.3.7, for example);

- **Number of instances of processes**: this is given in the process specification (see §2.15);

- **Response time for discrete data processes**: this is given in the process specification (see §2.15);

- **Minimum sample frequency for continuous processes**: this is given in the process specification (see §2.15);

- **Processing/activation for data processes**: this is given in the process specification (see §2.15);

- **Maximum transition time for bSTD transitions**: this is given in the bSTD (see §2.24.3.15);

- **Number of instances of terminators**: this is given in the terminator specification (see §2.13);

5.9.1.1 Major factors in determining processing requirement

The major types of processing that need to go on are:

- **functional calculations: as carried out by data processes;**

- **input and output to devices**: as shown in the context diagram;

- **accessing of stored information**: as shown as access flows;

- **control structure**: as shown by control processes. The processing requirement of control processes can usually be neglected as compared to that for data processes.

The first contribution is often much less than the other two items. This is not always true — some systems have a high proportion of their processing requirement in this area (for example, a system that solved a large number of equations in many variables). Some guidelines for estimating processing requirements for this are given in §5.9.1.2.

The amount of processing due to handling inputs and outputs is often considerable, particularly where displays are used, or rapidly changing real-world values are being controlled or monitored. Some guidelines for estimating processing requirements for this are given in §5.9.1.3.

Stored data access can often be very expensive in terms of the number of instructions required to access one item. Many commercial information storage systems have most available processing time used by the DBMS. Some guidelines for estimating this type of processing requirement are given in §5.9.1.4.

5.9.1.2 Estimating processing requirement for data processes
To determine processing requirement for system, the processing requirement for each of the data processes are added together. This leads to the algorithm:

5.9.1.2.1 *Continuous processes*
The above algorithm only works for discrete data processes. Continuous processes (including

```
Dops := 0 ;
For each event that the system deals with
    For each data process that is part of the response to that event
        Dops := Dops + frequency(event)*complexity(data process) ;
    end ;
end ;
```

event–detection processes) should have their minimum sampling rate estimated. This sampling rate is added to the process. If the process is likely to be allocated to an analog processor, this sampling rate is not appropriate.[1] The calculation is then carried out as for discrete processes activated with that same event frequency.

Choosing the sampling rate is outside the scope of this reference, but is treated in books on control theory.[2]

5.9.1.2.2 *Processing/activation*
This is defined as the required processing for one activation of the process. Conceptually, it corresponds to the number of machine instructions that need to be carried out each time the process is activated. There are two problem with this:

---

1 If a continuous process is likely to be allocated to a non-digital processor, then the above makes no attempt to estimate its likely cost.

2 See, for example [Pow78]

1. The path through the possible instructions will vary, depending on run-time values (for conditional selection of logic), so there is no one fixed value that can be used. To deal with this, either an average over all possible combinations of input parameters can be used, or the worst-case value can be used. The first is more optimistic and is probably more appropriate in most situations. In time-critical systems, the second, more pessimistic value gives a greater margin for any possible problems.

2. The effort involved in estimating the processing required by a data process may approach that of building the process and testing it. Furthermore, the estimate of complexity is required early in the life-cycle of the system — often before all the policy has been agreed. This has lead to the development of various pragmatic techniques for estimating the processing requirements from a limited set of key items. These are discussed in general terms below.

5.9.1.3 Estimating processing requirement for input-output

To estimate the processing requirement for a process that carries out input-output to the real-world, the event–response DFD is a good starting point. The flows from the event–response process to the environment can be immediately identified.

Each of these flows should be examined and the following factors determined:

- width in bytes of the data flow. This can be estimated from the data flow specification (including component flows), down to the ADT. Each data type will have an expected number of bytes, depending on the required precision, etc. In initial calculations, some guesstimates may have to be applied here.

- number of instances of the terminator for outputs. If a terminator is multi-instance, then sometimes the same flows have to be 'sent' to all instances of the terminator when an event occurs. Usually, though, when the event occurs, there will be one instance of the terminator generating the flow and one instance that requires output. Because the processing cost for one activation of the process is required, careful thought need to be given to these multi-instance terminators.

- expected type of interface. It is important to categorise the interfaces into a few major types. The processing required to output a real number to an analogue output device will be very different from that required to display it on the screen as ASCII characters. If displays are used, then whether they are used in text or graphic output mode is also likely to be significant — a rolling graph of sampled temperatures requires much more processing than a continuous display of the latest temperature.

For each type of interface used, the processing requirement can be estimated, taking the above factors into account. Each type of interface requires a 'typical processing requirement' parameter. For example, Displaying a real number on a text screen might be estimated as 3000 instructions. These estimates may be made from first principles or by benchmarking.

5.9.1.4 Estimating processing requirement for stored data access

In many commercial information systems, all other processing components are much less than that for accessing stored data. A very simple 'rough cut' estimate of the processing requirement for one activation of a data process may be calculated as the number of 'logical accesses/second' to the stored data. A create, read, update, or delete acting on a specific entity occurrence counts as one logical access. A create or delete acting on a relationship occurrence also counts as one logical access.

The processing requirement of match operations is a little more complex. If the match uses an identifier, then there is, at most, one occurrence that satisfies the test criterion. Most likely, this will be an access mechanism in the implemented system. This should be treated as one logical access. For match accesses that do not use the identifier, the implementation will require a 'search' or 'select', or their equivalent, depending on the database architecture. Often, this processing requirement will be proportional to the number of occurrences of the entity, with some proportionality constant.[1]

5.9.1.5 Calculating cost of processing

To calculate the cost of processing, a single average processing cost in terms of instructions/second per unit cost ('bangs per buck') is used. This factor may be taken as an average over a range of similar systems that have already been built, or it may be estimated from first principles.

A single factor is used, representing the concept that a single 'homogeneous' processing environment is used. It does not mean that a single processor is assumed for the chosen implementation, but rather that when the price/performance curve shows an 'upturn' the system will be distributed across several processors, with the same (roughly) price-performance ratio.

5.9.1.5.1 Obtaining better estimates

Better estimates of the processing requirements and cost may be obtained by carrying out preliminary allocation to possible processor types. This estimation is more complex, using several price-performance ratios and taking the inter-processor communication overhead into account. It will be covered in later versions of this reference.

5.9.2 Estimating storage cost of system

Strictly speaking, this should be determined at an enterprise, rather than system level. Stored information is a shared resource between several systems. This requires strategic planning and will be covered in later versions of this reference. The algorithm below should only be used 'when the system is the enterprise' (i.e. the system is truly 'stand-alone'). Because of the enterprise nature of stored data, the references in the algorithm are to 'enterprise' ERDs and specifications.

The storage requirement for an information model may be estimated in two parts — that for entities (including associative entities) and that for relationships. Usually, the storage requirements for relationships is significantly less than that for entities and their attributes. It also depends quite strongly on the data architecture chosen. It will be covered in later versions of this reference, but is neglected for this first-cut estimate.

---

[1] For some special hardware choices (e.g. associative disk drives, contents addressable file store etc.), the processing requirement will be proportional to the likely number of matches, or 'hits'.

1. The total storage is obtained to a good first approximation by estimating the storage requirement for each entity and then adding these together.

2. The storage requirement for one entity is obtained by estimating the storage for one occurrence of the entity and multiplying it by the expected number of occurrences of the entity.

3. The storage requirement for one occurrence of an entity is estimated by adding together the estimated storage for each attribute.

4. The storage requirement for each attribute is obtained be examining the ADT that it uses. For standard ADTs, there is a storage requirement that is typical for implementations in that enterprise; enumerated ADTs will only require 1 (or at most two) bytes; strings and related types require storage that may vary. For string types, the maximum expected length or the average expected length should be used, depending on whether it is expected that 'variable length records' will be used.

This may be expressed mathematically by:

$$\sum_{\text{entities}} (N_{\text{entity}}) \times \sum_{\text{attributes}} \quad (\text{number of bytes for attribute})$$

In algorithmic form, the total storage is calculated:

5.9.2.1 Calculating cost of storage

```
storage := 0 ;
For each entity that the enterprise deals with
    storage for occurrence := 0 ;
    For each attribute of the entity
        storage for occurrence := storage for occurrence + storage(ADT) ;
    end ;
    storage := storage + number of occurrences(entity) * storage for
        occurrence ;
end ;
```

To calculate the cost of storage, a single average cost in terms of 'bytes per buck' is used. This factor may be taken as an average over a range of similar systems that have already been built, or it may be estimated from first principles.

A single factor is used, representing the concept that a single 'homogeneous' storage environment is used. It does not mean that only one type of storage medium will be used in the chosen implementation, but rather that a characteristic average over all types of storage device and storage organisation is used.

5.9.2.1.1 Obtaining better estimates

Better estimates of the storage requirements may be obtained by carrying out preliminary allocation to possible storage media. This estimation is more complex, with more frequently used components being allocated to faster and more expensive technology. It will be covered in later versions of this reference.

5.9.3 Estimating cost of system interfaces

The cost of interfaces is often a significant part of the total system cost. This is particularly so for systems that deal with many instances of a user or device. Strictly speaking, it is not always possible to determine this cost on a system basis — resources of this type are often shared between several systems. This is therefore a strategic activity, supported by the Enterprise Resource Library. However, for systems that are truly 'stand-alone', the following algorithm may be used:

1. Categorize the terminators into types: people, devices.

2. For people, multiply the cost of providing one user interface by the number of occurrences of the terminator.

3. For devices, look at the number of separate signals to/from the device. Categorize these into types: digin, digout, anin, anout, stepper drive etc. Using a typical cost for one such channel, estimate the cost to interface to one occurrence of the terminator. Multiply this by the number of occurrences of the terminator.

In all cases, do not neglect the cost of cabling, modems, line drivers, cluster controllers, etc. All these costs are better calculated at an enterprise level, but for truly stand-alone systems, may be calculated on a system basis.

5.10 Quality Checks

The model must be checked for quality in various ways. Some of these are absolute require-ments, others are guidelines. For some views, components seen in the view must be specified or described in another view or views. The views, in effect, 'overlap'. These areas of overlap ensure that the functions, information, and dynamics all fit together so that the System Essential Model correctly specifies the system. In the diagrams given in §5.2.2, these inter-connections were shown as lines. The main checks are that the model should be:

- Complete: all required views must be present. For example, every data process must have a corresponding minispec; every process group must have a corresponding DFD.

- Consistent: because the different views overlap, there is the possibility that there is inconsistency between two different views showing the same component. For example, a minispec may only use stored information in a store it access on a DFD — the entities in this store are defined by a data store specification and the attributes in an entity specification.

- Correct: the two preceding criteria are of a structural, or syntactic nature. However, even if they are satisfied, there is still the possibility that the system fails to meet the user's true needs. Only if these needs have been correctly established and captured in the model is the model correct.

Obviously, these three criteria overlap to some extent. For example, the use of information in a store may seem inconsistent with the definition of the content of the store because the definition of the store's content is incomplete. However, it is convenient to regard these as three distinct types of criteria that can be applied to assure model quality.

5.10.1 Checking individual views

Each view must satisfy the rules for that modelling tool. For example, a bSTD must always satisfy behavioural consistency, have all states accessible, have one initial state etc. The rules for each of the modelling tools are given in the appropriate section of "Modelling Tools".

Each view must also be checked to ensure it correctly describes the required policy (see §5.10.4.2).

5.10.1.1 When to apply cross-checks

These cross-checks should not be thought of as 'only to be applied when that aspect is complete' — they should be applied as soon as possible in the model building process. For consistency checks, this will be possible as soon as the two views being compared are available. It may not be cost-effective to do this as soon as they exist (as they may both be very incomplete); the cross-check should be applied when they are more complete.

The cross-checks should always be applied before the System Essential Model is considered complete. The System Essential Model must pass all these criteria before it is considered of to be the required quality.

5.10.1.1.1 Cross-checks and automated support

For 'automated' support environments, these checks could either be carried out at all times, flagging individual problems, as and when they occur, or the checking could be switched 'off' and 'on', as required by the modeller. In either case, the tool should try to use the minimal information (as derived from input from the modeller) to determine what appears on each view. If this is correctly applied by the tool, then the consistency criteria described below will be automatically satisfied.

In pencil and paper environments, each view is logically independent and the checks described below must be carried out after any change to either view.

See §5.10.3.3.1 for a specific discussion relating to 'vertical balancing'.

5.10.2 Completeness of the System Essential Model

The tools used support each other by providing more detail on each aspect of the system. Full specification occurs when all views are provided.

It is also important that the model is minimal, in the sense that the system is not defined to have any requirement that is not really a requirement. This may also be viewed as a form of completeness. For example, there should be no functions other than those required as a response to an event. This type of quality check has been 'folded over' into the completeness check. For example, each function must have a corresponding event to which it is a response.

5.10.2.1 Single instance view types

1. There is a statement of purpose.
2. There exists a single context diagram. Alternatively, a set of partial context diagrams may be used — these are equivalent to a single, integrated context diagram.
3. There is an event list.

5.10.2.2 Environmental aspect completeness

5.10.2.2.1 Terminators

1. Each terminator on the context diagram has a corresponding terminator specification.
2. Each terminator specification has a corresponding terminator on the context diagram.

5.10.2.2.2 System inputs and outputs

1. Each data flow on the context diagram has a corresponding data flow specification.
2. Each event flow on the context diagram has a corresponding event flow specification.

5.10.2.2.3 System interface stores

1. Each data store on the context diagram has a corresponding data store specification.
2. Each named access flow on the context diagram has a corresponding access flow specification.

5.10.2.2.4 Event specification

1. Each event in the event list has a corresponding event specification. (The event specifications may be organised in the form of a table, if desired —see §2.22.1.2)
2. Each event specification has a corresponding event in the event list.

5.10.2.3 Behavioural aspect completeness

5.10.2.3.1 Data flow specification

1. Each data flow on a DFD has a corresponding data flow specification.
2. Each data flow specification has a corresponding data flow on at least one DFD.
3. Each ADT mentioned in a data flow specification has an ADT specification (this is part of the Enterprise Resource Library).

5.10.2.3.2 Event flow specification

1. Each event flow on a DFD has a corresponding event flow specification.
2. Each event flow generated (signalled, raised/lowered) by a minispec has a corresponding event flow specification.
3. Each event flow generated (signalled, raised/lowered) by a bSTD has a corresponding event flow specification.
4. Every event flow defined as part of a multiple evnt flow has a corresponding event flow specification.
5. Each event flow specification has a corresponding event flow on at least one DFD. It will also be referenced in at least one minispec or bSTD.

5.10.2.3.3 Data store specification

1. Each data store on a DFD has a corresponding data store specification.
2. Each named access flow on a DFD has a corresponding access flow specification.

5.10.2.3.4 Event store specification

1. Each event store on a DFD has an event store specification.
2. Each event store used (initialised or signalled to) by a minispec has a corresponding event store specification.
3. Each event store used (initialised or signalled to) by a bSTD has a corresponding event store specification.
4. Each event store specification has a corresponding event store that appears on at least one DFD.

5.10.2.3.5 Process specification

1. Each process seen on a DFD has a corresponding process specification.
2. The context process must be defined as a process group.
3. Each DFD (except the context diagram) has a process group on a higher-level DFD. he context diagram serves this overview purpose for figure 0.
4. Each minispec has at least one corresponding data (or, rarely, control) process on a DFD.
5. Each bSTD has a corresponding control process on a DFD.

5.10.2.4 Information aspect completeness

5.10.2.4.1 Data stores

1. Each entity or relationship mentioned in a minispec must be specified in a data store specification as being part of a store that appears on the DFD showing the data process. The process is linked to the store by a corresponding access flow, which may (or may not) be named.

2. If a data store is specified as containing another store, the included store must also be specified.

3. Each data store specification has a corresponding data store on the context diagram, on a DFD, or the data store must be defined as being included in a data store that is on a data flow or context diagram.

5.10.2.4.2 System entity and relationship specifications

1. Each entity on a system ERD that does not appear as an associative entity on any system ERDs has a corresponding system entity specification.

2. Each associative entity on a system ERD has a corresponding system associative entity specification.

3. Each relationship seen on a system ERD has a corresponding system relationship specification.

4. Each entity mentioned in a data store specification has a corresponding system entity specification.

5. Each relationship mentioned in a data store specification has a corresponding system relationship specification.

6. Each system entity specification has a corresponding entity on at least one system ERD.

7. Each system relationship specification has a corresponding relationship on at least one system ERD.

8. For every system associative entity specification, there is an associative entity on the system ERDs.

5.10.2.4.3 System eSTDs

1. Each entity with a state variable defined in the system entity specification has a system eSTD for each state variable.

2. Each system eSTD has a corresponding entity or associative entity seen on one or more system ERDs. The system entity specification lists this state variable.

5.10.3 Consistency of System Essential Model

Because the model is constructed in overlapping views, it is possible that two views may become inconsistent. Consistency checks exist for these reasons:

1. Where part of the system requirement appears in two views, it is stated consistently in both views. For example, if a bSTD states that it enables a particular process, it is important that the corresponding control process is shown with an enable to the controlled process in a DFD — if it were shown as a trigger, the System Essential Model would be inconsistent.

2. To ensure the same model component that appears in two different views is given the same name.

3. To ensure that model components are used correctly, according to their specification. For example, if a process is defined as continuous in a process specification, then it is illegal to trigger it — it may only be enabled or disabled.

The consistency checks listed here are grouped by aspects (environmental, behavioural and information). This is done for convenience of reference.

5.10.3.1 Consistency of the information aspect
These rules can be applied whenever the information aspect is complete:

1. Each access shown in a system eSTD must be permitted by the system entity or associative entity specification.

2. For any access action shown in a system eSTD, there is the same access on the system entity–event table for that event-entity combination.

3. Each entity seen on the entity–event table has an eSTD showing the event causing that entity to be accessed.

5.10.3.1.1 Creation of entities and relationships

1. Each entity, associative entity or relationship that the system has create access to (as defined in the system entity, associative entity or relationship specification) requires that the system has all available information to check that an occurrence of the entity *can* be created.

2. Each associative entity or relationship being created requires that occurrences of the entities that it refers to exist. The system must have create or match access to these entities. If the access is only match, then the system checks whether the entities exist before creating an occurrence of the relationship.

3. If an associative entity or relationship being created has a rule of association, then the system is responsible for checking that this rule is obeyed before creating the occurrence. The system must have read access to any entities and match access to any relationships involved in this rule of association. If attributes of entities are used in the rule of association, then the system entity specification for that entity must show the attribute with a read access.

4. Any entity that participates in a mandatory relationship must have an occurrence of the relationship created each time an occurrence of the entity is created. This means that the enterprise relationship specifications for each entity that the system has create access must be checked to determine these relationships. There must be a corresponding system relationship (or associative entity) specification for this relationship, showing the system access as "create".

5.10.3.1.2 Deletion of entities and relationships

1. If the system can delete an occurrence of an entity (as shown in a system entity or associative entity specification), then the system must also delete any relationships that refer to these entities. These relationships have system relationship specifications, showing the system's access as delete.

2. If the system can delete an occurrence of a relationship (as shown in a system relationship or associative entity specification), then the system must also delete any entities that have mandatory participation in this relationship. These entities must have system entity specifications, showing the system's access as "create".

5.10.3.2 Consistency of the environmental aspect

1. Every flow shown under system inputs in the event specification, is an input data flow to the context process.

2. Every output flow on the context diagram must be defined as a system output in the event specification (or table) for at least one event.

3. Every input flow on the context diagram must be defined as a system input in the event specification (or table) of at least one process listed as an event–response or event–detection process for the event.

4. Each event–response process has as an input any system input listed for the event.

5. Each event–response process has no other flows or flow components than those listed as stimuli, system inputs and outputs on the event/system interface table. Access flows do not come under this definition.

6. The only input data flow to an event–response process is a stimulus on the event/process table. All other inputs must come from data stores.

5.10.3.3 Consistency of the behavioural aspect

When upper level groups are defined or lower-level decompositions are created, it is important that the inputs and outputs are equivalent. (This is sometimes referred to as 'vertical balancing'.)

As flows and stores can be grouped, the views may be consistent without using the exact same names on the views.

5.10.3.3.1 Vertical balancing and automated support

For 'automated' support environments, if upper- and lower-level views have been created by the support tool under the control of the modeller, then vertical balancing should be automatically satisfied. It is a consequence of the grouping or partitioning used. There is no requirement that the cross-check is carried out in these cases — it is automatically true. However, the support tool must ensure that any changes to the levelled views are carefully controlled.

For example, adding an output to a process must automatically add that output to a higher-level diagram. This could be carried automatically. On the other hand, if an output is added to a higher-level diagram, it is not possible to determine which of the child process (or processes) it is an output from without other information.

Generally, therefore, an automated tool should allow modifications to 'lowest level' logic (minispec or bSTD), with the changes being automatically propagated upwards through the previously defined groups.

In pencil and paper environments, each view is logically independent and the checks described below must be carried out at least once and after any change to either view.

5.10.3.3.2 Data flow consistency in manual support environments
For each input data flow to the parent process, one of the following must be true:

- one or more of the processes on the child diagram must have the same input;
- the data flow is compound and all its elements are input to one or more processes on the child diagram.

Additionally, every data flow that is an input to a process on the child diagram, one of the following must be true:

- the flow must be an output from a process on the same diagram;
- the flow must appear as an input to the parent process;
- the flow must be a component of an input to the parent process.

Equivalent requirements hold for each output data flow.

5.10.3.3.3 Event flow consistency in manual support environments
An event flow input or output to one process is said to be input or output to another if any of the following are true:

- it is seen with the same name;
- it is included in another event flow;
- all the event flows it includes are input or output.

5.10.3.3.4 Data store access consistency in manual support environments
Data stores are accessed consistently if the entities and relationships they include are accessed consistently, regardless of which stores were used to include them. A data store access is consistent if:

- the same data store is accessed in exactly the same way
- the same included entity or relationship is accessed exactly the same even if in another store
- a data store including the store is accessed in exactly the same way

---

Event stores

1. Any event store updated by a process group must be updated by at least one child process.

2. Any event store written to (initialised or signalled) by a child process must be shown on the parent diagram, unless the store is only used by processes on this child diagram. On the parent diagram it is shown as being written to.

3. Any event store read (waited for) by a child process must be shown on the parent diagram, unless the store is only used by processes on this child diagram. On the parent diagram it is shown as being read.

Triggered process groups

1. If the parent process group is triggered, no child control process may enable a child process. (It is however, allowed to have a control process that triggers other processes in the group. This allows them to be 'sequenced', if required.

2. If the parent process is triggered, no process of the child diagram may generate a continuous flow.

3. If the parent process is triggered, no process of the child diagram may access an event store that does not appear on the parent diagram with an access to another process, other than the parent process.

5.10.3.3.5 Data flow diagram and process specification

1. A continuous process may be shown as "enabled" and "disabled" on a data flow diagram. It is not allowed to "trigger" it.

2. A continuous control process that has state behaviour must be specified with a bSTD, or with behavioural state transition and action tables.

3. A discrete control process must be specified with a minispec.

5.10.3.3.6 Data process/minispec consistency

The process specification for each specification on a data flow diagram defines whether it is specified by a minispec, control process, or data flow diagram. These checks are applied for all processes whose specification is by minispec.

1. Any data element that is an input to a data process must also be an input of its minispec.

2. Any data element that is an input of a minispec must also be an input to its data process.

3. Any data element that is an output by a data process must also be an output of its minispec.

4. Any data element that is an output of a minispec must be output from its data process.

5. If there is an access flow from a data store to a data process, there must be a match or read of an entity (or relationship) contained in that store by the minispec.

6. If there is an access to a data store from a data process, there must be a create, delete, or modify of an entity or relationship contained in that store by the minispec.

7. If a minispec reads or matches an entity (or relationship) there must be an access flow from the data process to a store containing that entity (or relationship).

8. If a minispec creates, deletes, or modifies an entity (or relationship) there must be an access to the data process from a store containing that entity (or relationship).

9. Any event store initialized by a minispec must be written to by the process which the minispec specifies.

10. Any event store signalled to by a minispec must be written to by the process which the minispec specifies.

11. Any event store written to by a process that is specified by a minispec must be initialized or signalled to by that minispec.

12. Any event flows output from a process that is specified by a minispec must be generated (raised, lowered, or signalled) in that minispec.

13. Any event flow generated (raised, lowered, or signalled) in a minispec must be output from the process it specifies.

5.10.3.3.7 Data Conservation

These rules can be applied to any data process. They may also be applied to the input and output data to a process group (including the context process).

1. Any input to a process should be used (required) by that process to derive an output element. (This includes the trivial case when the outputs are merely the inputs, 'routed through' the process.)

2. It should be *possible* to derive all the outputs of the process from its inputs. (This includes the trivial case when the outputs are merely the inputs, 'routed through' the process.) Note that when the specification of the system is complete, lower level details will show *how* this is achieved. The criterion that this generation of the output is *possible* is particularly useful before these details have been filled in. Note that it is unlikely that this criterion could be checked for automatically by any support tool in the immediate future – it requires human insight to apply it.

5.10.3.3.8 Control process/behavioural STD (or minispec) consistency

These rules should be applied when the control process is added to the event–response DFD, and any point thereafter if the bSTD is changed.

1. For any process triggered by a minispec or behavioural STD, or enabled/disabled by a bSTD, there is a process with the same name on the data flow diagram that contains the control process the behavioural STD is specifying.

2. Any event flow input to a control process must be used at least once within a condition on the bSTD or by the minispec for that control process.

3. Any discrete event flow output from a control process must be signalled from its bSTD or minispec.

4. Any continuous event flow output from a control process must be raised or lowered at least once by the bSTD or minispec.

5. Any process triggered by a control process must be triggered by its bSTD or minispec.

6. Any process enabled/disabled by a control process must be enabled/disabled by its bSTD or minispec.

7. Any process triggered by a bSTD or minispec must be triggered by its control process.

8. Any process enabled/disabled by a bSTD or minispec.must be enabled/disabled by its control process.

9. Any signal by a bSTD or minispec must be a discrete event flow output from its parent control process.

10. Any continuous event flow raised or lowered by a bSTD or minispec must be a continuous event flow output from its parent control process.

5.10.3.3.9 Activation consistency
These rules can be checked when control is added to the event–response DFD, and when the lower-level control processes are put in place.

1. Every process on a data flow diagram must either

    be explicitly enabled and disabled by a control process

    be explicitly triggered by a control process

    or, be permanently able to run and have a stimulus that will cause it to run.

2. If a control process triggers a process group all processes in the group has an activation mechanism of trigger.

3. If a control process enables/disables a process group no process in the group may have an activation mechanism of permanent.

5.10.3.3.10 Additional consistency rules
These can be applied after adding control to the DFD.

1. There is no more than one control process on any data flow diagram.

2. An event store must have at least one signal access and one wait access.

5.10.3.4 Consistency between the environmental and information aspects
A major check of consistency occurs when the environmental and information aspects are complete. Before spending any more time detailing system's requirements, it must be assured that all viewpoints represent the same system! These checks ensure the boundaries are the same. If some of the components of these views are built and not others, relevant checks should be made at that time. For example, if the information views and the context diagram are built, function/information viewpoint consistency can be checked. Events are not yet needed.

Extra entries, such as "stimulus", "system inputs", and "system outputs", exist on the event table and event specification to aid in checking scope consistency.

1. For each event in the event list, there is an event specification. (Note: This may be achieved by combining the two into an event table.)

2. Any store referenced in the "system inputs" or "system outputs" entries of the event specification has a corresponding entries in the entity-event table for each of the entity or relationships contained in the store.

3. Any attribute defined as used by the system in a system associative entity or entity specification should:

- be referenced by a data element specification for a data element that is (or is part of) an input to the system, as shown on the context process

or

- be derivable from inputs to the system (including inputs from stored data).

4. If the system only creates it, an entity or relationship must be included in a store accessed by the context process with a create access flow and by a terminator with an input or input/output access flow.

5. If a system uses but does not create it, an entity or relationship must be included in a data store accessed by the context process and by a terminator with a input/output or output access.

If any stimulus or system input in the event specification exists that is not input to the context process, it must either be added to a flow, or a flow created to bring it in from the appropriate terminator(s), or it must be removed from the event specification entries. All new flows must be fully specified.

If any system output does not exist as an output from the context process, it must be added to an output flow, or a flow must be added with that name, or the entry must be deleted for the system output. Fully specify all new flows.

If any inputs or outputs to the context process are not defined as a system stimulus, input or output, it must be determined which events need these flows. The table should be updated, adding and specifying events if needed. If no event within the systems scope is thought to need or produce these flows, they are not needed by the context process.

If the stimulus of a single event requires more than one discrete data input, and those data inputs do not occur at the same time at the same terminator, events have been missed. These evnts must be added to the event list and a corresponding event specification created (or a new row added to the evnt table).

5.10.3.4.1 Checking events against the system entity state transition diagram

If any of the events cause a response which accesses an entity, that event should occur as a condition in the corresponding system eSTD. If this is not the case, the enterprise eSTD should be checked. If the event is correctly shown there, the system eSTD should be amended to show this access. If the event is not shown in the enterprise eSTD, then the enterprise information modelling group should be notified that the event needs to be added to this enterprise eSTD. When this has been done, the system eSTD should be amended.

5.10.3.5 Consistency between the environmental and behavioural aspects

5.10.3.5.1 Consistency between the context diagram and DFDs

The context diagram acts as an overview data flow diagram. It may therefore be checked against the event–resonse DFD and/or figure 0 – see §6.10.3.3.1.

5.10.3.5.2 Consistency between events and the DFD

The event specification (or event table) acts as a link between the events and the behavioural views.

These can be checked when the event–response DFD is complete.

1. Any event–detection process either signals a control process or process group that the event has occurred, sends a data flow to an event–response process, or is also the event–response process for that event.

2. Any event–detection process has as an input any stimulus listed for the event.

3. Any event–response process has as an output any system output listed for the event.

4. Any event–response process has as an input any system input listed for the event.

5. Any event–response process has no other flows or flow components than those listed as stimuli, system inputs and outputs on the event/system interface table. Access flows do not come under this definition.

6. The only discrete input data flow to an event–response process is a stimulus on the event/process table. All other inputs must come from data stores or continuous data and event flows.

5.10.3.5.3 Consistency between events and DFDs

1. Compared to the event specifications, every process must be one of:

   - an event–response process;

   - an event–detection process;

   - a control process specified in the condition/control process entry;

   - a process group containing event–response or event–detection processes for one event (a child process);

   - a data process that acts as both an event–detection and event–response process for one event.

5.10.3.5.4 Consistency between events and bSTDs

1. For each event which may cause a change in state, there is a condition/control process entry of the event table or specification.

2. Each control process mentioned in the event table or event specification has a corresponding minispec or bSTD (or tabular equivalent).

3. Each condition in the event table or event specification is used by the minispec or bSTD (or tabular equivalent) of the control process paired with the condition.

5.10.3.6 Consistency between the behavioural and information aspects

Information that is shared between the system and other systems should be shown as shared on the context diagram as stores. As an alternative, these may be suppressed and this interface documented in other ways (see §2.12.3.8.1).

1. If there is an entity or relationship on the function–entity table that is not created by the system, it either must be accessed in some manner by the context process and by a create access by a terminator, or the creation must be added to the function–entity, system entity or relationship specification, systems eSTD, and entity–event table.

2. If there is an entity or relationship that is only created or only deleted by the system, but never matched, updated or read, a terminator has an input access to a store which includes the entity or relationship.

3. Each data access seen on the entity–event table should be seen in at least one data process that detects or responds to that event.

4. Each data access by an event–response process must be seen under that event in the entity–event table.

5. Each data access by an event–detection process must be seen under that event in the entity–event table.

6. Each process creating an entity, associative entity or relationship must ensure that the relationship meets all rules of association.

7. Each process accessing an entity must validate to ensure the entity is in the proper state for that event/access combination.

8. Each process deleting a relationship must validate that no entities which must participate in that relationship exist, or must also delete those occurrences.

9. Each process deleting an entity must validate that no relationships referring to that entity occurrence exist, or must also delete those relationship occurrences.

10. Each process creating an occurrence of an entity that must participate in a relationship must also create the occurrence of that relationship.

5.10.4 Correctness of the System Essential Model

The major test of quality is ensuring the model meets its objectives and can be used as needed. To trust the model, that is, to have faith that it is correct, even though it cannot be proven, requires extensive user review. This review should be done throughout the building of the model, one small view at a time.

Each view of the model should be reviewed by:

- subject matter experts to verify the policies specified;

- technical personnel not involved in the construction of the model to verify the ability of the model to communicate requirements.

5.10.4.1 Correct scope

To ensure the model represents the system:

1. Each output from the context process must exist to satisfy a responsibility listed in the statement of purpose.

2. Each responsibility listed in the statement of purpose must be associated with one or more inputs and/or outputs to the context process.

3. Each event detected by the system must be to meet one or more responsibilities in the statement of purpose.

4. Every entity the system uses (as shown on a system ERD) must be needed to support a responsibility in the statement of purpose.

5. All entities and relationships specifically mentioned in the statement of purpose should appear on the system entity relationship diagram.

5.10.4.2 Correct policy

1. To be certain the system will actually perform as needed, each minispec should be reviewed by two independent subject-matter experts. Given the same test information, they should be capable of determining the same outputs, using only the minispecs and 'hand calculation'. Two subject-matter experts should come up with the same resulting outputs.

2. Each bSTD must be verified as describing the required sequence of system behaviour.

5.10.5 Guidelines

5.10.5.1 For data flows between processes

5.10.5.1.1 Between processes in different event–response groups

There can never be a data flow between the responses to two different events — intermediate values must always be stored. This is because events can never happen at the same time.

5.10.5.1.2 Between processes in the same event–response group

Two different processes that are part of the response to the same event may be linked by a data flow. For example, the process group "Record customer deposit" might contain the processes "Validate customer deposit" and "Post customer deposit".

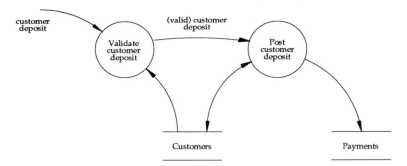

The process group "Record customer deposit" is the event–response. When it is enabled, "Validate customer deposit" and "Post customer deposit" are also enabled. The stimulus of "Record customer deposit" is "customer deposit". When it occurs, process "Record customer deposit" runs in zero time (top level view of what happens).

In more detail, when "customer deposit" occurs, "Validate customer deposit" runs in zero time, producing "(valid)customer deposit". This acts as the stimulus for "Post customer deposit" which updates the stored data.

5.10.5.1.3 Between event–detection and event–response processes

Data flows may also be used between an event–detection process and the response to that event if that response is already enabled. In other cases, a store must be used. This is true, even if the process will be immediately enabled or triggered when the control process receives the signal from the event–detection process. For example:

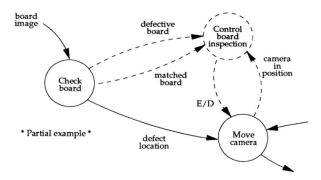

"Check board" (an event–detection process) determines that the event "Board is faulty" occurs by interpreting the data flow "board image". It generates a discrete event flow "defective board" to indicate that this event has occurred. The control process "Control board inspection" accepts this event flow and immediately enables the continuous process "Move camera" (as would be seen by examining the corresponding bSTD).

Even though "Move camera" is at a vanishingly small time interval after the data flow "defect location" is generated, it is *after* this time. It is the sequencing which requires a store. A correct version is:

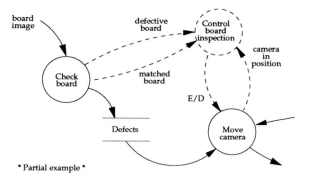

This shows a store "Defects". The value would be <Defect>.location for the <Defect> that is currently being recorded. This is indicated by the single <Defect> with <Defect>.status =

---

found. "status" is a state variable for this entity. After recording the information, the status of this <Defect> is changed to "recorded" by:

```
<Defect>.status = recorded
```

5.10.5.2 For grouping processes

1. An event–response process and event–detection process for the same event should reside on the same DFD.

2. There should be only one package or data event–response process for each event.

3. There should be only one control process detecting and responding to an event.

4. One guideline for grouping events might be that all the events in the group create the same output.

5. If there are many events which create the same output, or respond to the same input, it may be helpful to group them together. This should only be done if they constitute a single high-level function, which can be given a honest name.

6. If there are several processes that are always activated together, grouping them together allows them to be activated as a group.

7. If the change in states between two or more entities are almost always based on the same event (such as a car moving in and out of a tank) they may be combined into one control process.

5.10.5.3 For partitioning processes

1. The process could be broken to lower-level functions if:

    • it is too large to be easily reviewed (see §2.16.6.1);

    • a partial function may be reused in other processes.

2. If a process is decomposed into lower-level functions, the child DFD should not be too complex to understand. If this is a problem, then groups of functions should be chosen from the child diagram. These are shown as process groups on this child diagram, with the individual components of this group on a 'grand-child' diagram.

5.10.5.4 For defining stores

When choosing stores, they should always represent a cohesive collection of entities and relationships.

Stores at the lowest levels of DFDs should only contain one entity or relationship.

If there are too many entities or relationships in one store, several smaller stores should be created to allow meaningful real-world names of the stores.

6: Appendices

Table of Contents

6.1 Set Theory Basis of Information Model

This is not meant to be a complete discussion, but to show the correspondence between the ideas presented here and those in existing literature. See [Sto61], [Tha88], [Dow86] for further discussion of these concepts. The specification tools presented elsewhere in this manual are meant to be user-friendly equivalents of more rigorous, underlying theory. The intention is that we will eventually have CASE tools that provide user-friendly translations of this theory, but internally can be supported by the full rigour of theory. In the absence of automated support, this will not be possible, but the more user-friendly versions can still be used in pencil and paper environments.

Most of the concepts in this section relate to what are traditionally referred to as information modelling tools. However, this does not mean that YSM has 'gone rigorous' on information modelling and remained 'informal' on process modelling. The structured text grammar defined for minispecs is a Chomsky type 2 grammar (context-free grammar) [Cho59]. This grammar is defined in §6.6. The state machine used in YSM also has a very restricted and well-defined grammar (see §2.24). Together, these provide a rigorous formalism for defining and checking (including simulation of) essential models.

This section is organised as a series of notes, drawing the parallel between standard set theory and the YSM approach. YSM constants are shown in ordinary script and set constants in italics to distinguish between them.

6.1.1 Entities/sets

An entity is a set, with certain defining properties. An occurrence of an entity is an element of that set. In set notation, the symbols $\{instructor\}$ indicate the collection of all "instructors"s. This set would have a name, say "*Instructors*". To indicate that an element is a member of a set, the symbol $\in$ is used. Thus *john* $\in$ *Instructors*, or *john* $\in$ $\{instructor\}$

YSM uses an entity name to stand for the whole collection of occurrences and "<...>"s to indicate one occurrence of the set. The comparison between YSM and set theory is:

concept	YSM	set theory
one occurrence of entity	<Instructor>	*instructor*
set of all occurrences of entity (sometimes called entity-set)	Instructor	*Instructors* = $\{instructor\}$

6.1.2 Relationship/Relations

6.1.2.1 Binary relationships

A relationship involving two entities "X" and "Y" is defined as a set of ordered pairs "(x, y)", with $x \in X$ and $y \in Y$. This is equivalent to saying the relationship is the set of all lines on the instance diagram — each line on the instance diagram corresponds to an ordered pair (see §2.2.3.4). In set notation this set is written:

$$\{ (x, y) \mid x \in X, y \in Y \}$$

In set theory, this is referred to as a relation. A relation is a subset of the Cartesian product of two (or more) sets. The Cartesian product of two sets is the set of *all* possible ordered pairs from these sets and the relation is a specified subset of these. The Cartesian product is written $X \times Y$ and the relation is a subset of this, say R. In set notation, $R \subseteq X \times Y$.

A 'pure' relationship in YSM (one that is not an associative entity) may be regarded as being just this subset of the Cartesian product. Thus, if *Instructors* = {*instructor*} and *Courses* = {*course*}, then the relationship:

> <Instructor> can teach <Course>

is defined to be a set *Skills* $\subseteq$ *Instructors* $\times$ *Courses*
The comparison between YSM and set theory is:

concept	YSM	set theory
one occurrence of relationship	<Instructor> can teach <Course> (in minispec)	*(instructor, course)*
set of all occurrences of relationship (sometimes called relationship set)	can teach (on ERD)	*Skills*= {*(instructor, course)*}

6.1.2.1.1 Exceptions to YSM convention

The above convention is contravened by the use of the 'chevron' format is also used to identify the relationship in function–entity tables, entity–event tables and in the input-output list of minispecs. This is because the name of the relationship does not uniquely identify it — it is identified by its name, together with the entities it refers to (see §2.6.2.3.2).

The notation used in these cases makes it clear that it is a relationship (rather than an entity) as well as identify which specific relationship is being referred to.

This apparent dual use of the chevron notation should not cause problems. The sense in which it is being used can be inferred from the context in which it is being used. For example, in a minispec, the list of inputs might include "<Instructor> can teach <Course>", indicating that the data process needs to check for occurrences of this relationship. Within the body of the minispec, the expression "If (matching)<Instructor> can teach <Course>" might be found. However, this is only allowed if both "Instructor" and "Course" have been bound. It then gives a value which is TRUE if the specific instructor can teach that specific course (see §6.10.10).

6.1.2.1.2 Higher-order relationships/relations

A binary relationship is a set of ordered pairs; a higher-order relationship may involve ordered triples, ordered quadruples, etc., (the general term tuple is used to indicate an element of indefinite order).

Thus the relationship:

> <Course> is scheduled to run at <Location>, starting on <Day>

corresponds to the set of ordered triples:

> *Scheduled courses* = {*(c, l, d)* | *c* $\in$ *Courses*, *l* $\in$ *Locations*, *d* $\in$ *Days*}

6.1.2.2 Associative entity/defined sets

In YSM an associative entity is a relationship that can also act as an entity. It is equivalent to a set that is both a relation (a set of ordered tuples) and also used as one component of another relation. For example, the associative entity "Scheduled course" may be regarded as the set *Scheduled courses*, as defined above. The relationship <Customer> reserves place for <Delegate> on <Scheduled course> may be written as a set of ordered triples:

> *Bookings* = {(*k, d, s*) | *k* ∈ *Customers, d* ∈ *Delegates, s* ∈ *Scheduled courses*}

The entity "Scheduled course" acts as a relationship *and* entity in the YSM terminology; the set *Scheduled courses* is both a relation and set in set terminology.

6.1.3 Attributes/mappings

An attribute of an entity <E> is a mapping of the set E onto a set of values. (The set of possible values is referred to as the abstract data type in YSM and the co-domain or range in set theory.)

More formally, if the set of possible values V, then the attribute is a subset of $E \times V$ that satisfies the rule:

> if ($(x,y) \in E \times V$) and ($(x,z) \in E \times V$) then y = z.

In other words, there is a single value for each element of the set V for each element of the set E.

There may be several mappings of the same set S onto different ranges (or possibly the same range). Each is referred to as a named attribute in YSM terminology; in set terminology, each is a named subset of $E \times V$.

Note: both binary relationships and attributes are defined as being a subset of the cartesian product of two sets. The difference between the two concepts is a semantic one. In relationships, both sets a regarded as entities, in attributes, one set is regarded as a set of values.

6.1.3.1 Attributes of associative entities

An associative entity is a relationship and also has attributes. In set terms, this means that it is a subset of the Cartesian product of the sets that are involved in the relationship; it also has mappings onto value sets corresponding to the attributes.

6.1.3.2 Subtypes/subsets

In YSM, a subtype is a collection of occurrences of a more general entity, given a name to identify this grouping. This is exactly the same concept as a subset. In YSM, a subtyping is a way of breaking down an entity into exclusive groups. In set theory, this is just a named collection of subsets. To be consistent with the YSM interpretation, the intersection of any two of these subsets is null.

If a subtyping is complete, then every occurrence of the supertype is an occurrence of exactly one of the subtypes. In set theory, this is referred to as a partitioning.

6.1.3.3 Subtyping as a partial ordering

An entity may not be a subtype of itself (irreflexive property of subtyping). The supertype and subtype are different — although the supertype may not contain any occurrences that are not occurrences of the subtype at a specific point in time, it has the *potential* to do so.

More generally, if "A" is a subtype of "B", then it is illegal for "B" to be a subtype of "A". This is the antisymmetric property of subtyping. (It implies, by obvious logic, the irreflexive property.)

Furthermore, if "A", "B" and "C" are entities and "A" is a subtype of "B" and "B" is a subtype of C, then "A" is a subtype "C". This is referred to as the transitive property of subtyping. (This may be continued indefinitely — see below).

All of these are intuitively obvious and have very simple interpretations in terms of Venn diagrams.

These properties define subtyping to be an irreflexive partial ordering on the set of all entities. A more well-known irreflexive partial ordering is the "<" relation defined on the set of integers. For this reason, irreflexive orderings are often written using the "<" symbol.

Unlike the 'less than' relation, however, the subtyping relation is only partial. Given any two entities "A" and "B", it is *not* true that exactly one of A < B or B < A is true. The two entities are often noncomparable.

The reason for the above definition of subtyping as a partial ordering on the set of all entities is not that this is of direct significance to YSM. Rather, where 'inheritance' is concerned, the theory of partial orderings (which is well-known) may be applied. This leads to the rules laid down in YSM for subtyping and inheritance.

6.1.3.3.1 Network representation of subtypings

If a diagram is drawn showing all entities as nodes and subtypings as directed arcs, then a directed network (referred to as an inheritance net) is obtained. Because no entity can be a subtype of itself, there can be no closed cycles in this network. This is an example of a directed, acyclic graph.

Partial ordering and direct acyclic graphs are discussed in more detail in [Man87].

6.1.4 Entity existence, occurrence/predicate qualifiers

The named entity existence (see §6.3.3.1) "a is <E>" (read as "a is an occurrence of E") corresponds to the set statement "$\exists a \in E$" (read as "there exists an a, which is a member of the set E").

The universal named occurrence (see §6.3.3.2) "For all a<E>" (read as "for all occurrences of E") corresponds to the set statement "$\supset a \in E$" (read as "for all a's that are members of the set E").

Either of these predicates serve to bind the expression that follows. This is mostly outside the scope of this release of YSM, but will be relevant for developments in rule-based methods. However, this is consistent with the approach adopted in the rule clauses (see §6.3), existence tests (see §6.5), itemised loops (see §6.6.5.5), and the YSM data access grammar (see §6.10).

6.1.5 Rule clause grammar

Space precludes a full treatment (see [Dow86] or [Tha88]), but one simple example is shown below. In the YSM notation, the relationship:

> <Employee> reports to (manager)<Employee>

might have the rule:

> **RULES :** X reports to Y
>   satisfies:
>   1.   NOT(X = Y)

(see §6.3.4.1). In predicate form, this might be written:

> $\supset$x$\supseteq$y (reports(x,y) $\Rightarrow$ ~(x = y))

where reports(x,y) is the binary predicate constant corresponding to the relationship "reports to". This should be read as:

> for all x and all y, if x reports to y then it is not true that x = y.

In more theoretical treatments, the above would be cast into one of several standard forms (e.g. prenex normal form) (for further details consult [Tha88] or [Dow86]). This can be done automatically from the form given for YSM rules of association. It is intended that a system modeller can think in terms of the more natural "if the following preamble is true, then all the rule clauses must be true". Translation to the less friendly theoretical forms could be automated if desired.

6.1.6 Restrictions/subsets

In YSM a restriction is a limited set of values that may be used to define an attribute or an abstract data type. This is exactly the same as saying a specific subset of the values being restricted is chosen. The grammar used by YSM is exactly the normal predicate calculus used in set theory. For example, the ADT "cardinal" is defined as a restriction of the ADT "integer" by:

> **PARENT ADT :** integer
> **RESTRICTIONS :** cardinal ≥ 0

This is equivalent to the set definition:

cardinals = $\{ c \mid c \in Integers \wedge c \geq 0 \}$

A slightly more complex example is used to define the 'even number' as a restriction of the ADT "cardinal" by:

> **PARENT ADT :** cardinal
> **RESTRICTIONS :** f is a cardinal SUCH THAT 2*f = even

This is equivalent to the set definition:

evens = $\{ e \mid e \in cardinal \wedge [\exists f \in cardinals \ (2*f = e) \ ] \}$

As an example of the adequacy of the YSM notation, the ADT "odd-composite" (odd, non-prime numbers) is defined as a restriction of cardinal by:

> **RESTRICTIONS :**
>   odd-composite > 0
>   AND
>   (q is a cardinal, r is a cardinal
>   SUCH THAT
>   (odd-composite = q*r AND q$\neq$1 AND r$\neq$1))
>   AND
>   NOT(q is a cardinal SUCH THAT 2*q =odd-composite)

This is equivalent to the set definition::

$$\{x \mid x \in \mathbf{Z}^+$$
$$\wedge \; (\exists q, r \in \mathbf{Z}^+[q\neq 1 \; \wedge \; r\neq 1 \; \wedge \; x=q^*r] \;)$$
$$\wedge \sim (\exists q \in \mathbf{Z}^+[2^*q=x] \; \}$$

where $\mathbf{Z}^+$ is used to represent the set of cardinals.

6.2 Extended Backus–Naur Form

This is the notation used in defining syntax of grammars used by YSM. Note: EBNF is not part of YSM, but it is used to define YSM. (Although it should be noted that the DeMarco data dictionary notation was developed from it [DeM78]. For example, the structured text and data composition grammars are defined in terms of it.

The basic concept is that the set of legal expressions from the language are defined using production rules. A production rule shows how a named component of the language is defined in terms of a set of allowed ways of forming the item.

Each definition consists of a term, followed by the "::=" string (definition symbol), followed by legal productions of the item being defined. The "::=" is a meta-symbol — in other words, it is part of EBNF, rather than the grammar being defined. Where EBNF is used in this manual, defined terms are always shown in italics and meta-symbols are always shown 'sans-serif'. To delineate the beginning and end of a defined term, it is enclosed in "<" ">" (this is unlikely to give any confusion with the 'entity occurrence notation).

Each named item usually has several possible productions — each is separated from the next by a "|". This is again a meta-symbol. The productions themselves may include fixed text — this is always indicated in this manual by bold text.

As an example, consider the definition of a 'selection with alternative' in structured text (see §6.6):

> *<selection with alternative>* ::=
>   **if** *<condition>* **then**
>   *<statement>*
>   **else**
>   *<statement>*
>   **end**

This definition should be read as:

> *A selection with alternative is given by the characters "if", followed by a condition, followed by the characters "then", followed by a statement, followed by the characters "else", followed by another statement, followed by the characters "end".*

The above requires that "*<condition>*" and "*<statement>*" must also be defined.

The convention that a single space may be replaced by multiple spaces and/or line breaks, but no other line-breaks or spaces may occur in a legal production.

6.2.1 Iteration of construct

The above is the original BNF grammar, but it is 'extended' by allowing iteration.[1] Consider defining a 'multi-way selection with default' construct used in minispecs. This has several (one, two or many alternatives) that may be selected, depending on conditions. If none of these conditions hold, then a default statement is used. Although this *can* be defined using the meta-grammar introduced above, it is much more convenient to use an 'iteration' symbol. The

---

1 There are several 'extended' versions in use. Some use parentheses for grouping and a 'DeMarco-like' selection notation. The 'cehvrons' round defined terms is also sometimes dropped. The variant chosen for this manual was selected to avoid confusion for thos used to DeMarco data dictionary notation.

characters "{" and "}" (surprise!) are used for this. Using these symbols, the definition becomes:

```
<multi-way selection with default> ::=
    if <condition> then
        <statement>
    { elsif <condition> then <statement> }
    else
        <statement>
    end
```

The form of EBNF used in this manual assumes at least one iteration.

6.2.2 Literals

The above examples have 'hard-coded' lexical units such as **end** into the construct. This may not always be convenient. An alternative is to give each lexical unit a name and then define it as a literal. For example, the data flow composition grammar includes:

```
<left bracket> ::= '['
```

this defines this as a fixed lexical unit, consisting of the single character "[".

6.3 Rule Clause Definition.

The formulation of rules of association are based on first-order predicate calculus. This need not concern the practitioner of YSM, but for a more symbolic representation of the grammar described below, see [Tha88]. See also §6.1.5 which discusses rules in terms of set theory.

6.3.1 Structure of rule of association

A rule of association consists of two parts:

1. **Entity identifications:** there must always be a definition which identifies the participants in one specific occurrence of the relationship and gives them names by which they can be referred to in the rules. There may also be definitions of occurrences of any entities other than those involved in the relationship being specified (not all rules of association require this type of subsidiary definition). These are referred to as entity quantifications.

2. **Rule list:** this is a list of rules clauses, all of which must be true. There may be any number of these and the effective logic is that they must *all* be true.

6.3.2 Participant identification

A participant identification gives a name to the occurrence of each of the entities participating in the relationship. These names represent potential participants in one occurrence of the relationship. The relationship is allowed for these participants (or not), depending on the rule clauses.

Identification of each named participant is achieved by stating the relationship in a form with names replacing the 'entity slots' in the relationship frame. For example, in the relationship "<Employee> reports to <Employee>(manager)", names could be substituted:

```
RULES : X reports to Y(manager)
        satisfies:
        (here would be a list of rule cluses satisfied by X and Y)
```

Both "X" and "Y" are specific occurrences of the entity "Employee". The symbols "X" and "Y" are used to represent these specific occurrences in the rules. YSM refers to these as named occurrences of the entities, or bound variables. This reflects the fact that they cannot freely vary any more. Any name may be chosen for these bound variables, as long as it is different from any other name used in the rule of association. The role qualifier (see §2.2.3.4.7) may be retained, but it is more usual to drop it. For example:

```
RULES : X reports to Y
        satisfies:
        ...
```

is not ambiguous, because the second 'slot' in the relationship frame has already been designated as playing the manager role in the relationship (in the participating entities entry of the relationship specification).

6.3.2.1 Free occurrences

If the rules do not involve one of the participants, it may be left as an anonymous or free variable. For example, in the relationship "<Employee> reports to <Employee>(manager)", we might have:

```
RULES : <Employee> reports to X
       satisfies:
          ...
```

This indicates that there are going to be some rules about which occurrences of "Employee" can participate in the relationship as a manager, but there are no restrictions on which occurrences of "Employee" participate in the relationship in the reporting role. These rules will involve statements about "X", rather than "<Employee>". In effect, these free occurrences represent 'do not care' occurrences.

A free variable declared in this way does not represent a specific occurrence and cannot be used in the rule.

6.3.3 Entity quantifications

Many rules of associations only involve statements about the participants in the relationship. For other relationships, there are less 'local' constraints that must be checked. These may involve:

- comparison with other occurrences of the same entity;
- occurrences of other entities;
- different relationships involving entity occurrences.

There are two types of entity quantifications that may be used to introduce other entity occurrences into the rules. These are the existence of one named occurrence of an entity and the whole collection of possible occurrences. These will be referred to as 'named entity existence' and 'universal named occurrence'. Both of these set up bound variables for use in the rules.

6.3.3.1 Named entity existence

A named entity existence asserts that there is one (there may be more) occurrence of an entity. This specific occurrence is given a name, by which it can be referred to in the rule clauses. This entity occurrence may be of any entity, not just those that participate in the relationship whose rule is being defined. To assert such a named occurrence of an entity "E", the form:

```
A is <E>
```

is used. This should be read:

> *There is a particular occurrence of the entity E that satisfies the following rules. In the rules, it will be referred to as A.*

This is a named occurrence and is a bound variable, in the same way as named participants. The name chosen for this occurrence is not critical. Any name can be chosen, as long as it is different from any other named entity occurrences.

6.3.3.1.1 Conditional existence

A filter may also be applied to set of entity occurrences. For example:

```
A is <E> with A.age > 21
```

would restrict "A" to be an occurrence of the entity "E" which had a value for the attribute "age" greater than 21.

6.3.3.2 Universal named occurrence

A universal named occurrence asserts that there all occurrences of an entity, the rules hold. These occurrences are given a name, by which they can be referred to in the rule clauses. These entity occurrences may be of any entity, not just those that participate in the relationship whose rule is being defined. To assert such a universal named occurrence of an entity "E", the form:

```
For all A that are <E>
```

is used, which should be read:

> *For all occurrences of the entity E, the following rules are satisfied. In the rules, a single occurrence will be referred to as A.*

This is also a named occurrence and is a bound variable, in the same way as named participants and named entity existences. As with named entity existence, filters may also be applied to filter out a subset of the entity occurrences.

6.3.3.3 Multiple named occurrence declarations

There must be a participant identification clause and there may be one or more entity quantifications. These are then followed by the rule clauses that must be satisfied. In the EBNF notation used in this manual (see §6.2):

> *<entity identifications>* ::= *<participant identification>* { **AND** *<entity quantifications>* }
> **satisfies**
> *<rule clause>* { *<rule clause>* }

There must be at least one rule clause, but there may be more.

The syntax constants "AND" and "satisfies" are not critical, but are the lexical notation adopted in this manual. For example, "AND" which separates the participant identification from any entity quantifications and then subsequent entity quantifications from each other could replaced by any other convenient equivalent.

6.3.3.4 Reference to associative entities

If checks involve associative entities[1], a slightly different form of declaration is required. In their role as an entity, they are declared with a name that is used as an occurrence of that entity. In their role as a relationship, they are shown to be linking specific occurrences of the entities to which they refer. For example, the associative entity "Scheduled course" includes :

---

1 If the relationship whose rule is being defined is an associative entity, then it is traeted as a relationship in defining the bound variables. This section deals with situations where the rule may refer to other associative entities that are different from the relationship whose rule is being defined.

> **PARTICIPATING ENTITIES :**
> <Course> is scheduled to run at <Location>, starting on <Day>.

If the definition of another relationship includes a rule of association that needs to make references to occurrences of "Scheduled course", it might include:

> **RULES : ...**
>   AND S is C is scheduled to run at L, starting on D
>   satisfies:
>   ...

This defines an occurrence of the entity "Scheduled course" that can be referred as "S" in it's role as an entity. In it's role as a relationship, it refers to a Course ("C"), a Location ("L") and a Day ("D"). Any (or all) of these are now bound variables and can be used in the rule clauses. If some of the entities that the associative entity refers to are not required in the rule clauses, then they may be left as free variables. For example:

> **RULES : ...**
>   AND S is C is scheduled to run at <Location>, starting on <Day>
>   satisfies:
>   ...

This allows rules to be stated about the course, but not the location or day, which are free variables.

6.3.4 Rule list

The rule list is a list of rule clauses, each of which evaluate to TRUE or FALSE. For allowed occurrences of the relationship, all these rules must evaluate to TRUE. If they do not, the occurrence of the relationship being checked is not allowed.

A rule list can be combined into one compound rule, if desired, by the use of "AND"s. However, this usually decreases, rather than improves clarity.

Each rule clause is either a simple rule clause or a compound rule clause.

6.3.4.1 Simple rule clauses

The named occurrences of the entities are used in comparisons that give a boolean value — i.e. TRUE or FALSE (see §6.4.4). The simple rule clauses may be:

1. **Comparison of occurrences:** a test may be made to see whether they are the same or different occurrences of an entity. This is only allowed if they are occurrences of the same entity. For example, in the relationship "<Employee> reports to <Employee>(manager)", we might have a single rule clause:

> **RULES :** X reports to Y
>   satisfies:
>   1.   NOT(X = Y)

This is really a compound rule clause, because of the 'NOT' (see below), but rules more commonly demand that occurrences are different, demanding they are the same is rare. (See §6.1.5 for predicate form of this rule.)

2. **Expression involving values of attribute:** any expression in the attributes of the named occurrences may be used, providing the expression has a boolean value (see §6.4.4). For example, a single rule clause might include a comparison:

```
RULES : X is married to Y
     satisfies:
        1.      ABS(X.age – Y.age)  < 5
```

to disbar large age differences. (Sorry about 'ageist' example.)

3. **Expression involving value of a state variable:** for example:

```
RULES :
Course is scheduled to run at <Location>, starting on <Day>
     satisfies:
        1.<Course>.status = available
```

The value it is equated to must be one of the permissible states for that state variable, as shown in the eSTD. The only comparison that may be used is "=" (as above), or "≠".

4. **Relationship occurrence:** a relationship occurrence may be formed by using bound variables in place of the frame slots for any relationship in which they participate (not just the one being defined). For example:

```
RULES : X is scheduled to teach Y
                AND S is C is scheduled to run at <Location>, starting on <Day>
                        with S = Y
        satisfies:
            1.        X can teach C
```

6.3.4.2 Compound rule clauses

Rule clauses may be combined using the usual boolean operators (see §6.4.4). The following are the production rules for legal combinations:

1. Each rule clause must evaluate to TRUE or FALSE for any specific occurrences used in it.

2. A rule clause may be formed as "NOT(rule clause a)".

3. A rule clause may consist of "rule clause a AND rule clause b".

4. A rule clause may consist of "rule clause a OR rule clause b".

In the above, "rule clause a", "rule clause b" indicate previously defined rule clauses. Either may be simple or compound. Quite complex rules may be built up using this grammar. For example, "<Person> is married to <Person>" might include:

> **RULES :** A is married to B
> AND For all C that are <Person>
> satisfies:
> 1. A.age ≥ 18
> 2. B.age ≥ 18
> 3. A.gender ≠ B.gender
> 4. NOT ((C is mother of A AND C is mother of B)
> OR
> (C is father of A AND C is father of B))

has four rule clauses. The first three are simple and the fourth is compound.

6.3.4.2.1 *Example of using other relationships and entities*

As an example where the rule involves other entities, the relationship <Instructor> is allocated to teach <Scheduled course> might have:

> **RULES :** X is scheduled to teach Y
> AND S is C is scheduled to run at <Location>, starting on <Day>
> with S = Y
> satisfies:
> 1.   X can teach C OR ((X.age > 28) AND (C.status = in testing))

This uses a relationship <Instructor> can teach <Course> and a state variable of the entity "Course". It makes the rather curious demand that instructors can only be allocated to teach courses if they have been certified as qualified to teach it (via the relationship) or the course is being tested and they are so old they can take the aggravation of an untested course (another ageist example!). The ERD fragment is shown below:

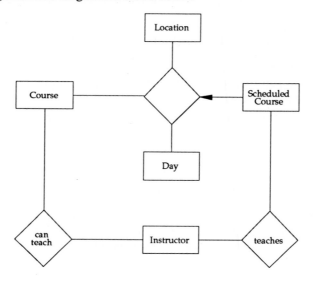

6.4 Standard YSM Abstract Data Types

The standard YSM ADTs are provided as part of the method (user-defined ADTs are also allowed — see §2.10). They may be used to specify attributes and data elements. Each type has certain characteristics which are described here. The way in which these characteristics will be implemented is 'transparent' to the user.

Each ADT has certain pre-defined operation associated with it (user-defined operations are also allowed — see §2.17). These operations may be used for any item declared to be of that type. For example, an attribute specified as an integer can be added to another integer, giving an integer result. This operation may be accessed in several different formats, for example:

> "add x to y and store the result in a",
> "assign the value of x + y to a",
> "a := x + y ;",

but it is the same operation, whichever of these formats is used. These are referred to as different dialects. An enterprise may use a single dialect, or it may use several dialects, depending on different audience backgrounds and requirements.

No specific dialect forms are given in this manual for operations associated with ADTS. This is to save space.

6.4.1 alphabetic

ABSTRACT DATA TYPE : alphabetic

MEANING : This is any sequence of alphabetic characters.

STRUCTURE : simple

NUMBER OF VALUES : discrete

VALUE DEFINITION : inheritance

PARENT ADT : string

RESTRICTIONS : Each character contained in the string must be one of A, B, …,Z, a, b, …, z.

ORDERING : linear

PARAMETERS : —

6.4.1.1 Operations

6.4.1.1.1 Inherited operations

The alphabetic ADT inherits all operations defined for the string ADT:

1. Concatenation with a string returns a string;
   with a text returns a text;
   with an alphabetic returns an alphabetic;
   with an alphanumeric returns an alphanumeric.

2. Substring or slice to select part of alphabetic, returns an alphabetic.

3. char(x, n) returns the nth character of x as an alphabetic character.

4. length(x) returns the length of alphabetic x as a cardinal.

6.4.1.1.2 Specific operations

1. The operations $<, \leq, =, \neq, \geq, >$ are used to test lexical ordering of two alphabetics, returning a boolean value.

6.4.1.2 Comments

1. The minimum length and maximum length parameters are inherited from string.

2. The standard alphabet used in YSM is:

> A, B, C, D, E, F, G, H, I, J, K, L, M, N, O, P, Q, R, S, T, U, V, W, X, Y, Z,
> a, b, c, d, e, f, g, h, i, j, k, l, m, n, o, p, q, r, s, t, u, v, w, x, y, z

If any other alphabet is required, it could be given a distinct name.

6.4.2 alphanumeric

ABSTRACT DATA TYPE : alphanumeric

MEANING : This is any sequence of alphabetic or numeric characters.

STRUCTURE : simple

NUMBER OF VALUES : discrete

VALUE DEFINITION : inheritance

PARENT ADT : string

RESTRICTIONS : Each character that is part of the alphanumeric is either an alphabetic character (see alphabetic ADT) or 0, 1, 2, 3, 4, 5, 6, 7, 8, 9.

ORDERING : none

PARAMETERS : —

6.4.2.1 Operations

6.4.2.1.1 Inherited operations

The alphanumeric ADT inherits all operations defined for the string ADT:

1. Concatenation with a string returns a string;
   Concatenation with a text returns a text;
   Concatenation with an alphanumeric or alphabetic returns an alphanumeric.

2. substring or slice to select part of alphanumeric, returns an alphanumeric.
   where s is an alphanumeric, start and n are cardinals, returns the n characters, starting at the ith character.

3. char(x, n) returns the n'th character of x as a character.

4. length(x) returns the length of alphanumeric x as a cardinal.

6.4.2.1.2 Specific operations

1. The operations = and ≠ are used to test lexical equivalence of two alphanumerics, returning a boolean value.

6.4.2.2 Comments

1. The minimum length and maximum length parameters are inherited from string.

6.4.3 angle

ABSTRACT DATA TYPE : angle

MEANING : Used to measure rotations and directions.

STRUCTURE : Simple

NUMBER OF VALUES : continuous

VALUE DEFINITION : inheritance

PARENT ADT : real

RESTRICTIONS : 0 <= angle < 2π

ORDERING : cyclic

DIMENSIONS : 1

UNITS : radians,
          degrees,
          minutes,
          seconds

PARAMETERS :

parameter	: type	: default
accuracy	: angle	: 0
resolution	: angle	: 0

6.4.3.1 Operations

1. Angles may be added or subtracted. Negative angles are equivalent to the same angle modulo 'one rotation' (360 degrees, or 2π radians).

2. The operations "sin", "cos" and "tan" all have a single input of type "angle" and return a value of type "real". These functions have the usual mathematical meaning.

3. The operations "arcsin", "arccos" and "arctan" all have a single input of type "real" and return a value of type "angle". These functions have the usual mathematical meaning ($\sin^{-1}$, $\cos^{-1}$, $\tan^{-1}$).

6.4.3.2 Comments

6.4.4 boolean

ABSTRACT DATA TYPE : boolean

MEANING : The boolean type has two values TRUE and FALSE. It is sometimes referred to as the "logical" data type, but that should be avoided because of overuse of the term "logical". Boolean types are used to describe whether something is true or not. Note: the boolean ADT cannot be restricted.

STRUCTURE : Simple

NUMBER OF VALUES : finite

VALUE DEFINITION : value list

VALUES : TRUE, FALSE

ORDERING : none

6.4.4.1 Operations
The three standard operations are NOT, AND and OR, discussed in more detail below.

6.4.4.2 Boolean expressions
Using the above, complex expressions may be defined. The following gives the standard grammar used for this in YSM. This is used in several modelling tools.

If X is a boolean item, then "NOT(X)" is a legal boolean expression whose value is: FALSE when X is TRUE; TRUE when X is FALSE. This may also be written "~X".

If "X" and "Y" are boolean items, then "X AND Y", "X OR Y" are both legal boolean expressions. They may also be written "X $\wedge$ Y" and "X $\vee$ Y" respectively. Their values are given by the following table:

X	Y	X OR Y	X AND Y
TRUE	TRUE	TRUE	TRUE
TRUE	FALSE	TRUE	FALSE
FALSE	TRUE	TRUE	FALSE
FALSE	FALSE	FALSE	FALSE

6.4.4.2.1 Parentheses in boolean expressions

If "X" is a boolean item, then "(X)" is an allowed boolean expression with the same value as X. The value of an expression in parentheses must always be evaluated before combining it with any other expression. If there are several "OR"s, "AND"s at the same level, then all "AND"s must be evaluated first. For example:

> waiting AND NOT(oktosend OR failed)

or it's equivalent:

> waiting $\wedge$ ~(oktosend $\vee$ failed)

is a legal boolean expression providing "waiting", "oktosend" and "failed" are all boolean items. In this case the expression is only TRUE if all three of the following conditions hold:

"waiting" has the value TRUE
"oktosend" has the value FALSE
"failed" has the value FALSE

ABSTRACT DATA TYPE : cardinal

MEANING : The cardinal data type is used to represent a "natural number" i.e. one of 0, 1, 2, 3, ... In parameter definitions, the term "infinity" is used to represent "an indefinitely large value".

STRUCTURE : Simple

NUMBER OF VALUES : discrete

VALUE DEFINITION : inheritance

PARENT ADT : integer

RESTRICTIONS : cardinal ≥ 0

ORDERING : linear

PARAMETERS :

parameter	: type	: default
minimum	: cardinal	: 0
maximum	: cardinal	: infinity

6.4.5 cardinal

6.4.5.1 Operations

1. first(n) gives true if n=0.

2. next(n) gives the next cardinal. (This is more usually written "n + 1".)

3. If n > 0 true, then prior(n) gives previous cardinal. If n = 0, then prior(n) is undefined. (This is more usually written "n – 1".)

4. Addition and multiplication are allowed.

5. Subtraction is allowed, as long as the result is non-negative.

6.4.5.2 Comments

6.4.6 character

ABSTRACT DATA TYPE : character

MEANING : Any single printing or display character.

STRUCTURE : Simple

NUMBER OF VALUES : finite

VALUE DEFINITION : inheritance

PARENT ADT : string, minimum length=1, maximum length = 1

RESTRICTIONS : —

ORDERING : none

PARAMETERS : —

6.4.6.1 Operations

1. none

6.4.6.2 Comments

1. The character is the 'atomic' component of a string. There are the following general classifications of characters:

    - alphabetic: chosen from an 'alphabet' (YSM does not prescribe what this alphabet is, but any enterprise using the string and related ADTs should standardise on an alphabet);

    - digits: the characters 0, 1, …, 9;

    - printing characters: the digits, alphabetic characters and symbols such as ", ", ', " [,], !, ., #, $, %, …;

    - non-printing characters: for example 'escape', 'bell', 'DLE';

2. Other subtypes (for example 'graphic' characters) may be defined, as required.

3. The number of characters is finite, though it may be very large. For example, an alphabet such as Kanji requires many more characters than western alphabets. Note: the number of strings is *not* finite, but discrete.

6.4.7 currency

ABSTRACT DATA TYPE : currency

MEANING : This represents an amount of "money", without any commitment to a particular currency or storage representation for it. For any currency that needs to be supported, a suitable "units" term should be defined.

STRUCTURE : Simple

NUMBER OF VALUES : discrete

VALUE DEFINITION : inheritance

PARENT ADT : integer

RESTRICTIONS : —

ORDERING : linear

UNITS : Dollars, Sterling, Mark, franc, Yen (for example)

PARAMETERS :

parameter	: type	: default
smallest unit/base	: cardinal	: 100

6.4.7.1 Operations

add, subtract, "<", "≤", "=", "≠", "≥", ">".

6.4.7.2 Comments

1. This ADT is suitable for storing information about any credit, value or debt that may need to be stored and reported on in one or more base currencies. However, systems that trade international currency usually regard one specific currency as their base storage unit. This ADT should not be used for such systems without carefully checking that the policy of conversion to this base currency is carefully captured. ADTs can still be used, but abstract data typing does not absolve the analyst from establishing the exchange rate rules.

2. Other currencies can also be defined as separate units.

3. More than one unit may be defined for the same country. However, the smallest unit/base parameter is used to set the minimum unit of currency that can be used. All storage and calculations will be to that precision. For example, in the UK and US there are 100 of the smallest units in the base unit of currency. Respectively, there are 100 pence in one pound and 100 cents in one dollar.

Its default of 100 is quite common, but should a greater precision be required, this optional parameter should be declared for any attribute that requires greater precision. For example, in the UK, the smallest unit used to be a farthing (one quarter of an old penny). There were 240 pennies in one pound. For such a financial system, the parameter should be given a value of 960.

4. Currency constants are defined using the units defined as being available. For example:

```
3.25 Dollars
```

Care must be taken that any such constant can be represented in exact numbers of the smallest unit. For example:

```
2.03 Pounds
```

is sensible in UK currency. However, before the UK currency was converted to decimal umits, it would not have been a sensible constant, as it did not equate to an exact number of the (old) penny. In those cases where the base unit is not a decimal equivalent of the smallest unit, fractions should be used.

5. Currency constants may also be written using special characters, so:

```
$3.25
```

is also allowed.

6.4.8 date

```
ABSTRACT DATA TYPE : date

MEANING : Corresponds to a particular day. The point from which the "first" day is reckoned
          does not matter, but may be considered to be "an indefinite time in the past".
          Functions are provided (see below) that convert this type into character strings in
          several formats, or "day of month", "month", etc.

STRUCTURE : Simple

NUMBER OF VALUES : discrete

VALUE DEFINITION : inheritance

PARENT ADT : cardinal

RESTRICTIONS : —

ORDERING : linear
```

6.4.8.1 Operations

1. next(d) gives the day following day d.

2. day of month(d) gives a cardinal in the range 1–31

3. month(d) gives a cardinal in the range 1–12

4. year(d) gives an integer (negative numbers being 'BC').

5. julian(d) gives the julian day as an integer.

6. usasci(d) gives the date as text in US format (mm:dd:yy).

7. internl(d) gives the date as text in international format (dd:mm:yy).

6.4.8.2 Comments

1. This is the 'local date'. The date ADT should not be used to 'time stamp' information, unless all enterprise operations take place within a single time zone. Two operations could occur at the same time, yet be on different local days. See "time" for further discussion.

6.4.9 integer

ABSTRACT DATA TYPE : integer

MEANING : The set of 'whole numbers' — ..., –3, –2, –1, 0, +1, +2, +3, ... Note that the "+" may omitted from positive integers.

STRUCTURE : Simple

NUMBER OF VALUES : discrete

VALUE DEFINITION : defined

DEFINITION : Any sequence (terminating or non-terminating) of the digits 0, 1, 2, 3, 4, 5, 6, 7, 8, 9, (optionally) preceded by a "+" or "–" character (the first digit must not be a 0.) In parameter definitions, the term "infinity" is used to represent "an indefinitely large positive value" (a non-terminating sequence preceded by an optional "+"). The term "negative infinity" is used to indicate "an indefinitely large negative number" (a non-terminating sequence preceded by a "–" character).

ORDERING : linear

PARAMETERS :	**parameter**	**: type**	**: default**
	minimum	: integer	: negative infinity
	maximum	: integer	: infinity

6.4.9.1 Operations

1. The operations next and prior are always available.

2. Also supports $<, \leq, >, \geq, \neq, =$, as a comparison between two integers and giving a boolean result. If a and b are integers:

   ```
   a = b
   ```

   gives a value which is TRUE or FALSE.

3. Negation is by a preceding "–" sign. For example:

   ```
   a := --b
   ```

   where "a" and "b" are integers, assigns the negative value of b to a.

4. Add, subtract and multiply are all available, giving an integer result. Each may be used in their normal infix form, for example:

   ```
   a := b + c
   ```

   or in a prefix form, for example:

```
a := add(b,c)
```

whichever is more convenient..

5. The operator "div" is defined as the quotient when one integer is divided by another. It is truncated towards zero. For example, if "a" had the value 13 and "b" had the value 4:

```
a div b
```

would have the value 3. It may also be written as "a / b", if desired, or in the prefix form "div(a, b)".

6. The operator mod is defined as the remainder when one integer is divided by another. If "a" had the value 13 and "b" had the value 4:

```
a mod b
```

has the value 1. The prefix form is "mod(a, b)".

7. The operation "real" converts an integer to a real number. For example, if "X" is an integer and has the value 1, then:

```
real(X)
```

is a real number with the value 1.0. This is an explicit type conversion. As a matter of style, an integer may be used in a place where a real is expected and then an implicit conversion of this integer to a real takes place.

6.4.10 real

```
ABSTRACT DATA TYPE : real

MEANING : This represents a position on a continuous line that is extended indefinitely in
          either direction.

STRUCTURE : Simple

NUMBER OF VALUES : continuous

VALUE DEFINITION : defined

DEFINITION : Any sequence of decimal characters, including at most one "." character and
             (optionally) preceded by a "+" or a "−" (the first digit cannot be 0). In parameter
             definitions, the term "infinity" is used to represent "an indefinitely large value".
             Note that this definition includes rational numbers, which have a finite or
             recurring decimal expansion, as well as irrational numbers, which are
             represented by an infinite non-recurring expansion.

ORDERING : linear
```

PARAMETERS :	parameter	: type	: default
	minimum	: real	: minus infinity
	maximum	: real	: infinity

6.4.10.1 OPERATIONS :

1. Supports $<, \leq, >, \geq, \neq, =$, as a comparison between two reals and giving a boolean result. For example:

   ```
   a = b
   ```

   where "a" and "b" are reals gives a value which is TRUE or FALSE.

2. Negation (by using a "−" sign in front of a real) is also allowed. For example:

   ```
   a := −b
   ```

   where "a" and "b" are reals would assign the negative value of b to a.

3. The binary operators add, subtract, multiply, divide are all available, returning an real value. Each of these may be used in their normal infix form, for example:

   ```
   a := b + c
   ```

   or in a prefix form, for example:

   ```
   a := add(b,c)
   ```

   whichever is more convenient.

4. Exponentiation is also allowed. With a general base this is denoted:

```
power(a, N)
```

or the infix form:

```
a**N
```

for a^N. For e^x (where $e = 2.71828...$), the special form:

```
exp(x)
```

is used.

5. If "x" is real and $x \geq 0$, then sqrt(x) is also a real.

6. The operation "integer" converts a real to an integer number. For example, if X is a real and has the value 1.6, then:

```
integer(X)
```

is an integer with the value 1. This is an explicit type conversion. As a matter of style, a real may be used in a place where an integer is expected and then an implicit conversion of this real to an integer takes place.

6.4.10.2 Comments

1. Real numbers include:
the integers ... 0, +1, −1, +2, −2, ..., decimal fractions ... E.g. −17.8,..., recurring decimals ... (e.g. $0.\overline{142857}$) and irrational numbers (e.g. $\sqrt{2}$ =1.4142135623..., e = 2.718...)

6.4.11 string

ABSTRACT DATA TYPE : string

MEANING : Informally: any sequence of characters. An abstract definition of this involves recursion and the operations, concatenation, length and slice. It is beyond the scope of this manual to give it.

STRUCTURE : simple

NUMBER OF VALUES : discrete

VALUE DEFINITION : defined

VALUES : (see meaning)

ORDERING : none

PARAMETERS :

parameter	: type	: default
minimum length	: cardinal	: 0
maximum length	: cardinal	: infinity

6.4.11.1 Operations

1. Supports $\neq$, = as a comparison, returning a boolean value. For example:

    ```
    a := b
    ```

    where "a" and "b" are strings gives a value which is TRUE or FALSE.

2. Concatenation with another string is allowed, returning a string.

3. substring or slice to select part of string, returns a string. This operation requires the start position and number of characters to be extracted.

4. char(x, n) returns the nth character of x as a character. (It is a special case of slice.)

5. length(x) returns the length of string x as a cardinal.

6.4.11.2 Comments

1. See "character" for a discussion of the different types of characters that the string can contain.

2. In some implementations this is regarded as an array of characters, but that is a way of getting around the fact that those implementations do not support this type. In fact, YSM regards a character as a specific type of string (with length 1).

6.4.12 text

ABSTRACT DATA TYPE : text

MEANING : This is any sequence of printing characters

STRUCTURE : simple

NUMBER OF VALUES : discrete

VALUE DEFINITION : inheritance

PARENT ADT : string

RESTRICTIONS : See character for a description of what characters may print.

ORDERING : none

PARAMETERS : —

6.4.12.1 Operations

6.4.12.1.1 Inherited operations

1. Supports ≠, = as a comparison, returning a boolean value.

2. Concatenation with another text, alphabetic, alphanumeric is allowed, returning a text; Concatenation with a string returns a string.

3. length(x) returns the length of text x as a cardinal.

4. substring (or slice) to select part of text, returns a text.

5. char(x, n) returns the nth character of x as a character.

6.4.13 time

ABSTRACT DATA TYPE : time

MEANING : This represents the continuously changing time since some arbitrary point in the past.

STRUCTURE : Simple

NUMBER OF VALUES : continuous

VALUE DEFINITION : inheritance

PARENT ADT : real

RESTRICTIONS : —

ORDERING : linear

DIMENSIONS : T

UNITS : secs, mins, hours, days (for example)

PARAMETERS :	parameter	: type	: default
	accuracy	: time	: 0
	resolution	: time	: 0

6.4.13.1 Operations

1. julian(t) gives a real number, which is the julian day, including a fractional part.

2. date(t, difference) gives the equivalent date ADT for a location where the 'time difference' is difference. Note: the date can be different at two different locations, at a given time instant, because of time zones.

3. time of day(t, difference) gives the equivalent time of day for a location where the 'time difference' is difference. (The time of day equivalent to a given time depends on the time zone.)

4. time(d, td, difference), where "d" is a given "date", "td" is a "time of day" and "difference" is a "time difference" gives the equivalent time. This may be used to convert from local time and date to 'universal time'. For example, for US Eastern standard time, this might be:

```
time(d, td, 5 hours)
```

but in Holland (for most parts of the year):

```
time(d, td, -1 hour)
```

6.4.13.2 Comments

1. This time does *not* reset … it continues to 'run on'.

2. This type is used to 'time-stamp' information, particularly where an enterprise spans several time zones.

3. The exact 'offset' used in the definition of this ADT is not important. However, it may be convenient to use tiem scales such as those already defined for scientific purposes (see [TAA89]).

4. In a more obvious scientific application, a possible attribute defined as a "time" might be "time of closest approach" for a 'flyby' of a planet by a space probe.

6.4.14 time interval

> **ABSTRACT DATA TYPE :** time interval
>
> **MEANING :** This gives the difference in time between two instants of time.
>
> **STRUCTURE :** Simple
>
> **NUMBER OF VALUES :** continuous
>
> **VALUE DEFINITION :** inheritance
>
> **PARENT ADT :** real
>
> **RESTRICTIONS :** —
>
> **ORDERING :** linear
>
> **DIMENSIONS :** T
>
> **UNITS :** secs, hours, mins, msecs (for example)
>
> **PARAMETERS :**
>
> | parameter | : type | : default |
> |-----------|--------|-----------|
> | accuracy | : time | : 0 |
> | resolution | : time | : 0 |

6.4.14.1 Operations

1. $<, \leq, =, \neq, \geq, >$ as usual.

2. Also supports addition and subtraction (as long as the result is non-negative).

6.4.14.2 Comments

1. The time difference may be positive or negative, depending on which event occurs first.

2. Other time units may be defined, as required.

6.4.15 time of day

ABSTRACT DATA TYPE : time of day
MEANING : This represents the time since midnight on a particular day. Functions are provided (see below) to convert this abstract type into commonly used representations of the time of day.
STRUCTURE : Simple
NUMBER OF VALUES : continuous
VALUE DEFINITION : inheritance
PARENT ADT : time
RESTRICTIONS : $0 \leq$ time < length of day
ORDERING : cyclic

6.4.15.1 Operations

1. hour(x) gives a value between 0 and 23 (cardinal).
2. min(x) gives a value between 0 and 59 (cardinal).
3. sec(x) gives a value between 0 and 59 (cardinal).
4. minute(x) gives a value between 0 and 1439 (cardinal).
5. second(x) gives a value between 0 and 86399 (cardinal).
6. internl(x) gives a text in international format (hh:mm:ss).

6.4.15.2 Comments

1. This ADT should not be used to 'time stamp' information across time zones. The "time" ADT should be used..
2. A possible attribute that used this ADT might be "start time" for the entity "shift".

6.5 Restriction Grammar

This grammar is used to define a restriction of a parent ADT. This 'smaller' set of values may be:

- **Defined as a new named ADT:** this ADT may then be used for declaring many different data items. These can then be legally compared and used as typed inputs and outputs to operations.

- **Used as a set of values for an attribute:** the attribute is then restricted to the values thus defined. Note: a set of values defined in this way should be considered 'anonymous' — no other data item may be compared with it (even if it has the same set of values).

The definition below is in a style that is consistent with other frame specification styles. For the formal set theory equivalent, see §6.1.6.

1. A restriction may consist of a comparison of parent ADT values with values from that ADT. The operators allowed are those for the parent ADT. Values may include units if those units are supported for the parent ADT. For example, the ADT "cardinal" is defined as a restriction of the ADT "integer" by:

   PARENT ADT : integer
   RESTRICTIONS : cardinal $\geq$ 0

   An attribute "reaction temperature" might be typed as a restriction of the ADT "real" by:

   PARENT ADT : real
   RESTRICTIONS : reaction temperature $\geq$ 50

   which would indicate that the lowest possible value for this specific attribute would be 50. The operators "=" and "$\neq$" are always allowed in such comparisons. For ordered ADTs, the operators "<", "$\leq$", "$\geq$" and ">" are also allowed.

2. A restriction may consist of a comparison of a function defined on the parent ADT and a given value. For example, the ADT "fairly small" might be defined as a restriction of "real" ADT by use of the exponential function (see §6.4.10):

   PARENT ADT : real
   RESTRICTIONS : exp(fairly small) < 1.1

   An attribute "term" might be defined as a restriction of the ADT "cardinal" by:

   PARENT ADT : cardinal
   RESTRICTIONS : factors(term) = 0

   where "factors" is an operation taking a cardinal as an input and returning a cardinal value. See §2.17 for a discussion of how to define such operations.

3. A restriction may consist of "NOT(restriction a)". For example, the ADT "non-zero integer" can be defined as a restriction of the ADT "integer" by:

> **PARENT ADT :** integer
> **RESTRICTIONS :** NOT(non-zero integer = 0)

4. A restriction may consist of "restriction a AND restriction b". Either restriction a or restriction b may be enclosed in parentheses to reduce ambiguity. If this is done, then the expression in parentheses is evaluated first, in the usual way.

   For example, the ADT "rank" (used as the "row" in chess games) is defined as a restriction of the ADT "cardinal" by:

> **PARENT ADT :** cardinal
> **RESTRICTIONS :** rank > 0 AND rank ≤ 8

5. A restriction may consist of "restriction a OR restriction b". If either restriction a or restriction b involves OR or AND, then it may be enclosed in parentheses to reduce ambiguity. If this is done, then the expression in parentheses is evaluated first, in the usual way.

6. A restriction may be in the form:

   existence list SUCH THAT existence test

   where the existence list contains one or more assertions about the existence of named values (not necessarily from the ADT that is being restricted) and the existence test is any expression in the ADT being defined and these named values. This expression must evaluate to TRUE or FALSE and the values in the ADT being defined as a restriction of the parent ADT are those that give the value TRUE. For example, to define the ADT "even" as a restriction of the ADT "cardinal":

> **PARENT ADT :** cardinal
> **RESTRICTIONS :** f is a cardinal SUCH THAT 2*f = even

An example that shows how each of the above may be combined is to define the ADT "odd-composite" (odd, non-prime numbers) as a restriction of cardinal by combining three restrictions with "AND":

> **RESTRICTIONS :**
> odd-composite > 0
> AND
>   (q is a cardinal, r is a cardinal
>   SUCH THAT
>   (odd-composite = q*r AND q≠1 AND r≠1))
> AND
>   NOT(q is a cardinal SUCH THAT 2*q =odd-composite)

See §6.1.6 for the set-theoretic equivalent to this.

6.6 Structured Text Grammar

Structured text builds the function up out of simple statements using standard constructs. A simple statement is one of the following:

- standard YSM operations;
- data access operations;
- standard operations 'owned' by abstract data types;
- enterprise operations;
- assignment statements.

From these 'building blocks', the required function is built up using the following constructs:

- **sequence**: one statement, followed by another, followed by another, ... ;
- **parallel**: several statements used at the same time;
- **selection**: one or more than one condition is tested and, depending on which is true, different statements used;
- **iteration**: the same statements are used repeatedly. The number of repeats may be fixed (for example "do 6 times") or it may be variable (termination of the iteration depending on the satisfaction of a test condition).

Any of these is referred to as a compound statement. In EBNF notation (see §6.2), the above is formalised as:

```
<statement> ::=
        <simple statement>
    |   <compound statement>

<compound statement> ::=
        <sequence statement>
    |   <parallel statement>
    |   <selection statement>
    |   <iteration statement>
```

6.6.1 Syntax and semantics

The allowed statements are defined in abstract form using EBNF notation (see §6.2). This gives a basic syntax, containing all the required elements. This syntax may be varied, but the same elements must be present.

The semantics of each construct is then given. This describes how it should be interpreted.

Two possible lexical styles are then shown for the construct. These are those described in this manual as 'procedural' (rather 'programese') and numbered text (nearer to 'ordinary' text, with data flows delimited by "%" characters). For 'sequence' and 'parallel', actual staements are give, but for the later examples, these are just referred to as statement, to avoid overloading with multiple examples.

Many of the constructs require that the end of the construct is denoted in some way. Here it is indicated by and "end" or the 'count' of lines in numbered text. Other possibilities might be the use of a "horizontal line" or "left-hand bracket" (as in the 'days of month' example used in §2.16.3.13.6).

6.6.2 Sequence

6.6.2.1 Abstract form

> *<sequence statement>* ::=
>> *<statement>* { ; *<statement>* }
> | **in sequence do** *<statement>* **end**

6.6.2.2 Semantics

There are two forms of the statement. The first is the normal default for structured text, with statements executed in the sequence in which they are listed. The second form is required if the construct is nested within a parallel construct.

6.6.2.3 Procedural

A sequence of statements is of the form:

```
a := length(v1);
b := length(v2) ;
```

6.6.2.4 Numbered text

Numbered text numbers each statement sequentially. If the sequence is regarded as a compound statement, then each component is numbered by adding a single digit to the number of that statement. For example, suppose the sequence were numbered "6". This would give:

```
6. Perform the following 2 operations:
      6.1. calculate %a% as the length of %v1%.
      6.2. calculate %b% as the length of %v2%.
```

If the sequence construct appears as component 3 (say) in a higher-level construct, which is numbered 5.2. (say) then the components are numbered 5.2.3.1., 5.2.3.2., 5.2.3.3. .... For example:

```
5.2.3. Perform the following 2 operations:
      5.2.3.1. calculate %a% as the length of %v1%.
      5.2.3.2. calculate %b% as the length of %v2%.
```

6.6.3 Parallel

6.6.3.1 Abstract form

> *<parallel statement>* ::=
>> **in parallel do** *<statement>* **end**

6.6.3.2 Semantics

In a parallel construct, there are several statements that need to be carried out, but the order in which they are carried out in is immaterial. This construct is used when the fact that there is no required sequencing needs to be stressed. Where this is not particularly significant, or the sequencing is important, the standard sequence construct (see above) is used.

6.6.3.3 Procedural

```
In parallel do
        a := length(v1);
        b := length(v2) ;
end
```

6.6.3.4 Numbered text

```
6. Perform the following 2 statements in parallel:
        6.1. calculate %a% as the length of %v1%.
        6.2. calculate %b% as the length of %v2%.
```

6.6.3.5 Nesting parallel and sequence constructs

Where the statement is nested, care must be taken to distinguish between operations that can be carried out in parallel and those that must be carried out in sequence. For example:

```
In parallel do
        d := inner product(v1, v2) ;
        In sequence do
                In parallel do
                        a := length(v1) ;
                        b := length(v2) ;
                end ;
                c := a*b ;
        end ;
end ;
angle := arccos(d/c) ;
```

In the numbered text style, this would be:

```
1. Perform the following two statements in parallel:
        1.1. Use the values in %v1% and %v2% to calculate the
                inner product and store the result in %d%.
        1.2. Perform the following two statements in sequence:
                1.2.1. Perform the following two statements in parallel:
                        1.2.1.1. Calculate %a% as the length of %v1%.
                        1.2.1.1. Calculate %b% as the length of %v2%.
                1.2.2. Multiply %a% and %b% together, giving %c%.
2. %angle% is computed as the inverse cosine of %d% divided by %c%.
```

6.6.4 Selection statement

6.6.4.1 Abstract form

```
<selection statement> ::=
        <simple selection>
    |   <selection with alternative>
    |   <multi-way selection>
    |   <multi-way selection with default>
```

6.6.4.2 Semantics

Selections are ways of choosing operations to be performed, depending on the values of conditions.

6.6.4.3 Condition

For all these constructs, there are one or more conditions that are tersted for. These are boolean expressions, in other words, they evaluate to TRUE or FALSE. This expression can involve any data item that is available and operations and comparisons on such items, as long as the result is TRUE or FALSE.

6.6.4.3.1 Examples of conditions

1. A continuous event flow may be used as a boolean item. For example:

```
If  auto_available  then ...
```

2. If "X" is a real input to a data process, it could be compared with a real constant, for example:

```
If  X > 10.0  then ...
```

3. Values of attributes can also be compared with test values. For example:

```
If <Patient>.lean body weight > 50 kg  then ...
```

which would carry out statements only if the particular, identified patient(s) were heavier than this limit. The attribute "lean body weight" is accessed (this example requires that <Patient> has been bound — see §6.10.1.4).

4. The values of state variables for bound entities may also be used. For example:

```
If <Scheduled course>.status = open  then ...
```

which would carry out statements only if the particular, identified Scheduled course is open for booking.

6.6.4.4 Simple selection

6.6.4.4.1 Abstract form

```
<simple selection> ::=
        if <condition> then <statement>
    |   if <condition> then <statement> end
```

6.6.4.4.2 Semantics

This contains a condition and one or more statements that are carried out, depending on the value of the condition.

The test is carried out and if the result of the test is TRUE then the statement(s) carried out. If the result of the test is FALSE, then these statements are not carried out.

If more than one statement is to be selected when the condition is is true, the end of the selection is indicated in some way (e.g. by an "end").

6.6.4.4.3 Procedural

For example:

```
if month=February then number of days := 28 ;
```

and:

```
if x then
        statement ;
        statement ;
        statement ;
end ;
```

6.6.4.4.4 Numbered text

```
6. if x then carry out the following 3 operations:
        6.1. statement.
        6.2. statement.
        6.3. statement.
```

6.6.4.5 Selection with alternative

6.6.4.5.1 Abstract form

```
<selection with alternative> ::=
        if <condition> then <statement> else <statement> end
```

6.6.4.5.2 Semantics

For a selection with alternative, there are two possible sets of actions.

The test is carried out and it the result of the test is TRUE then the first statement (or set of statements) is carried out. If the result of the test is FALSE, then the second statement(s) are carried out. In other words, either (but never both) set of statements are carried out.

As with a simple selection, the selection criterion is any expression that evaluates to TRUE or FALSE.

6.6.4.5.3 Procedural

```
if x then
        statement ;
        statement ;
else
        statement ;
end ;
```

6.6.4.5.4 Numbered text

```
6. if x then  carry out the two operations:
        6.1.1. statement.
        6.1.2. statement.
otherwise, carry out the operation:
        6.2. statement.
```

6.6.4.6 Multi-way selection

6.6.4.6.1 Abstract form

```
<multi-way selection> ::=
    if <condition> then
        <statement>
    { elsif <condition> then <statement> }
    end
```

6.6.4.7 Semantics

A multi-way selection consists of a sequence of conditions, each with associated statements. One (and exactly one) of these alternative statements will be carried out based on which conditions are TRUE.

The conditions and associated statements are listed in pairs. If the first condition is true, then the first statement is carried out;
If the first condition is not true, but the second is true, then the second statement is carried out;
If neither the first condition nor the second condition is true, but the third is true, then the third statement is carried out; …and so on …

6.6.4.7.1 Procedural

```
if  x  then
        statement ;
        statement ;
        statement ;
elsif  y  then
        statement ;
        statement ;
elsif  z  then
        statement ;
end ;
```

6.6.4.7.2 Numbered text

```
6. if x then  carry out the following three operations:
        6.1.1. statement.
        6.1.2. statement.
        6.1.3. statement.
otherwise, if y is true, then carry out the following two operations:
        6.2.1. statement.
        6.2.2. statement.
        otherwise, if z is true then carry out the following operation:
        6.3. statement.
```

6.6.4.8 Multi-way selection with default

6.6.4.8.1 Abstract form

```
<multi-way selection with default> ::=
    if <condition> then
        <statement>
    { elsif <condition> then <statement> }
    else
        <statement>
    end
```

6.6.4.8.2 Semantics

As a variant of a multi-way selection, an alternative action may be supplied. In this case, if none of the conditions are true, then the alternate statement is carried out.

6.6.4.8.3 Procedural

```
if  x  then
        statement ;
        statement ;
        statement ;
elsif  y  then
        statement ;
        statement ;
elsif  z  then
        statement ;
else
        statement ;
        statement ;
end ;
```

6.6.4.8.4 Numbered text

> 6. if x then carry out the following three operations:
>   6.1.1. *statement.*
>   6.1.2. *statement.*
>   6.1.3. *statement.*
> otherwise, if y is true, then carry out the following two operations:
>   6.2.1. *statement.*
>   6.2.2. *statement.*
>   otherwise, if z is true then carry out the following operation:
>   6.3. *statement.*
> and in all other cases, carry out the following three operations:
>   6.4.1. *statement.*
>   6.4.2. *statement.*
>   6.4.3. *statement.*

6.6.5 Iteration

6.6.5.1 Abstract form

<iteration statement> ::=
   <fixed loop>
| *<counted loop>*
| *<itemised loop>*
| *<pre-test iteration>*
| *<post-test iteration>*

6.6.5.2 Semantics

An iteration of an statement indicates that the statement is to be carried out repeatedly. There are several variants of this that are allowed, as shown above. Each is listed separately below.

6.6.5.3 Fixed Loop

6.6.5.3.1 Abstract form

<fixed loop> ::=
   repeat *<repeat number>* **times**
      <statement>
   end

6.6.5.3.2 Semantics

This type of iteration is always carried out the same fixed number of times (for example, "7").

6.6.5.3.3 Procedural

> repeat 7 times
>   *statement* ;
>   *statement* ;
> end ;

6.6.5.3.4 Numbered text

> 6. Perform the following two operations 7 times:
>   6.1. statement.
>   6.2. statement.

6.6.5.4 Counted Loop

6.6.5.4.1 Abstract form

```
<counted loop>
        for <loopvar> := <loopstart> to <loopfinish>
                <statement>
        end
    |   for <loopvar> := <loopstart> to <loopfinish> increment <loopstep>
                <statement>
        end
```

6.6.5.4.2 Semantics

This type of iteration uses a counter which is incremented each time the iteration is carried out. This type of iteration is useful if repeated logic needs a 'counter'. The iteration uses an initial value, a loop increment and a final value. These are likely to be temporary data items (see §2.16.3.9).

These work as follows:

1. The counter is set to the initial value of "loop1".

2. The counter is compared to the upper limit of "loop2" and if it is greater than this value, the iteration is exited. If it is not greater than the upper limit, then the following two steps are carried out:

3. Perform statement(s) inside the iteration construct.

4. Increment counter by "loop step" and go to step 2.

Notes

1. At the point where the iteration is commenced, the initial, loop increment and final values of "loop1", "loop step" and "loop2" must be well-defined.

2. All the loop parameters must be of the same type, which must have a linear ordering (see §2.10.3.10) and support the addition operation. (This is likely, therefore, to be "cardinal", "integer" or "real".)

3. If "loop increment" is omitted, for example:

> for loop := 3 to 7
>   statement ;
>   statement ;
> end ;

then it is assumed to be 1.

6.6.5.4.3 Procedural

```
for  loop := loop1 to loop2 increment loopstep
        statement  ;
        statement  ;
end ;
```

6.6.5.4.4 Numbered text

```
6. Using %loop% with an initial value of %loop1% and increasing it by %loopstep%
up to the value %loop2%, perform the following two operations:
        6.1.  statement.
        6.2.  statement.
```

1. Any of "loop1", "loop2", "loopstep" may be literal constants of the same type (usually integer or cardinal). For example:

```
For  %i% = 1 to 32 increment 2
```

6.6.5.4.5 Possibility of zero iterations

The iteration may be performed zero times. For example:

```
For i := 5 to 3
```

6.6.5.5 Itemised Loop

6.6.5.5.1 Abstract form

```
<itemised loop> ::=
        for each <entity selection>
            <statement>
        end
        | for each <data flow>
            <statement>
        end
```

6.6.5.5.2 Semantics

This may only be used where there is more than one identical input that the process operates on. This is most commonly seen in repeating a set of statements on certain selected occurrences of an entity.

In this case, the first iteration refers to the first occurrence, the next to the second .... There may be zero iterations, if there are no occurrences of the entity. Each occurrence may constitute:

- **all of an entity in sequence**: for example:

```
For each <Patient>
```

- **selected occurrences in sequence**: for example:

```
For each <Patient> with <Patient>.lean body weight > 100 kg"
```

Note: state variables may be used as 'pseudo-attributes' (see §6.10.16).

6.6.5.5.3 Procedural

```
for each <Patient>
        statement ;
        statement ;
end ;
```

6.6.5.5.4 Numbered text

```
6. For each <Patient>, perform the following two operations:
        6.1. statement.
        6.2. statement.
```

An alternative, less 'pedantic' version of this is:

```
For each <Patient> with lean body weight > 100kg".
```

6.6.5.5.5 Binding in itemised loop
When an itemised loop is used with an entity, it binds that entity within the loop. On exit from the loop, the entity is no longer bound.

6.6.5.6 Pre-test iteration

6.6.5.6.1 Abstract form

```
<pre-test iteration> ::=
    while <condition>
        <statement>
    end
```

6.6.5.6.2 Semantics
In this type of iteration a test is carried out before each iteration. This pre-test must evaluate to TRUE or FALSE.

If the test gives the value TRUE the iteration is carried out. This is continuously repeated until the test gives the value FALSE. This type of iteration may be carried out zero times if the condition is never true.

6.6.5.6.3 Procedural

```
while x
        statement ;
        statement ;
end ;
```

6.6.5.6.4 Numbered text

```
6. While x is true, perform the following two operations:
        6.1. statement.
        6.2. statement.
```

6.6.5.7 Post-test iteration

6.6.5.7.1 Abstract form

```
<post-test iteration> ::=
    repeat
        <statement>
    until <condition>
```

6.6.5.7.2 Semantics

In this type of iteration a test is carried out after each iteration. This post-test must evaluate to TRUE or FALSE.

If the test gives the value FALSE another iteration is carried out. This is continuously repeated, until the test gives the value TRUE. This type of iteration must be carried out at least once.

6.6.5.7.3 Procedural

```
repeat
        statement  ;
        statement  ;
until x ;
```

6.6.5.7.4 Numbered text

```
6. Perform the following two operations:
        6.1. statement.
        6.2. statement.
until x is true.
```

6.7 Standard YSM Dialects

The exact form of the way in which an operation is used is not at issue — if there is a standard way of multiplying two real numbers together, whether it is invoked by:

```
:= multiply(x, y)
```

```
multiply x by y
```

or

```
x*y
```

or any other variant is not a method issue. YSM describes this situation as being three versions of the same operation in different dialects. Within this manual, enterprise operations are used in the following named dialects:

- **procedural**: one form used in minispecs. This is given by using the operation name, followed by the arguments enclosed in parentheses.

- **textual**: an alternative, more 'wordy' form used in minispecs. The operation is part of a standard phrase, with operands in particular places in the phrase. In a more relaxed form, this is often used with delimiters around the operands (see §2.16.3.3.1).

- **algebraic**: used to describe forms such as "x*y". This also includes all boolean expressions — see §6.4.4.

- **bSTD**: the form of the operation when shown on a bSTD.

A general discussion of the dialect concept is given in §1.2.3.2.1 and the use of dialects in minispecs is described in §2.16.3.13.4.

Note: the terms "procedural", "textual", "algebraic" and "bSTD" are not critical — any other preferred name may be used. These are the names chosen for this reference manual.

6.7.1 Dialect conversion

One dialect could be unambiguously converted into another if the above prescription is followed. This allows verification of minispec logic in one dialect and then conversion to another for execution, emulation etc.

6.7.2 Use of named dialects

The use of named dialects is an optional part of YSM. If it is to be used in its full form, it requires that each minispec is declared to use one (or more) dialects. All inputs and outputs for minispecs must be typed and then the correct enterprise operation can be identified and checked for correct type usage.

6.8 Data Flow Composition Grammar

This appendix gives the grammar used for defining composition of data flows.

<data flow composition> ::=

    <data flow specification> { *<together with>* *<data flow specification>* }

    | *<data flow specification>* {*<together with>* *<left paren>*
                           <data flow specification> *<right paren>* }

    | *<left brace>* *<data flow specification>* *<right brace>*

    | *<lower limit>* *<left brace>* *<data flow specification>* *<right brace>*

    | *<lower limit>* *<left brace>* *<data flow specification>* *<right brace>* *<upper limit>*

    | *<left brace>* *<data flow specification>* *<right brace>* *<upper limit>*

    | *<left bracket>* *<data flow specification>* { *<selection sep>*
                            <data flow specification> } *<right bracket>*

and

    <data flow specification>::= *<data flow composition>* | *<data flow name>*

    <lower limit> ::= cardinal   *<upper limit>* ::= cardinal

where cardinal has its usual interpretation.

Note: the above definitions imply:

1. At least one component of a compound data flow must be non-optional.
2. There must be at least two selections in a selection list.
3. A selection cannot be optional.

6.8.1 Lexical conventions

The following are the usual lexical conventions in these constructs:

    <together with> ::= '+'

    <left paren> ::= '('   *<right paren>* ::= ')'

    <left brace> ::= '{'   *<right brace>* ::= '}'

    <left bracket> ::= '['   *<right bracket>* ::= ']'

    <selection sep> ::= '|'

6.9 Standard YSM Operations

This appendix gives the standard YSM operations in alphabetical order. See also data access operations in §6.10.

Each has several variants, or dialect forms. For example, the operation that generates a signal has:

```
USE :
        procedural:     signal(human win)
        textual:        signal %human win%
        bSTD:           S: human win
```

in three different dialects. The standard YSM dialects are listed in §6.7.

6.9.1 Availability in data and control processes

In general, these operations may be used to carry out standard operations in data processes or control processes. However, there are some restrictions detailed for each operation. For example, operations that deal with data are not allowed in control processes.

If an operation is not available in a control process, there is no entry for the bSTD dialect.

If an operation is not available in a data process, then there is no entry for the procedural or textual dialects.

6.9.2 Model component identifiers

In some operations, it is necessary to identify which model component the operation is to act on. For example, "enable" always acts on a specific named process. In the specification of the operations, this is referred to as the "process identifier". Other operations use "flow identifiers" and "event store identifier".

These identifiers are a different level of abstraction from the identifiers used within the model (to identify occurrences of entities) — they are *meta-identifiers*. In practice, the examples given will make the use of the operations clear. The formal specification frame is not needed for most practitioners of the method.

6.9.2.1 Clock identifiers
In these operations "clock identifier" is the name of a clock used in a bSTD. For example, "secs" always gives the time by a given clock.

6.9.2.2 Data flow identifiers
In these operations, a "data flow identifier" is the name of the data flow. For example, "issue" always acts on a named data flow.

6.9.2.3 Event flow identifiers
In these operations, an "event flow identifier" is the name of an event flow. For example, "signal" always generates a named event flow.

6.9.2.4 Event store identifiers
In these operations, an "event store identifier" is the name of an event store. For example, "initialise" always initialises a named event store.

6.9.2.5 Flow identifiers

A "flow identifier" is the name of a data or event flow. For example, "available" tests for availability of a named continuous flow.

6.9.2.6 Process identifiers

In these operations, a "process identifier" is the name of the process. For example, "enable" always acts on a named process.

6.9.3 Available

OPERATION : available

MEANING : Used to detect whether a given continuous flow is available at that point in time.

ARGUMENTS :

argument	: type	: direction
flow name	: identifier	: in
available	: boolean	: out

DEFINITION :

    pre-

        there is an active process that generates the flow

    post-

        available = TRUE

    pre-

        all processes that generate this flow are inactive (disabled).

    post-

        available = FALSE

USE :

    procedural: available(flow name)

    textual: %flow name% is available

    bSTD: available(flow name)

6.9.3.1 Comments

Note: flow name must be the name of a continuous event or data flow. Although this tests for a continuous flow, it may be used in continuous or discrete processes. The use of available in testing for continuous data flows is discussed in §2.14.3.4.4 and continuous event flows in §2.14.3.5.3.

6.9.3.2 Instance extensions

In minispecs, there may be several instances of the same flow generated by a multi-instance process. To cater for this, the forms:

available(flow name, instance = N)

(where N is any available data item of the same type as the identifier for the process instances) may be used. As an alternative:

instance %N% of %flow name% is available

may be used.

6.9.4 Begin

OPERATION : begin

MEANING : Used to reset a named clock to zero.

ARGUMENTS :

argument	**: type**	**: direction**
clock name	: clock identifier	: in

DEFINITION :   pre-

post-
$$msecs(clock\ name) = 0$$

USE :
    bSTD: B: clock name

6.9.4.1 Comments
The way this is used in bSTDs is discussed in §2.24.3.4.5.

6.9.5 Clock functions (secs, msecs, etc.)

OPERATION : secs

MEANING : This returns the number of seconds since a given clock was started.

ARGUMENTS :

argument	**: type**	**: direction**
clock name	: clock identifier	: in
secs	: cardinal	: out

DEFINITION : See meaning.

USE :

bSTD:	"secs(x) = n" causes an event to be generated at a given time. In general the value of the function is compared with a cardinal constant.

6.9.5.1 Comments

Other similar functions — "microsecs", "msecs", "mins", "hours", "days" are also available. All of these may only be used within a bSTD (or tabular equivalent).

The way these are used in bSTDs is discussed in §2.24.3.4.

6.9.6 Current time

OPERATION : current time

MEANING : This is used to find the current clock time.

ARGUMENTS :

argument	: type	: direction
current time	: time	: out

DEFINITION : See meaning

USE :
      procedural: "current time ( )"

6.9.6.1 Comments

This is used as a standard way of generating a 'time stamp'. If an attribute is assigned a value using this, it may be checked later using the same function. For example:

```
If <Scheduled picture>.start time = current time( ) then
        ...
```

would check the attribute "start time" of the entity "Scheduled picture" and if it was 'now', carry out some actions.

6.9.7 Disable

OPERATION : disable

MEANING : This is used in a control process to disable a given process. The process is either a data process or a process group.

ARGUMENTS :

argument	**: type**	**: direction**
process name	: process identifier	: in

DEFINITION : pre-
        A process with this name exists
    post-
        The named process is not active

USE :
    bSTD: "D: process name"

6.9.7.1 Comments

This is used in a control process to stop a continuous process or prevent a discrete process from running when its stimulus occurs. See §2.14.3.17 for further discussion of the semantics of "disable".

6.9.8 Enable

OPERATION : enable

MEANING : This is used to enable a particular, named process. The process may be a process group or data process.

ARGUMENTS :

argument	: type	: direction
process name	: process identifier	: in

DEFINITION : pre-

A process with this name exists

post-

The named process is active

USE :

bSTD: "E: process name"

6.9.8.1 Comments

This is used in a control process to start a continuous process or allow a discrete process to run when its stimulus occurs. See §2.14.3.17 for further discussion of the semantics of "enable".

6.9.9 Falling

```
┌─────────────────────────────────────────────────────────────────┐
│ ┌───────────────────────────────────────────────────────────┐   │
│ │ **OPERATION** : falling                                    │   │
│ └───────────────────────────────────────────────────────────┘   │
│ ┌───────────────────────────────────────────────────────────┐   │
│ │ **MEANING** : Generates an unnamed discrete event at the   │   │
│ │   time a particular named continuous                       │   │
│ │   event flow changes from TRUE to FALSE.                   │   │
│ └───────────────────────────────────────────────────────────┘   │
│ ┌───────────────────────────────────────────────────────────┐   │
│ │ **ARGUMENTS** :                                            │   │
│ │           **argument**      **: type**      **: direction**│   │
│ │           flow name         : flow identifier   : in       │   │
│ └───────────────────────────────────────────────────────────┘   │
│ ┌───────────────────────────────────────────────────────────┐   │
│ │ **DEFINITION** : pre-                                      │   │
│ │           continuous named flow has value FALSE at time T  │   │
│ │                 AND                                        │   │
│ │                 was TRUE at (T – delta),                   │   │
│ │                      (where T= 'now' and delta →0)         │   │
│ │       post-                                                │   │
│ │           un-named discrete event signal exists            │   │
│ └───────────────────────────────────────────────────────────┘   │
│ ┌───────────────────────────────────────────────────────────┐   │
│ │ **USE** :                                                  │   │
│ │     bSTD:       falling(x)                                 │   │
│ └───────────────────────────────────────────────────────────┘   │
└─────────────────────────────────────────────────────────────────┘
```

6.9.9.1 Comments

This operation is not allowed in data processes, but the procedural version is allowed in continuous processes specified by minispecs.

The use of this operation in bSTDs is discussed in §2.24.3.9.

6.9.10 Initialise

OPERATION : initialise

MEANING : Creates a named event store. If the event store already exist, any existing events held in the store are deleted.

ARGUMENTS :

<u>argument</u>	<u>: type</u>	<u>: direction</u>
event store name	: event store identifier	: in

DEFINITION :
         pre-
             none
        post-
             event store exists AND 'wait' on this event store would cause the waiting process to suspend

USE :
    bSTD: I: event store name

6.9.10.1 Comments

This creates an event store with the name of its argument. It may only be used with an event store. After the initialise, the event store exists and contains no events (until a 'signal' to the event store occurs). For example:

```
I: track available
```

creates an event store "track available" with no occurrences of "track available" in it. If initialise is used with an existing event store, then any events in that store are destroyed, but existing waits are retained.

6.9.11 Instance

OPERATION : instance

MEANING : Used to detect which instance of a multi-instance process generated a flow.

ARGUMENTS :

argument	: type	: direction
flow name	: data flow identifier	: in
instance	: data element identifier	: out

DEFINITION :

      pre-

          flow exists

     post-

          instance has given value

USE : procedural: instance(flow name)

       textual: instance of %flow name%

6.9.11.1 Comments

This function is used to determine which instance of a multi-instance process generated a given flow. (In effect, this provides access to a hidden field whose value is automatically assigned when the flow is issued from a multi-instance process.) For example:

instance(position update)

used in "Schedule lifts" would give the values of the identifier corresponding to which instance of "Control lift" generated the flow. This example is discussed in §2.16.3.1.2.

Note: the value of this function is not defined for single-instance processes.

See also "issue" — §6.9.12.

6.9.12 Issue

OPERATION : issue

MEANING : Used to highlight outputs from a minispec. These outputs are data flows only (signals and data access operations have their own operations).

ARGUMENTS :

argument	: type	: direction
flow name	: data flow identifier	: in

DEFINITION :

pre-
> flow has all component values assigned

post-
> components of flow are available to all processes that use this data flow as an input (as specified on DFDs)

USE :

procedural: issue(flow name)
textual: issue %flow name%

6.9.12.1 Comments

In a procedural minispec (see §2.16.3.13) this is used to issue a data flow from a data process. It is optional in structured text. Omitting it makes no difference to the function of the process, but some like to include it explicitly. As a matter of style, all "issue" statements may be collected together 'at the end' of a minispec to highlight the production of outputs.

6.9.12.2 Flows to multiple instance processes

For multiple instance processes, the destination of the data flow must be identified. To identify which instance the flow should be 'sent to' an extension of the "issue" operation is used. For example:

```
issue(lift allocation, instance(control lift) = N)
```

generates the data flow "lift allocation", containing instance identification taken from the data item "N" (which must contain an item of the same type as the identifier declared for the multi-instance process). This causes the flow to be received only by the instance of "Control lift" with this identifier. This example is discussed in §2.16.3.1.2.

Note: if a multi-instance process is enabled/disabled or triggered, then it will have a corresponding control process. This group will be multi-instance, so that within the group, only that instance is dealt with. As a consequence, no instance identification is needed for enable, disable or trigger. However, this group must be sent signals using this extension of issue.

See also "instance" §6.9.11.

6.9.13 Lower

```
┌─────────────────────────────────────────────────────────────────┐
│ ┌───────────────────────────────────────────────────────────┐   │
│ │ OPERATION : lower                                         │   │
│ └───────────────────────────────────────────────────────────┘   │
│ ┌───────────────────────────────────────────────────────────┐   │
│ │ MEANING : Used to set a continuous event flow to FALSE. This may only be used in │
│ │           continuous processes.                           │   │
│ └───────────────────────────────────────────────────────────┘   │
│ ┌───────────────────────────────────────────────────────────┐   │
│ │ ARGUMENTS :                                               │   │
│ │              argument      : type            : direction  │   │
│ │              flow name     : event flow identifier   : in │   │
│ └───────────────────────────────────────────────────────────┘   │
│ ┌───────────────────────────────────────────────────────────┐   │
│ │ DEFINITION :    pre-                                      │   │
│ │                       flow exists                         │   │
│ │                 post-                                     │   │
│ │                       flow has value FALSE                │   │
│ └───────────────────────────────────────────────────────────┘   │
│ ┌───────────────────────────────────────────────────────────┐   │
│ │ USE :                                                     │   │
│ │      procedural:  lower(flow name)                        │   │
│ │      textual:     lower %flow name%                       │   │
│ │      bSTD:        L: flow name                            │   │
│ └───────────────────────────────────────────────────────────┘   │
└─────────────────────────────────────────────────────────────────┘
```

6.9.13.1 Comments

This sets the value of its argument to false. It may only be used with an output continuous event flow. For example:

```
lower(auto_OK);
```

would make the output event flow become false.

6.9.14 Present

OPERATION : present

MEANING : Used to check whether an optional component of a data flow is present.

ARGUMENTS :

argument	: type	: direction
flow name	: data flow identifier	: in

DEFINITION :   pre-

flow is present as a component of a flow that the data
process has as an input

post-

result = TRUE

pre-

flow is not present as a component of a flow that the data
process has as an input

post-

result = FALSE

USE :
procedural: present(flow name)
textual: %flow name% is present

6.9.14.1 Comments

This is used to check whether a component of a data flow is present. For example if a flow "A"
is defined to have the following composition:

    STRUCTURE : B + (C) + [D | E | F + {G}]

this operation could be used to check whether "C" is present by:

    if present(C) then ...

where ... is used to indicate operations to be carried out if "C" is present. These will not be
carried out if "C" is not present. We might also use:

    if present(F) then
            for each G carry out the following operations
                    ...
            end ;
    end ;

which would carry out some operations on each of the "G"s, provided they are present.

As an alternative form:

    If %F% is present ...

may be used.

6.9.15 Raise

OPERATION : raise

MEANING : Used to set a continuous event flow to TRUE. This may only be used in continuous processes.

ARGUMENTS :

argument	: type	: direction
flow name	: event flow identifier	: in

DEFINITION :

pre-
      flow exists
post-
      flow has value TRUE

USE :

procedural: raise(flow name)
textual: raise %flow name%
bSTD: R: flow name

6.9.15.1 Comments

This sets the value of its argument to true. It may only be used with an output continuous event flow. For example:

```
raise(auto_OK);
```

would make the output event flow become true.

6.9.16 Rising

OPERATION : rising

MEANING : Generates an unnamed discrete event at the time a particular named continuous event flow changes from FALSE to TRUE.

ARGUMENTS :

argument	**: type**	**: direction**
flow name	: flow identifier	: in

DEFINITION : pre-
    continuous named flow has value TRUE at time T
       AND
       was FALSE at (T – delta),
          (where T= 'now' and delta →0)
   post-
    un-named discrete event signal exists

USE :
   procedural: rising(flow name)
   bSTD: rising(flow name)

6.9.16.1 Comments

This operation is not allowed in data processes, but the procedural version is allowed in continuous processes specified by minispecs.

The use of this operation in bSTDs is discussed in §2.24.3.9.

6.9.17 Signal

```
OPERATION : signal

MEANING : Generates a given discrete event flow.

ARGUMENTS :
            argument      : type                    : direction
            flow name     : flow name identifier    : in

DEFINITION : See meaning

USE :
       procedural:  signal(flow name)
       textual:     signal %flow name%
       bSTD:        S: flow name
```

6.9.17.1 Comments
This generates a discrete event flow with the name of its argument. For example:

```
signal(transmission ready);
```

would generate a discrete event flow "transmission ready".

6.9.17.2 Signals to multiple instance processes
For multiple instance processes, the destination of the signal must be identified. An extension of the "signal" operation is used. For example:

```
signal(stop at next floor, instance(control lift) = N)
```

generates the signal "stop at next floor", containing instance identification taken from the data item "N" (which must contain an item of the same type as the identifier declared for the multi-instance process). This causes the signal to be received only by the instance of "Control lift" with this identifier. This example is discussed in §2.16.3.1.2.

6.9.18 Time of day

OPERATION : time of day
MEANING : This returns the local time.
ARGUMENTS :

argument	**: type**	**: direction**
time	:time of day	: out

DEFINITION :

USE :
procedural: "time of day( )" returns time of day.

6.9.18.1 Comments

1. Returns local time (see ADT "time of day").

2. This should not be used to "time stamp", except within a single time zone.

6.9.19 Todays date

```
┌──────────────────────────────────────────────────────────┐
│ ┌────────────────────────────────────────────────────┐   │
│ │ OPERATION : todays date                            │   │
│ └────────────────────────────────────────────────────┘   │
│ ┌────────────────────────────────────────────────────┐   │
│ │ MEANING : This returns the current date.           │   │
│ └────────────────────────────────────────────────────┘   │
│ ┌────────────────────────────────────────────────────┐   │
│ │ ARGUMENTS :                                        │   │
│ │            argument           : type  : direction  │   │
│ │            todays date        : date  : out        │   │
│ └────────────────────────────────────────────────────┘   │
│ ┌────────────────────────────────────────────────────┐   │
│ │ DEFINITION :                                       │   │
│ └────────────────────────────────────────────────────┘   │
│ ┌────────────────────────────────────────────────────┐   │
│ │ USE :                                              │   │
│ │      procedural: "todays date( )" returns date.    │   │
│ └────────────────────────────────────────────────────┘   │
└──────────────────────────────────────────────────────────┘
```

6.9.19.1 Comments

This is used as a standard way of generating a 'date stamp'. If an attribute is assigned a value using this, it may later be checked using the same function. For example:

```
If <Scheduled picture>.day = todays date( ) then
      ...
```

would check the attribute "day" of the entity "Scheduled picture" and if it was 'today', carry out some actions.

6.9.20 Trigger

OPERATION : trigger

MEANING : This causes a given discrete process to run.

ARGUMENTS :

argument	: type	: direction
process name	: process identifier	: in

DEFINITION : See meaning

USE :

bSTD:	T: process name

6.9.20.1 Comments

This is used in a control process to activate a process. See §2.14.3.23 for further discussion of the semantics of "trigger".

6.10 Standard YSM Data Access Grammar

6.10.1 Introduction

6.10.1.1 Basic data operations

Data access operations act on stored data, which takes the form of entities, associative entities and relationships. These operations are only allowed in data processes. The basic operations are:

1. **create**: this creates a new occurrence of an entity or relationship. For entities, some or all the attributes may be assigned values.

2. **delete**: this deletes one or more occurrences of an entity or relationship.

3. **match:** this allows a check of whether any occurrences of an entity or relationship satisfying a criterion exist. It returns a boolean value (TRUE or FALSE) which can be used in conditional logic within the process. In addition, it identifies those occurrences satisfying this criterion (see §6.10.1.4). Where required, the number of occurrences may be accessed (see §6.10.2).

4. **update**: this changes one or more attributes of an identified entity to a new value. The old value (if present), in effect is 'overwritten'. The new value must be of the correct type.

5. **read**: this allows the value of an attribute of a given entity to be obtained. This value may be used in comparisons in conditional logic or assigned to any accessible data item of the correct type.

Note: where entity is referred to in the above, this includes associative entity.

6.10.1.2 Inputs and outputs to data access operations

This section gives a general description of the allowed data access operations. Where these require parameters (attributes, identifiers, etc.) there are many possible syntaxes for the way in which this is done. Rather than insist on one fixed syntax, YSM allows any form of these instructions to be used, provided they include the required parameters. The required parameters for each operation are given inthe following table, together with a definition of how these parameters should be interpreted. A possible (rather formal) version of such a grammar is given in the sections §6.10.3 onwards.

	OPERATION	INPUTS	OUTPUTS
Entity	create	entity identification	
	delete	entity selection	
	match	entity selection	TRUE or FALSE
	update	entity selection, attribute assignment	
	read	entity identification, attribute name	value of attribute
Relationship	create	relationship identification	
	delete	relationship selection	
	match	relationship selection	TRUE or FALSE
Associative entity	create	associative entity identification	
	delete	associative entity selection	
	match	associative entity selection	TRUE or FALSE
	update	associative entity selection, attribute assignment	
	read	associative entity identification, attribute name	value of attribute

6.10.1.3 Identification and selection

Identifications and selections are ways of identifying occurrences of an entity or relationship. Often we need to be sure that only one occurrence is selected for a subsequent operation. This can be achieved by using an:

- an identification, which guarantees at most one 'hit';

- a selection, which may give more than one 'hit' (there is an implied "for each" in the application of the operation).

The terms in the table are:

1. **entity identification**: any way of identifying the unique occurrence of the entity. This must be by one of the identifiers of the entity. Entity identifiers are attributes or combinations of attributes.

2. **relationship identification**: the unique way of identifying an occurrence of the relationship, which must always be an entity identification for each of the entities that participate in the relationship.

3. **associative entity identification**: any way of identifying the unique occurrence of the associative entity. This must include at least one of the identifiers of the associative entity. These identifiers may be attributes or combinations of attributes (as for any entity). Associative entity identifiers may also be combinations of entity references (as for any relationship).

4. **entity selection**: any way of selecting 0, 1 or more occurrences of an entity.

5. **relationship selection**: any way of selecting 0, 1 or more occurrences of the relationship. This must always correspond to a selection of entities that participate in the relationship.

6. **associative entity selection**: any way of selecting 0, 1 or more occurrences of an associative entity.

7. **attribute assignment**: the name of the attribute and a value to be assigned to it. This value may either be an available data item, the result of an enterprise operation or a constant, provided the value is of the same type as the attribute.

Several attribute assignments may be carried out in a single operation. If more than one entity occurrence is selected, then assignments to each of these can occur in parallel.

6.10.1.4 Binding

Fundamental to the use of these data access operations is the concept of binding (or selection).

In its simplest form binding means 'identifying an occurrence of'. (This corresponds to an identification.) Thus for an entity (say <Patient>), one occurrence of patient can be identified by a "match". (This can be guaranteed if the match is carried using the identifier.) The operation "delete <Patient>" would then delete that one occurrence of patient.

In more general cases, a match might be satisfied by several occurrences of <Patient> — for example "find if there are any patients whose names are Mike" (this is a selection). There are now several occurrences of <Patient> identified. The entity "Patient" is now bound and a subsequent "delete <Patient>" would delete all these selected patients.

Some of the operations described below require the entities that they involve to be bound before they are used and some set up the binding as part of their action. Sometimes, an operation has a different interpretation, depending on whether the entity is bound or not.

1. **binding entities**: to bind an entity, use any operation that identifies or selects occurrences of that entity.

2. **binding relationships**: to bind a relationship, use operations that identify or select occurrences of each of the entities that participate in the relationship. If any of these entities are not bound, the relationship is not bound.

3. **binding associative entities**: an associative entity may be bound as an entity (by an entity selection), or as a relationship (by selections for each of the entities that participate in the relationship).

6.10.1.4.1 Bindings and data processes

Bindings are local to a particular data process — in other words, what one process does has no effect on the bindings in another process. When a binding is set up by a data process, it remains in force until another binding over-rides it. Each time the data process is activated, all entities and relationships it uses are initially unbound.

6.10.1.4.2 Multiple bindings

Within a data process, it is possible to have two bindings in force for a given entity at one time, but only by using local terms. For example, we could use a local term ("very ill patient" perhaps) that is defined:

```
<Very ill patient> ::= <Patient> with <Patient>.infection level > 6 ;
```

This local term is now an alternative name for a selected subset of the <Patient> entity (see §2.16.3.10.2). Selections can now be carried out on "Patient" and "Very ill patient" independently. See §6.10.8.1 for a further discussion and example of this.

6.10.2 Matched and unmatched counts

The operation match determines how many occurrences of an entity or relationship exist (usually, there is some selection criterion that they must satisfy). This is indicated by an operation of the form "match", followed by the specification of the item(s) to be checked. This returns a value of TRUE or FALSE. It also has the side effect of changing the bindings of the items being tested.

In some cases, the number of matching occurrences is required. This is obtained by using a parallel operation "number of matching", followed by the specification of the item(s) to be checked. This returns a cardinal number, which are the number of occurrences satisfying this test.

In all other respects, these two operations are identical. In particular, if an entity (or relationship) is unbound, then the number of occurrences of it may be determined. For example, if the entity "Instructor" is not bound, then:

```
number of manuals required := number of matching(<Instructor>)
```

would set a data item to "number of manuals required" to be the same as the number of occurrences of the entity "Instructor".

For convenience, an operation "number of" has the same effect. This counts the number of occurrences of an entity or relationship, irrespective of binding.

6.10.3 Create entity

This has the effect of creating exactly one new occurrence of the entity. At least one identifier must be assigned a value (or set of values, if it is a compound identifier) that is distinct from any other occurrence of the entity. (It is therefore almost obligatory to precede a "create" by a "match" to ensure that this is true.) In addition, a list of attributes may be assigned values at the same time as this occurrence is created. This list is of indefinite length, with each attribute assignment separated from the next by a ",". The values for these attributes are any available items of compatible type. For example:

```
create <Patient> with
        number := %admission number%,
        date admitted := todays date( ),
        name := %interview name%
end ;
```

creates a new occurrence of "Patient" (assuming that "number" is the identifier).

6.10.3.1 Shortened forms

A shortened form is allowed, using the typing of available data items (see §2.18.3.6). The assignments of values to the corresponding entity is then automatic. For example, suppose, %admission number% was typed as <Patient>.number and %interview name% typed as <Patient>.name. The above could then be shortened to:

```
create <Patient> with
        date admitted := todays date( ),
end ;
```

This shortened form is only allowed if at least one of the available data items is typed as the identifier for the entity. If all attributes that are to be assigned values have corresponding typings, then an even shorter form of create is available. For the above example, if the "date admitted" attribute did not have to be set, we could use:

```
Create <Patient> ;
```

6.10.3.2 Effect on binding

The "create" operation is unaffected by the binding of the entity when it occurs. After the "create", this single occurrence of the entity is bound. As a consequence, either of the two cases above may be replaced by:

```
create <Patient> ;
<Patient>.date admitted := todays date( ) ;
```

6.10.4 Delete entity

This has the effect of deleting any occurrences of an entity that satisfy a criterion. It has several forms:

6.10.4.1 Deleting all bound occurrences

If the entity is bound, then the operation is merely "delete", followed by the name of the entity. This deletes all bound occurrences of the entity. For example:

```
delete <Patient> ;
```

6.10.4.2 Deleting selected occurrences

If the entity is not bound, then the operation must include a selection that defines one or more occurrences of the entity which are the ones to be deleted. For example:

```
delete <Patient> with <Patient>.age < 18 years ;
```

deletes all occurrences of patient who had the attribute "age" currently greater than 18 years.

6.10.4.3 Shortened forms

A shortened form "delete matching" is available if there is no ambiguity about what the "match" test is. This will be the case if exactly one of the "available data items" is typed as an identifier of the entity (see §2.18.3.6). For example, if the data process had a stimulus "%patient removal%, containing %patient id% (together with other items), with %patient id% typed as <Patient>.number, then:

```
delete matching <Patient> ;
```

would delete the single instance of <Patient> with this number (supposing "number" to be the identifier for "Patient).

The full form could also have been used, which would have been:

```
delete <Patient> with <Patient>.number = %patient id% ;
```

6.10.5 Match entity

This returns a value TRUE or FALSE depending on whether any matching occurrences of the entity exist. For example:

```
If <Patient>.name = %required name% then
```

could be used to see if there were any patients of the required name. The value could also be assigned to an accessible boolean item. For example:

```
%patients to be deleted% := Match(<Patient>.name = %required name%)
```

This would require %patients to be deleted% to be a temporary boolean item.

The match operation always 'starts from scratch' — it assumes no occurrences are bound. In other words, it always checks *all* occurrences. When the operation is invoked, then the result is true if *any* patient has that name and these occurrences are bound.

6.10.5.1 Unspecified match entity

If exactly one available data item (see §2.16.3.1.1) is typed as an identifier of the entity (see §2.18.3.6), then a reduced form of the operation may be used. For example:

```
If matching <Drug> ...
```

could be used if only one available data item was typed as an identifier of the entity "Drug". See §2.16.3.12.2 and §2.16.3.13.11 for examples of this.

6.10.6 Update entity

This is referenced by using the name of the entity in chevrons, followed by the attribute name, together with a value that needs to be assigned to the attribute. This value may be a constant, expression or available item of compatible type to the attribute. For example:

```
<Patient>.narrative history := %doctors notes%;
```

sets the value of the attribute "narrative history" of the entity "Patient" to the value of a component of a data flow, assuming they are of compatible type. They are of compatible type if the data item is typed as that attribute of the entity (see §2.18.3.6) or if they have the same underlying abstract data type (in this case, probably "text"). The ":=" symbol should be read:

is assigned the value

Note: the entity <Patient> must be bound when the statement is invoked. Each occurrence of <Patient> that is bound will be updated.

6.10.7 Read entity

This uses the name of the entity in chevrons, followed by a "." and then the attribute name. The result is a value of the same type as the attribute is declared as (in attribute specification). This value may be assigned to any accessible item with a compatible type. For example:

```
%patient to be weighed% := <Patient>.name ;
```

The value may also be used in expressions (for iterations, selections, etc.). For example:

```
If <Patient>.name = %required name%  then
```

could be used to select some operations to be performed or not, depending on whether the attribute "name" and the data element "required name" had the same value. (For this to be a legal statement they must have compatible types.)

Note: the entity <Patient> *must* be bound when the read statement is used.

6.10.8 Create relationship

To define an occurrence of a relationship, each of the entities must have a unique occurrence defined as taking part in this occurrence of the relationship. Thus, an operation of the form:

```
create "<Patient> is assigned to <Doctor>" ;
```

would create an occurrence of the relationship "is assigned to", that refers to an occurrence of a "doctor" and an occurrence of a "patient". Both <Patient> and <Doctor> must be bound at the point at which this statement occurs. If more than one "Patient" or "Doctor" is bound, then an occurrence of the relationship is set up for each possible pair. (If there were two bound occurrences of "Patient" and four bound occurrences of "Doctor", eight occurrences of the relationship would be created.)

6.10.8.1 Example, using multiple binding of an entity

As already stated, binding an entity destroys any binding that was in place prior to that new binding. This may cause problems, particularly when a relationship is recursive. Consider a requirement that each patient be assigned a "buddy" — another patient who counselled and helped him/her. If this were the case, a data process might be needed to record the relationship. Assume that it has an incoming data flow, containing %patient1% and %patient2% as the identifiers for the two patients. A first attempt at the minispec might be:

```
If <Patient> exists with <Patient>.id = %patient1% then
        If <patient> exists with <Patient>.id = %patient2% then
                create "<Patient> is buddy of <Patient>" ;
        end ;
end ;
```

Unfortunately, the second test destroys the binding set up in the first test. The solution is to create local terms which can each be bound separately. (Only one local term is required, but the example is easier to understand with two). The terms are defined:

```
<Patient 1> ::= a <Patient> with <Patient>.id = %patient id 1% ;

<Patient 2> ::= a <Patient> with <Patient>.id = %patient id 2%
```

and then the statements:

```
If <Patient 1> exists then
        If <Patient 2> exists then
                create "<Patient 1> is buddy of <Patient 2>";
        end ;
end ;
```

will work as required. Note: even though <Patient 1> and <Patient 2> are alternative names for <Patient> there is no problem — each can have a binding independently of the other (see §6.10.1.4).

6.10.9 Delete relationship

6.10.9.1 Specified deletion

To destroy an occurrence of a relationship, each of the entities taking part in the relationship must be bound. After the "delete", no occurrences of the relationship will refer to those combinations of occurrences. For example, if the "Doctor" and "Patient" entity were both bound, the statement:

```
delete "<Patient> is assigned to <Doctor>" ;
```

would ensure that no occurrence of the relationship "is assigned to", would refer to any of the selected occurrences of "Doctor" and "Patient".

6.10.9.2 Unspecified deletion

If one of the entities is unbound, then the delete operation is still allowed. It is not allowed if more than one entity is unbound. For example, suppose that one specific occurrence of "Doctor" is identified, but <Patient> is not bound. The operation:

```
delete "<Patient> is assigned to <Doctor>" ;
```

would then delete *all* occurrences of "Patient" allocated to that "Doctor".

6.10.10 Match relationship

6.10.10.1 Bound form

To check whether an occurrence of a relationship exists between specific occurrences of the entities, each of the entities must be bound. One or more occurrences may be selected for each entity, as required. For example, the operation:

```
<Patient> is assigned to <Doctor>
```

returns the value TRUE or FALSE, depending on whether any of the selected occurrences of "Patient" are assigned to any of the selected "Doctor"s. Both <Patient> and <Doctor> must be bound at the point at which this statement occurs.

6.10.10.2 Unbound form

If one of the entities participating in the relationship is unbound, then the "match" operation may be used to check whether there any occurrences of the unbound entity that satisfy the match. For example, suppose one <Doctor> had been selected. The operation:

```
<Patient> is assigned to <Doctor>
```

would check whether any occurrences of "Patient" were assigned to that specific Doctor. After the match, these occurrences of <Patient> are now bound.

If neither of the entities are bound, then match determines whether there are *any* occurrences of the relationship. After using match in this way, all occurrences are bound.

6.10.10.2.1 Higher-order relationships

For higher-order relationships, leaving any of the entities unbound and then using match will determine whether there are any occurrences of the relationship that refer to the bound occurrences of the relationship. In addition, these occurrences are bound after the match.

6.10.10.3 Use of relationship match

As with entity match (see §6.10.5), this operation may be used in two ways. It may be used to return a boolean item used in a test, for example:

```
If <Patient> is assigned to <Doctor> then ...
```

or it may be assigned to a boolean data item, for example:

```
%matching patients% := <Patient> is assigned to <Doctor> ;
```

6.10.11 Create associative entity

An associative entity is a relationship and as for any relationship, to define an occurrence of it, each of the entities must have a unique occurrence defined as taking part in this occurrence of the relationship. Thus, an operation of the form:

```
create <Scheduled treatment>
```

is equivalent to:

```
create "<Patient> is scheduled to receive <Drug>"
```

if "Scheduled treatment" is the relationship "<Patient> is scheduled to receive <Drug>". (Either form of the operation may be used.)

All entities that participate in the relationship must be bound where this statement occurs.

6.10.12 Delete associative entity

This has the effect of deleting any occurrences of an associative entity that satisfy a criterion. It has two forms:

- Treating the associative entity as an entity, then any of the ways of deleting an entity (see §6.10.4) may be used.

- Treating the associative entity as a relationship, then any of the ways of deleting a relationship (see §6.10.9) may be used. In this version, the "relationship frame" is used. For example, if the associative entity "Scheduled treatment" had a relationship frame "<Patient> is scheduled to receive <Drug>", then:

```
delete "<Patient> is scheduled to receive <Drug>"
```

would delete all scheduled treatments for a patient with a particular drug if they were both bound. If the <Patient> is not bound, then all treatments with a specific drug could be deleted; if <Drug> is not bound, but <Patient> is, then all treatments for a specific Patient would be deleted.

6.10.13 Match associative entity

The associative entity is both an entity and a relationship. The form of this statement may reflect either use. For its use as an entity, see "match entity". For its use as a relationship, the "relationship frame" is used. For example, suppose the associative entity <Scheduled treatment> had a relationship frame "<Patient> is scheduled to receive <Drug>". To check whether an occurrence of this exists, the statement:

```
<Patient> is scheduled to receive <Drug>
```

returns the value TRUE or FALSE, depending on whether that specific "patient" has been scheduled to receive that "drug". If both <Patient> and <Drug> are bound at the point this statement occurs, then the check is carried out for those selected occurrences. If one or either (but not both) are unbound, then the check is for *any* occurrences that satisfy the criterion.

6.10.13.1 Use of associative entity match

As with entity match (see §6.10.5), this operation may be used in two ways. It may be used to return a boolean item used in a test, for example:

```
If <Patient> is scheduled to receive <Drug> then ...
```

or it may be assigned to a boolean data item, for example:

```
%treated patients% := <Patient> is scheduled to receive <Drug> ;
```

6.10.14 Update associative entity

The associative entity is treated as an entity for this operation. For example:

```
<Scheduled treatment>.treatment time := %required time% ;
```

could be used to allocate a value to the attribute "treatment time" of the associative entity "Scheduled treatment".

Note: the associative entity <Scheduled treatment> must be bound when the statement is invoked.

6.10.15 Read associative entity

The associative entity is treated as an entity for this operation. For example:

```
%required time% := current time( ) +
                <Recommended treatment>.treatment period ;
```

could be used to allocate a value to "required time".

Note: the associative entity <Recommended treatment> must be bound when the statement is invoked.

6.10.16 Use of state variables

State variables of an entity may be used as if they were attributes in some respects.

Specifically, they may be used to select matching instances (one or many) of the entity that are in this state). For example:

```
For each <Scheduled course> with status = open do:
    ...
end
```

which would carry out some predefined operations on all "Scheduled course"s that had "open" as the value of the state variable"status".

The value of a state variable for bound occurrences may also be changed to reflect a new state in those occurrences of the entity. For example:

```
For each <Scheduled course> with start date = todays date( ) do:
        <Scheduled course>.status := running ;
end
```

which would change all courses starting today to "running".

State variables can not be used in any other way. In particular, the 'value' of the state can not be assigned to any data item.

6.11 Internal System Essential Model

This section is a preliminary description of the internal model of the System Essential Model and automated support for it. There is no pretence that this is a complete or rigorous description. It is included to justify the way YSM is organised around the concept of a internal model and multiple views into it.

With the advent of automated support tools, this view of the Essential Model will predominate. The multiplicity of view cross-checking rules will be replaced by 'visibility' control and filters, as described below.

6.11.1 Internal model components

The internal model contains the following:

- a set of functions, each with a minispec;

- a set of data processes, each an instantiation of a function and with a defined stimulus;

- a set of abstract state machines, each with a specification (usually a bSTD);

- a set of control processes, each an instantiation of an abstract state machine;

- a set of terminators, each with a list of inputs and outputs;

- a set of data items, which may be an attribute, non-attributed data item or grouping of other data items;

- a set of data flows. Each uses a data item, with an origin and destination;

- a set of event flows. Each has a meaning and persistence, with an origin and destination;

- a set of events, each with a specification;

6 11.2 The distinction between a function and the data process

A function is a required component of the YSM internal model. Its characteristic behaviour is to transform inputs to outputs according to a predetermined definition. A function is an abstraction. To be used, it has to be realised as a data process.[1]

In YSM, enterprise operations are functions — they can potentially be used as and when required (maybe many times). A data process is the realisation of the function. This is analogous to the concept of an entity and 'occurrences of the entity'. A legal Essential Model may have several data processes all of which are realisations of a single function.

YSM also uses the term 'data process ' for the graphic icon on the DFD. This is a little confusing. The situation with state machines is rather more logical.

A state machine is a machine driven by events. YSM allows these to be used. They are shown on a DFD using a 'control process' icon and often loosely referred to as control processes. In

---

1 The literature refers to this as instantiation.

fact, there can be several state machines with the same logic in a model — they are instances of an 'abstract state machine'.

6.11.3 Deriving views from the internal model

All views into the Essential Model may be derived from this.

Some are obvious. For example, there may be several instantiations of an abstract state machine, possibly with changes on input and output event flow names. Each would have a bSTD that is derived from that of the abstract state machine with the names changed.

Some are not so obvious, but none-the-less, derivable automatically from the internal model. Two examples are given below:

6.11.3.1 Deriving system entity specifications
The entities accessed by the system can be derived from each of the minispecs. (Recall that the entities used by the enterprise are defined in the Enterprise Essential Model. The System Essential Model defines the visibility of these components to the system.)

The accesses (create, read, update, delete) to each of these can also be identified from the minispecs. The attributes accessed by the system are those accessed by any of the minispecs. The identifiers used by the system are any of the entity's identifiers that are used in a match by any of the minispecs.

6.11.3.2 Deriving the 'flat' DFD
For example a form of the flat DFD (control processes are omitted, to simplify the example) can be derived automatically as follows:

1. Each data process is shown.
2. For each data process, there is a store for each entity or relationship accessed by it (as stated in the corresponding minispec). The names of each store are arbitrary (but see below).
3. Accesses from the data process to the data store may be derived from the minispec.
4. All data flow outputs from the data process can be derived from the minispec. These are connected to all other data processes and terminators for which there is a data flow with this name.
5. All data flow outputs from terminators are connected to all other data processes and terminators for which there is a data flow with this name.

6.11.4 Types of view

There are several different types of view into the internal model:

1. Required views: these are ones that must be present and verified. Examples include minispecs and state machine specifications.

2. Retained views: these are ones that are not required to capture the internal model. All logical information in them may be derived from the internal model. However, they provide useful views into the model and may be retained from one 'session' to the next. (A session is used in the sense of one session of use of an automated support environment.) To do this, the support tool has to store such information as positional information for each derived graphic view, user-friendly names for stores and groups of data flows.

3. Temporary views: these may be produced by the system modeller (or generated automatically) during one session, but not retained at the end of a session. Information entered into these views may update the internal model, but the view itself is not retained.

Bibliography

[Bac69] C. W. Bachman, "Data Structure Diagrams", *Data Base*, Vol. 1, No. 2 (Summer 1969), pp. 4-10.

[Boe76] B. W. Boehm, "Software Engineering", *IEEE transactions on Computers*, Vol. C-25, No. 12 (December 1976), pp. 1226-41.

[Boe81] B. W. Boehm, *Software Engineering Economics*, Prentice-Hall, 1981.

[Böh66] C. Böhm and G. Jacopini, "Flow diagrams, Turing Machines and Languages with Only Two Formation Rules", *Communications of the ACM*, Vol. 9, No. 5 (May 1966), pp. 366-71.

[Bro75] F. P. Brooks, *The Mythical Man-Month*, Addison-Wesley, 1975.

[Che76] P. Chen, "The Entity-Relationship Model - Towards a Unified View of Data", *ACM transactions on Database Systems*, Vol. 1, No. 1 (March 1976), pp. 9-36.

[Cod70] E.F. Codd, "A Relational Model for Large Shared Data Banks", *Communications of the ACM*, Vol. 13, No. 6 (June 1970), pp. 377-387.

[COD71] *Data Base Task Group Report*, ACM, April 1971.

[DeM78] T. DeMarco, *Systems Analysis and Specification*, YOURDON Press, 1978.

[Dij65] E. Dijkstra, "Programming Considered as a Human Activity", *Proceedings of the 1965 IFIP Congress*, North-Holland Publishing Co., 1965, pp. 213-17.

[Dij68a] E. Dijkstra, "Go To Statement Considered Harmful", *Communications of the ACM*, Vol. 11, No. 3 (March 1968), pp. 147-48.

[Dij68b] E. Dijkstra, "Structure of the THE – Multiprogramming System", *Communications of the ACM*, Vol. 11, No. 5 (May 1968), pp. 341-346.

[Dij69] E. Dijkstra, "Structured Programming", Conference report of a NATO sponsored conference in Rome, October 1969. It is reprinted in *Software Engineering, Concepts and Techniques*, edited by Buxton, Naur, Randell, Litton Educational Publishing Inc., 1976.

[Gan77] C. Gane and T. Sarson, *Structured System Analysis: Tools and Techniques*, Improved System Technologies, 1977.

[Ker74]   B. W. Kernighan and P. J. Plauger, *The Elements of Programming Style*, McGraw-Hill, 1974.

[McL67]   M. McLuhan and Q. Fiore, *The Medium Is the Message*, Bantam, 1967.

[McM84]   S. McMenamin and J. Palmer, *Essential Systems Analysis*, YOURDON Press, 1978.

[Mil56]   G.A. Miller, "The Magical Number Seven, Plus or Minus Two: Some Limits on Our Capacity for Processing Information", *Psychological Review*, Vol. 63, No. 2 (March 1956), pp. 81-97.

[Mye75]   G.J. Myers, *Reliable Software Through Composite Design*, Petrocelli/Charter, 1975.

[Oll78]   T. W. Olle, *The Codasyl Approach to Data Base Management*, Wiley, 1978.

[Par72]   D. L. Parnas, "On the Criteria to Be Used in Decomposing Systems into Modules", *Communications of the ACM*, Vol. 5, No. 12 (December 1972), pp. 1053-58.

[Rei75]   D. J. Reifer, "Automated Aids for Reliable Software", *Proceedings of the 1975 International Conference on Reliable Software* (April 1975), pp. 131-42.

[Ros77]   D. T. Ross and K. E. Schoman, "Structured Analysis for Requirements Definition", *IEEE Transactions on Software Engineering*, Vol. SE-3, No. 5 (January 1977), pp. 6-15.

[Ste74]   W. P. Stevens, G.J. Myers and L. L. Constantine, "Structured Design", *IBM Systems Journal*, 1974, Vol. 13, No. 2, pp. 115-139.

[War85]   P. T. Ward and S. J. Mellor, *Structured Development for Real-Time Systems*, YOURDON Press, 1985 (volumes 1 and 2), 1986 (volume 3).

[Wei75]   J. Weinberg, *An Introduction to General Systems Thinking*, Wiley, 1975.

[Wir71]   N. Wirth, "Program Development by Stepwise Refinement", *Communications of the ACM*, Vol. 14, No. 4 (April 1971), pp. 221-227.

[You75]   E. Yourdon and L.L. Constantine, *Structured Design: Fundamentals of a Discipline of Computer Program and System Design*, YOURDON Press, 1975. Second edition was published in 1978.

[You76]   E. Yourdon, "The Emergence of Structured Analysis", *Computer Decisions*, Vol. 8, No. 4 (April 1976), pp. 58-59.

[You77]   E. Yourdon, *Structured Walkthroughs*, YOURDON Press, 1977.

[YOU84]   YOURDON Inc. public seminar material, 1984 onwards.

Index

References (see page 654)

Bac69 4
Boe76 8
Boe81 8
Böh66 3
Bro75 8, 71
Che76 7, 9, 377
Cho59 565
Cod70 5
COD71 4
DeM78 6 - 7, 15, 42, 571
Dij65 3
Dij68a 4
Dij68b 5
Dij69 4
Dow86 565, 568 - 569
Fla81 54, 377
Gan77 6
Gor88 247
Ken83 387
Ker74 4
Kor86 54

Len83 116
Man87 568
McL67 41
McM84 7, 9, 496
Mea55 333
Mil56 3, 70
Mye75 5
Oll78 4
Par72 5
Pow78 225, 540
Rei75 9
Ros77 6
Roy88 185
Shi87 333, 356
Ste74 4
Sto61 60, 565
TAA89 597
Tha88 565, 568, 573
War85 9
Wea68 328
Wei75 15, 385
Wir71 5
You75 5
You76 6
You77 6
YOU84 9

BUSINESS REPLY MAIL

FIRST-CLASS MAIL   PERMIT NO. 3518   RALEIGH, NORTH CAROLINA

POSTAGE WILL BE PAID BY ADDRESSEE

Yourdon™
Forum II
8521 Six Forks Road
Raleigh, N.C. 27690-1530

YOURDON™

Training • Consulting • Software

YOURDON is the world leader in systems development training. Now we also offer proven method implementation and method evaluation services. As a Division of CGI Systems, Inc., YOURDON can provide you and your organization with the most cost effective method options to meet all your systems development requirements. Give us a call at the YOURDON office nearest you:

YOURDON North American Offices
YOURDON **Worldwide Headquarters**
Forum II
8521 Six Forks Road
Suite 400
Raleigh, NC 27615
(800) 326-0799
FAX (919) 847-2457

YOURDON **Western Office**
90 New Montgomery Street
Suite 420
San Francisco, CA 94105
(800) 234-8083
FAX (415) 546-6539

YOURDON **Canadian Office**
2015 Peel, bureau 425
Montreal, Quebec
Canada H3A 1T8
(514) 849-2766
FAX (514) 849-5766

YOURDON **European Office**
Central Business Park Utrecht - Bldg 3
Atoomweg 350 - P.O. Box 8476
NL-3503 RL Utrecht
Netherlands
(31) 30 41 45 11
FAX (31) 30 41 14 38

YOURDON™, A Division of CGI Systems, Inc.

For more information, please fill out the attached Postage-paid reply card and return to YOURDON. **All replies will receive a 20% training discount voucher.** ✂

Please tell me more about YOURDON™ world class products and services. I am particularly interested in the following areas:

☐ YOURDON™ Systems development **training seminars**
  ☐ Business/Information Systems
  ☐ Control/Engineering Systems

☐ YOURDON™ Method implementation and **consulting services**

☐ YOURDON™ CASE software tools
  ☐PC ☐Workstations/Mainframe

☐ YOURDON™ Technical publications/manuals

Your Name: _____

Title: _____

Company: _____

Address: _____

City: _____ Zip: _____

State: _____

Phone: _____

FAX: _____